BRITISH CERAMIC PROCEEDINGS
NO.55

21st CENTURY
CERAMICS

Recent British Ceramic Proceedings

Stoke—on—Trent Libraries

Please return or renew by last date shown

HANLEY LIBRARY
PHONE No. 01782 238455

THOMPSON, D. P. (ed)
21st century ceramics

6.10.97

If not required by other readers this item may be renewed, in
person, by post or by telephone. Please quote details above and
date due for return. If issued by computer the numbers on the
barcode label on your ticket and on each item, together with date
of return are required.

DYPE145

21st CENTURY CERAMICS

Edited by

D. P. Thompson and H. Mandal
*Department of Mechanical, Materials and
Manufacturing Engineering (Materials)
The University of Newcastle
Newcastle upon Tyne, UK*

**British Ceramic Proceedings
No.55**

THE INSTITUTE OF MATERIALS

30R1L

Book 616
Published 1996 by
The Institute of Materials
1 Carlton House Terrace
London SW1Y 5DB

ISSN 0268–4373
ISBN 0–901716–85–5

The papers contained in this fifty-fifth volume
of the British Ceramic Proceedings were presented
at a meeting of the Ceramic Science Section,
during the Annual Convention of
the Ceramic Industry Division
held at Keele University,
5–6 April 1995.

Typeset by
Dorwyn Ltd,
Rowlands Castle, UK

Printed and bound at
The University Press
Cambridge, UK

Contents

Foreword

The last half of the 20th Century has seen an enormous increase in interest in the more specialised branches of ceramics, originally described collectively as *Special Ceramics* or subsequently as *Technical Ceramics* or *Fine Ceramics* (the latter especially in Japan). These materials are now commonly sub-divided for convenience into *Structural Ceramics* – describing materials in which the main characteristic is excellence of mechanical properties, with high temperature capability often associated as an additional benefit – and *Functional Ceramics* – originally coined to describe materials with a specific additional property of excellence, but now almost exclusively used to refer to the ever-increasing family of *Electrical Ceramics*. More recently, interest in *Bioceramics* has increased out of all proportion, in response to the pressing current need for the development of bio-compatible implants and other medical and dental materials. All these types of ceramics are distinguished from *Classical* or *Traditional Ceramics* in that the starting materials are not clay-based; furthermore, more careful tailoring of the chemistry of the starting mix has usually been the key factor in controlling the phases produced and their resulting properties. In some cases this has meant departing from conventional oxide-based systems into nitride, carbide, boride or silicide based systems.

Inevitably the signficant developments in these materials in the 1960s and 1970s was accompanied by journalistic euphoria, and predictions based on linear (or even more radical) extrapolations of expected growth rates were in many cases shown in the 1980s and 1990s to be gross overestimates of the real situation. This has been noticeable in the case of Structural Ceramics, whereas in both Functional Ceramics and Bioceramics, impressive increases in market sales have accompanied the increased levels of research and development dedicated to these materials.

Clearly, at the end of one century and the start of another, it is appropriate to consider the status of this group of ceramics as a whole, and to evaluate which materials will continue on an upward curve of development and market exploitation into the next century, and which materials will saturate or decrease in importance in response to competition from other materials, or excessively high production costs. The present volume is a compilation of the papers presented at the symposium: *21st Century Ceramics*, organised by the Ceramics Science Committee of Institute of Materials at the 1995 Annual Ceramics Convention, held at the University of Keele, UK on 5–6th April 1995. The programme was subdivided into the topics: Bioceramics, Nanoceramics, Novel Processes, Developments in Processing Technology, Electroceramics, Nitrogen Ceramics and Composite Ceramics. Authors were asked to focus their papers in the context of current developments, and to look forward into the next century and comment on the likely evolution (or otherwise) of the materials which formed the basis of their papers. The resulting compilation of papers constitutes an excellent state-of-the-art summary of current developments in this continuously emerging field, and provides a good overview for industrialists considering investment in this area in the immediate future.

D. P. Thompson
H. Mandal

University of Newcastle
August 1995

Ceramic Processing

Recent Advances and Applications of Thermal Sprayed Ceramic Coatings

A. J. STURGEON

TWI, Abington, Cambridge, CB1 6AL, UK

ABSTRACT

Thermal spraying is increasingly being used to deposit high quality ceramic coatings for applications ranging from the well established zirconia thermal barriers used in aero engines to more recent uses in the automotive, electronic and biomedical industries. New thermal spraying processes for preparing high quality ceramic coatings are also becoming available, with high velocity oxyfuel (HVOF) being a good example. A review of recent developments and applications of thermal sprayed ceramic coatings will be given. Recent work on the use of HVOF to deposit coatings of alumina and hydroxyapatite will also be presented.

1. INTRODUCTION

Thermal sprayed ceramic coatings offer an attractive and cost effective route to enhance the surface characteristics of many engineering components. This is in part due to their excellent resistance to extreme temperature, corrosion and wear. They provide, for example, thermal barriers in aero-engine gas turbines, hard coatings to increase the life of dies and cutting tools, and wear resistant coatings on rolls for use in the paper and steel industries. Today the largest applications for thermal sprayed ceramic coatings are in zirconia thermal barriers for aero-engine gas turbines and to a lesser extent for automotive, diesel and land-based turbines. Coatings of alumina, alumina with titanina, and chromia do not form part of this application area, but are used primarily for wear parts and are preferred coatings for most general wear problems when damage due to impact is absent. Thermal sprayed ceramic coatings are also considered for applications where the electrical or chemical nature of the coating is essential to the function of a component. Good exmaples are dielectric coatings for power hybrid packaging and coatings on medical implants to promote compatibility with body tissue. These and other current applications are summarised in Table 1.

2. THE HVOF PROCESS

The traditional thermal spraying processes for preparing high quality ceramic coatings are air plasma spraying, low pressure plasma spraying and detonation flame spraying. High velocity oxyfuel (HVOF) is a more recent thermal spraying process

Table 1. Current applications for thermal sprayed ceramic coatings

Industry	Application
Petrochemical	Wear resistant coatings on pump shafts, plungers, turbine rotor shafts, mechanical seals, compressor rods
Plastic Industry	Extruder barrels, screws, cutter shafts, extruder die plates
Textile Industry	Wear resistant coatings on draw rolls, finish applicator rolls, take up rolls, heater plates, thread guides
Aerospace	Combustor can thermal barriers, after burner spray bars
Steel Industry	Wear resistant coatings on tinning and chromium plating lines, melt coat rolls, chemical treatment rolls
Pulp and Paper Industry	Wear resistant coatings on pump sleeves, seals, cylinder liners
Medical	Bioactive coatings on orthopaedic and orthodontic implants
Electrical Insulation, General Industry	Induction coils, brazing fixtures, alternator plates, corona rolls

which has become established as a very competitive process for the deposition of highly quality metallic and metal bonded carbide coatings, such as tungsten carbide-cobalt. There are now several variants of the HVOF process commercially available, including Jet-Kote II (Thermadyne Stellite), HV 2000 (Miller Thermal), Top Gun (UTP), Diamond Jet (Sulzer Metco), CDS (Sulzer Metco) and the JP-5000 (Eutectic Tafa). Each one has differences in design, but all are based on the same fundamental principles.

A schematic diagram of the HVOF process is given in Fig. 1. An internal combustion process rapidly heats and accelerates a powder consumable to high velocities. The combination of a high chamber pressure of over 4 bar and gas flow rates of seveal hundred litres per minute generate hypersonic gas velocities of typically 1800m.s^{-1}, and combustion temperatures of above 2800°C. Some features of the HVOF process are compared in Table 2 with those for competing thermal spraying processes. Suitable combustion fuel gases include propylene, propane, hydrogen, MAPP and for a few systems, acetylene or liquid fuels such as kerosene. The velocity and temperature reached by the particles is a complex function of both combustion conditions and parameter settings. Recent work[1] has measured velocities of 600 to 1000m.s^{-1} for HVOF sprayed alumina particles of a size of 5–15μm. The temperatures reached by the particles during their rapid passage through the gun and onto the moment of impact will depend on the particular type of HVOF equipment, fuel and spraying conditions used. The lower temperature of the HVOF process, relative to plasma and detonation spraying, has resulted in uncertainty over its ability to deposit high quality ceramic coatings. The results reported here will demonstrate that the HVOF process can be used to prepare ceramic coatings which compare well to those deposited by air plasma spraying.

Table 2 Thermal spraying processes

Deposition technique	Heat source	Propellant	Typical temperature of heating environment	Typical particle velocity m.s.$^{-1}$	Average spray rate kg.hr^{-1}	Coating porosity % by volume	Relative bond strength
Flame spraying	Oxyacetylene	Air	3000	40	2–6	10–20	Fair
Plasma spraying	Plasma arc	Inert gas	>5000	200–400	4–9	5–8	Good
Low pressure plasma spraying	Plasma arc	Inert gas	>5000	400–600	4–9	<5	Very good
Detonation gun spraying	Oxygen/acetylene/ nitrogen gas detonation	Detonation shock waves	4500	800	0.5	<5	Very good
High velocity oxyfuel	Fuel gase	Combustion jet	3000	400–800	2–4	–	–

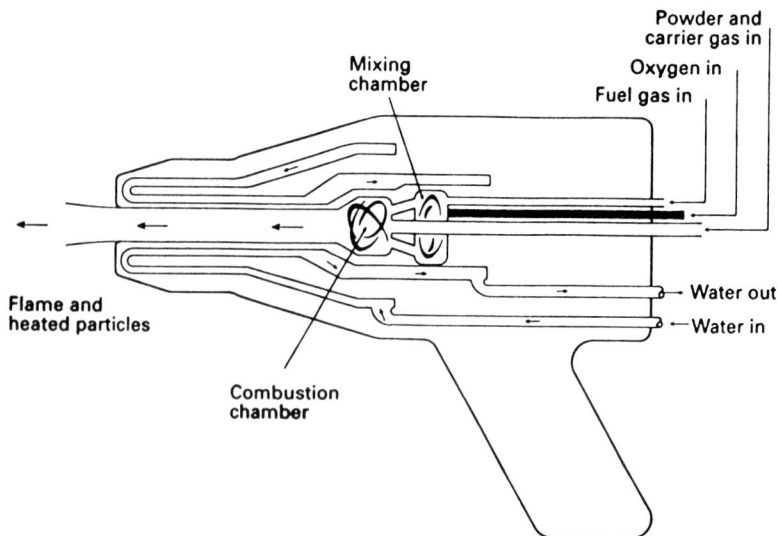

Figure 1.   Schematic diagram of HVOF process

3. ALUMINA COATINGS

Alumina coatings with a nominal thickness of 200μm have been prepared using the
Top Gun HVOF process and both acetylene and hydrogen fuel gases.[2] The powder
consumable was a 99.5% purity α-alumina (Plasmalloy 1110TG) with a mean particle
size of 15μm and with 95% of the powders under 29μm in size. For comparison, an air
plasma alumina coating was also prepared using a Miller Thermal SG100 gun with an
Ar/He 40kw plasma and with an α-alumina powder (Plasmalloy 1010) routinely used
for this system.

The measured values for Vickers microhardness (HV 300g), porosity, elastic
modulus, crystalline structure and surface roughness of the HVOF and air plasma
sprayed alumina coatings are reproduced in Table 3. Scanning electron images of
cross-sections for the HVOF coatings are shown in Figs 2 and 3. The HVOF
alumina coating sprayed using hydrogen fuel has a porosity measured at 3.9%. This
is similar to the 3.5% porosity measured for the air plasma sprayed coating. The use
of acetylene with the HVOF process gave a much higher porosity content of about
12.6%. The alumina coating prepared using hydrogen fuel consists mostly of
γ-alumina, with α-alumina present as a minor phase. This is again similar to the air
plasma sprayed alumina coating. The α-alumina starting powder has been converted to a primarily γ-alumina coating. The coating prepared using acetylene as
the fuel has a higher α-alumina content, with less γ-alumina being formed during
the spraying process.

The difference in microstructure results from the extent each of the two fuel gases
are melting and accelerating the alumina powder during the spraying process. The
coatings are formed by the build up of lamella as individual particles impact, deform

Table 3. Coating characteristics

Process	Porosity volume %	Elastic modulus, GPa	Alpha alumina content, %	Micro-hardness Vickers, 300g	Surface roughness um.Ra
HVOF hydrogen fuel	3.9	95	12	1215	1.4
HVOF acetylene fuel	12.6	43	36	879	4.4
Air plasma, Ar/He	3.5	40–50	9	1431	4.3

and then rapidly freeze. A higher level of melting and a higher particle velocity prior to impact is expected to give better deformation of the particle, leading to improved bonding between lamella and to lower porosity. As the same alumina powder 1110TG was used for both fuel gases, the temperature reached by the particles will, depend upon both the temperature and heat content of the surrounding combustion environment. Theoretical calculated values for combustion power and flame temperature for the two fuel gases are given in Table 4 using data available in the literature.[3] The values are for the oxygen to fuel gas ratios used in this work. Also given in Table 4 are mean values of particle velocities recently measured for the 1110TG alumina powder and same two fuel gases.[1]

The use of hydrogen fuel rather than acetylene can be expected to give a lower temperature environment of 2834°C compared to 3160°C, but with a higher combustion power, estimated at 120kW compared to 68kW, together with a faster particle velocity of typically 540m.s^{-1} compared to 280m.s^{-1}. The melting point of

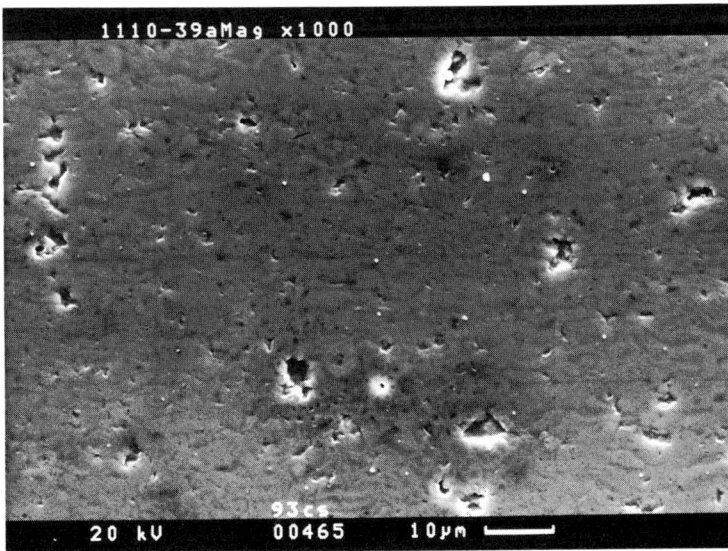

Figure 2.   Scanning electron image of an alumina coating sprayed using hydrogen fuel.

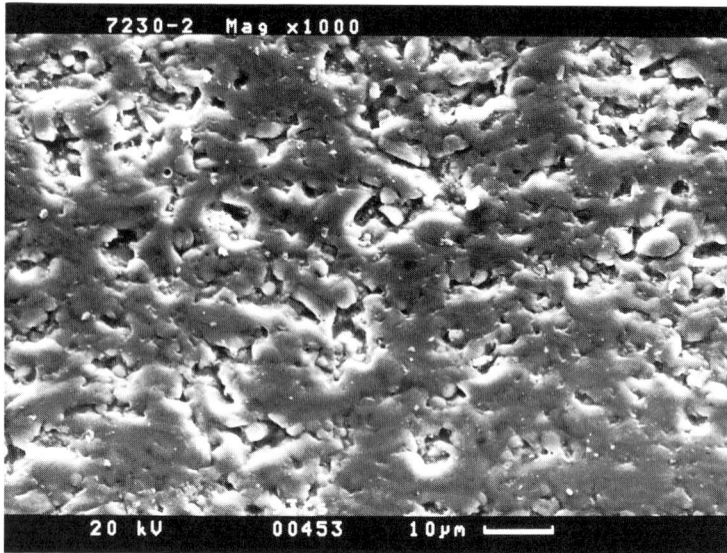

Figure 3.   Scanning electron image of an alumina coating sprayed using acetylene fuel.

alumina is 2015°C, so both fuel gases are capable of giving combustion temperatures sufficiently high to melt the powder.

An indication that a higher proportion of powder particles are melted by the hydrogen fuel is given by the relative α-alumina and γ-alumina ratios present in Table 3. It is generally accepted that during solidification on impact, γ-alumina is homogeneously nucleated if the initial droplet is completely molten.[4] This is a consequence of the energy barrier to nucleations for γ-alumina being less than that for α-alumina. Particles which are not completely molten are believed to crystallise from pre-existing nuclei to α-alumina. A higher observed γ-alumina to α-alumina ratio in the coating suggests a higher proportion of the powder particles were completely melted just prior to impact.

Measured abrasive wear rates[2] for the HVOF alumina coatings are reproduced in Fig. 4. The alumina HVOF coating prepared using hydrogen fuel has a significantly lower wear rate than the coating sprayed using acetylene, and is also noticeably lower than the air plasma sprayed alumina coating. The plasma sprayed coating acts as a reference to which the HVOF coatings can be compared. The improvement in wear

Table 4.   Typical combustion characteristics for different fuel gases

Characteristic	Acetylene fuel gas	Hydrogen fuel gas
Oxyfuel ratio	1.63	0.35
Flame temperature, °C	3160	2835
Combustion power, kW	68	120
Mean particle velocity, m.s.$^{-1}$	280	540

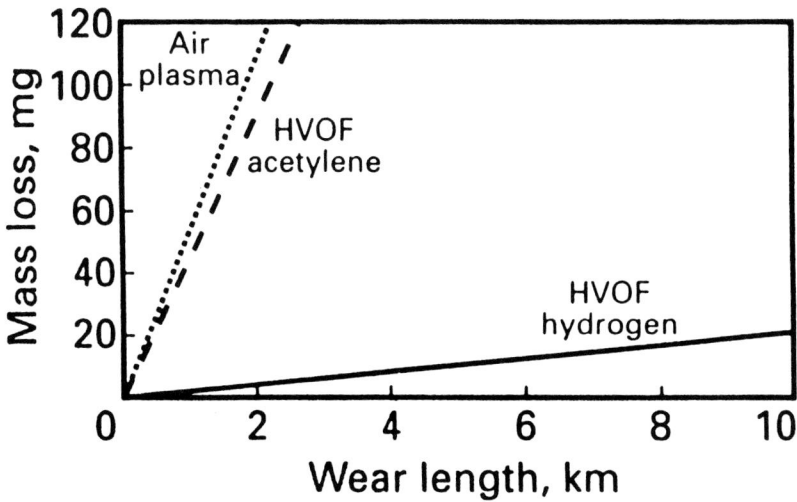

Figure 4.   Abrasive wear test.

performance is quite considerable and is an unexpected and surprising result. Until now, the consensus has been that the HVOF process is not suitable for the deposition of high quality alumina coatings, and that plasma spraying is the preferred process. This work has clearly demonstrated that HVOF must be considered as a competing thermal spraying process for the preparation of high quality alumina coatings.

The reasons for the improved wear performance obtained using the HVOF process with hydrogen fuel are not clear. A higher hardness for the plasma sprayed coating has not resulted in the best wear performance. Both the HVOF (using hydrogen) and plasma coatings consist of primarily γ-alumina with low α-alumina contents and porosity levels. The higher value of elastic modulus measured for the HVOF coating does however suggest that there are structural differences between the coatings. Higher impact velocities associated with HVOF spraying using hydrogen fuel may be giving better inter-lamella bonding, contributing to the higher measured elastic modulus and improved wear performance.

4. DIELECTRIC COATINGS

Work at TWI has demonstrated that the HVOF process can deposit coatings of alumina that have dielectric strengths of 50 to 70kV.mm^{-1} for a nominal 100µm thickness. Some results are summarised in Table 5. These values match if not exceed the values obtained for air plasma sprayed coatings. The surface roughness of the HVOF alumina coatings is typically 1.5µm Ra. This value indicates a surface finish that is suitable for thick film metallisation, and is significantly better than plasma sprayed coatings where 4µm Ra is more typical.

Traditional hybrid power substrates consist of alumina sheet joined to a metallic heat sink often using the direct bonded copper (DBC) route. The alumina dielectric

Table 5. Electrical properties of porcelain enamelled steel compared to those for HVOF deposited coatings.

Property	Alumina 96%	Enamelled steel	HVOF alumina	Plasma alumina
Surface roughness μm Ra	<0.5	<1	1.5	4
Dielectric strength (dc) kV.mm^{-1}	>23	>12 (40μm)	51–74 (160–53μm)	40 (160μm)
Dielectric constant	9.7	7.5–9.0	8.1–8.9	–
Dissipation factor	0.0004	0.003	0.006–0.009	–
Volume resistivity Ωcm	10^{14}	10^{13}	10^{12}–10^{13}	–

has a thickness usually in excess of 600μm. Thermal spraying technology allows new designs for power hybrid substrates to be considered. An example is shown in Fig. 5. An electrical insulating alumina layer can be sprayed directly onto a metal heat sink, with the copper interconnection tracks sprayed onto the alumina coating. HVOF deposited coatings with thicknesses of only 200μm will still give a suitably high dc breakdown voltage for power hybrid substrate applications.

These new designs offer the potential for improvements in thermal dissipation capacity of hybrid substrates by reducing the number of interfaces between power chip and heat sink, and allowing for a reduction in dielectric thickness.

5. HYDROXYAPATITE COATINGS

Coatings of hydroxyapatite, a form of calcium phosphate [$Ca_{10}(PO_4)_6(OH)_2$] with a similar composition to the mineral phase found in bone, are now routinely deposited onto the stems of surgical implants to achieve fixation to bone tissue.[5] Hydroxyapatite

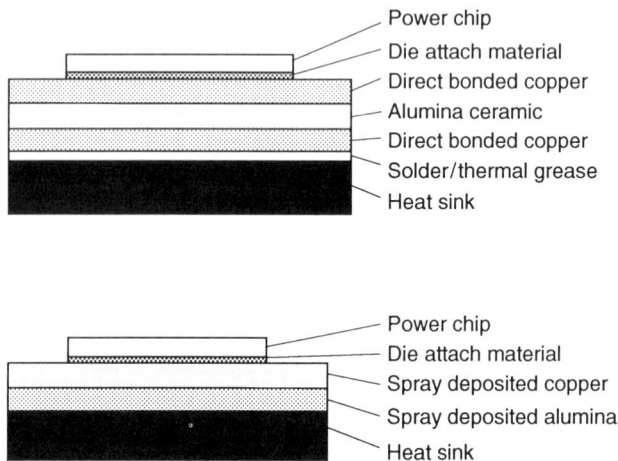

Figure 5.   Thermal sprayed hybrid substrate design.

is classed as a biological active material which encourages bone growth on its surface. Such coatings increase the implant to bone joint integrity and give long term stability. Many processes have been investigated for depositing hydroxyapatite coatings, including electrophoretic deposition, physical vapour deposition, glass enamelling, sol-gel and the most successful to date, air plasma spraying.[6] Hydroxyapatite coatings of thickness 50 to 400μm with a range of porosity levels are prepared by plasma spraying.

Plasma sprayed coatings with predetermined levels of crystallinity can be achieved with correct selection of parameter settings, but the range of these settings is limited. As a consequence, it is difficult to give both control over the level of porosity and degree of crystallinity in the coating. This is largely due to the high temperature of the plasma environment (typically >5000°C) which can lead to decomposition and excessive melting of the hydroxyapatite powder.

An alternative approach is to use a process such as HVOF spraying which offers a lower temperature heating environment. HVOF spraying should allow control of coating crystallinity for a wider range of parameter settings to give more flexibility in controlling other characteristics such as porosity.

Coatings of hydroxyapatite have been prepared[7] with the Top Gun HVOF system using a plasma spraying grade of powder with a nominal particle size of 25–45μm. The two fuel gases acetylene and hydrogen were again used. X-ray diffraction spectrum collected from the powder shows it to be highly crystalline with no amorphous content. Coatings prepared with both acetylene and hydrogen fuel form adherent layers on titanium alloy test pieces. X-ray diffraction spectra for the two coatings reveal that hydroxyapatite is the only crystalline phase present. A difference in the degree of crystallinity between the two coating exists, with the coating prepared using hydrogen fuel having a noticeable amorphous content, whereas acetylene fuel gives a coating with very little amorphous content.

Further work is now under way to establish the extent porosity and structure in the coating can be varied in a controlled manner whilst still retaining the high level of crystallinity.

6. SUMMARY

High velocity oxyfuel thermal spraying offers a competitive process for preparing ceramic coatings. Fuel gas type plays an important role and must be considered in any coating application. The higher impact velocity is believed to give better inter lamella bonding, leading to higher elastic modulus and improved wear performance for alumina coatings. Two newer applications where HVOF spraying is being considered are alumina dialectric coatings for new designs of power hybrid packages and hydroxyapatite coatings for improved biological compatibility of orthopaedic implants.

REFERENCES

1. A. J. Sturgeon, D. A. J. Ramm: *Proc 7th Nat Thermal Spray Conf*, Boston, USA, 20–24 June 1994, 239–244.

2. A. J. Sturgeon, M. D. F. Harvey: *Proc 14th Int Thermal Spray Conf*, Kobe, Japan, May 1995, to be published.
3. A. D. Hewitt: Welding and Metal Fabrication, November 1972, 382–389.
4. R. J. McPherson: *J. Mat. Sci.,* 1980, **15**, 3141–3149.
5. P. J. Ducheyne: *Biomed. Mater. Res. Appl. Biomat.*, 1987, **21**A, 219–236.
6. C. C. Berndt *et al*: *Materials Forum*, 1990, **14**, 161–173.
7. A. J. Sturgeon, M. D. F. Harvey: *Proc 14th Int Thermal Spray Conf*, Kobe, Japan, May 1995, to be published.

Pyrolysis of Al_2O_3-SiC Nanocomposite Powder Using Preceramic Polymer as SiC Precursor

B. SU*, M. STERNITZKE, C. E. BORSA and R. J. BROOK

*Department of Materials, University of Oxford, Parks Road, Oxford,
OX1 3PH, UK*

ABSTRACT

An approach to fabricate Al_2O_3/SiC nanocomposite powder is exploited by mixing ultrafine alumina powder with Si-polymeric precursor and converting the polymer to fine grained SiC particle at high temperature. Three SiC preceramic polymers were used in this study, including polycarbosilane, polysilane and polysilazane. The structural development of polymeric precursors and polymer derived Al_2O_3/SiC nanocomposite powders during pyrolysis process have been investigated by means of FT-IR, XRD and TEM. The results show that the pyrolytic behaviour of the polymers is different depending on polymer type and pyrolysis conditions. Controlling the polymers from oxidation during pyrolysis is important to get a uniform, fine-grained Al_2O_3/SiC nanocomposite powder. Pyrolysis under reducing atmosphere (H_2–Ar) is preferred to avoid mullite formation which is due to the polymer oxidation. The preceramic polymers which were coated on alumina powder will undergo a transformation from amorphous Si–C–O or Si–C–N–O network to crystalline SiC particles with the increase of pyrolysis temperature. β-SiC whiskers were observed in the pyrolysis products at 13000°C under H_2–Ar, while Al_2O_3/SiC nanocomposite powder with SiC particle size below 20 nm were obtained at 1500°C.

1. INTRODUCTION

Nanocomposites are materials in which at least one of the phases has grain size at the nanometer scale. Materials with such nanosized grains are reported to have some unusual and interesting properties, for example, enhanced mechanical properties.[1,2] However, ceramic composites which have grain size less than 200 nm are difficult to prepare by conventional powder processing methods due to the problem of agglomeration. Various attempts have been made to fabricate ceramic nanocomposites with finer grains and better homogeneity. Among them, sol-gel and polymer pyrolysis methods offer the potential for improved compositional homogeneity due to the preparation from liquid solutions and the homogeneous mixing of the precursor at the molecular level, and have been used to produce Al_2O_3/SiC nanocomposites.[3,4] But it is difficult to get fully densified ceramic bodies.[5]

---

* Now at IRC in Materials for High Performance Applications and School of Metallurgy and Materials, University of Birmingham, Edgbaston, Birmingham, B15 2TT, UK.

An alternative method has been exploited to fabricate Al_2O_3/SiC nanocomposites by the polymer precursor route. The method involves blending commercial ultrafine alumina powder with preceramic polymer which provides the SiC source and converting the mixture to Al_2O_3/SiC nanocomposite powder by subsequent pyrolysis and fabricating the ceramic from such a powder using hot-pressing. Fully dense nanocomposites with well dispersed SiC particles have been fabricated.[6] In this paper, we utilise three different preceramic polymers (polycarbosilane, polysilane and polysilazane) as the SiC precursors to produce nanosized SiC particles. The structural evolution of the polymers during pyrolysis was investigated by FT-IR using both diffuse-reflectance and transmittance mode. X-ray diffraction was used to characterise the crystallisation sequence of the polymer derived SiC and Al_2O_3/SiC nanocomposite powders. The structural change of the nanocomposite powders was investigated by TEM in combination with EDX.

2. EXPERIMENTAL PROCEDURE

2.1 Materials

Three preceramic polymers were used in our study. Polycarbosilane (PCS, Changsha Institute of Technology, China), synthesised using a similar method to Yajima,[7] is widely used to produce SiC fibers. Polysilastyrene (PSS, ABCR GmbH & Co KG, Germany), which is a copolymerised alkyl/aryl-substituted polysilane, is reported as generating particulate SiC.[8] Polyvinylmethylsilyhydrazine (PV, Elf Atochem, France), which is a kind of liquid polysilazane, can be converted to an Si–C–N ceramic and eventually to SiC at high temperatures.[9] Their representative structures are shown in Fig. 1. Commercial ultrafine alumina powder (AKP53, Sumitomo Co., Japan) with average particle size of 200 nm was used.

2.2 Nanocomposite Powder Preparation

The alumina powder was first mixed with polymer solutions dissolved in xylene by ultrasonic methods. The weight ratio of alumina powder to polymer is 15 so the

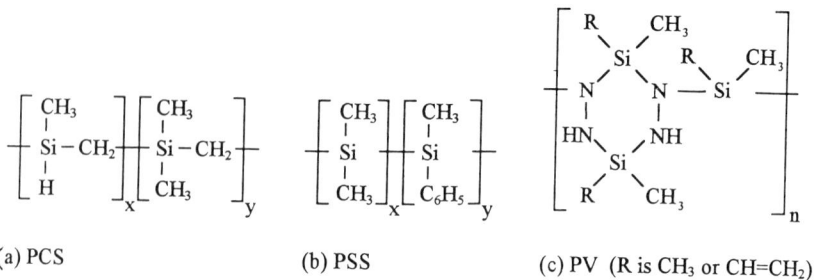

(a) PCS   (b) PSS   (c) PV (R is CH_3 or $CH=CH_2$)

Figure 1.   Typical structure of the polymer precursor used for the SiC source: (a) PCS, (b) PSS, (c) PV.

volume fraction of SiC residue is 5% assuming 60 wt% yield of the SiC precursors. The organic solvent was evaporated and the powder/polymer mixtures and the polymers were put into a silicate tube for crosslinking at 450°C for 2h under argon separately. After crosslinking, both mixtures and polymers were broken and sieved with a 250 micron mesh. Pyrolysis of these powders was carried out in an alumina tube under either inert (argon and nitrogen) or reducing atmosphere (1% hydrogen–argon). The heating rate was typically 5°C min^{-1} and the pyrolysis temperatures were from 1000 to 1700°C. The holding time was 2h. Weight loss was measured for three polymers during crosslinking and pyrolysis. The pyrolysed powders were then used for further characterisation.

2.3 Characterisation

Both diffuse-reflectance and transmittance FT-IR spectra were obtained on a Fourier transform spectrophotometer using powder and KBr pellet samples respectively. X-ray diffraction (XRD) data were obtained using monochromatic CuKα radiation on a Philips PW 1150/70 X-ray diffractometer. The mean grain size (L) of the β-SiC crystalline phase was calculated from the Sherrer equation: L=0.9λ/Bcosθ, where B is the diffraction line-broadening measured as the line width of half the maximum intensity (in radians) and λ is the Cukα wavelength (i.e. λ=0.154 nm) and θ is the Bragg angle (θ=17.8° for β–SiC(111)). Microstructural features of pyrolysed powders were characterised by transmission electron microscopy (Philips CM20) with an X-ray energy dispersive spectroscopy (EDX) analyser attachment (Link AN10000).

3. RESULTS AND DISCUSSION

The IR transmitance spectra changes of PCS, PSS and PV during crosslinking and pyrolysis are shown in Fig. 2. Each polymer exhibits a different structural change. For PCS, the peak at 2100 cm^{-1} (Si–H stretching) decreased slightly after crosslinking and only peaks in the bands of 800 cm^{-1} (Si–C stretching), 1080 cm^{-1} (Si–O stretching) and 470 cm^{-1} (Si–O–Si deformation) existed in the pyrolysis products. For PSS, the change is quite different during the crosslinking stage. A strong peak at 1020 cm^{-1} (CH$_2$ wagging in Si–CH$_2$–Si) and an increase in 2100 cm^{-1} indicated a transformation reaction from PSS to PCS structurer with a formal insertion of methylene into an Si–Si bond during crosslinking, which agrees with the experimental observation that large amounts of gaseous reflux formed in the tube resulting in the weight loss of 74.7% in comparison with PCS of 28.4%. When pyrolysed at temperatures below 1300°C, the peak corresponding to Si–O stretching (1080 cm^{-1}) was very strong, probably due to the oxygen contained in the polymer which exhibited high sensitivity towards oxidation during PSS to PCS transformation in the 100 to 450°C temperature range.[10] It was noticed that a peak around 3500 cm^{-1} (H$_2$O) existed in the as-received PSS, which would form silanol groups, Si–OH during crosslinking, as evident from the IR spectra of crosslinked PSS in which a broad peak appeared around 3200–3700 cm^{-1} (Si–OH stretching). However, when the pyrolysis temperature was raised to

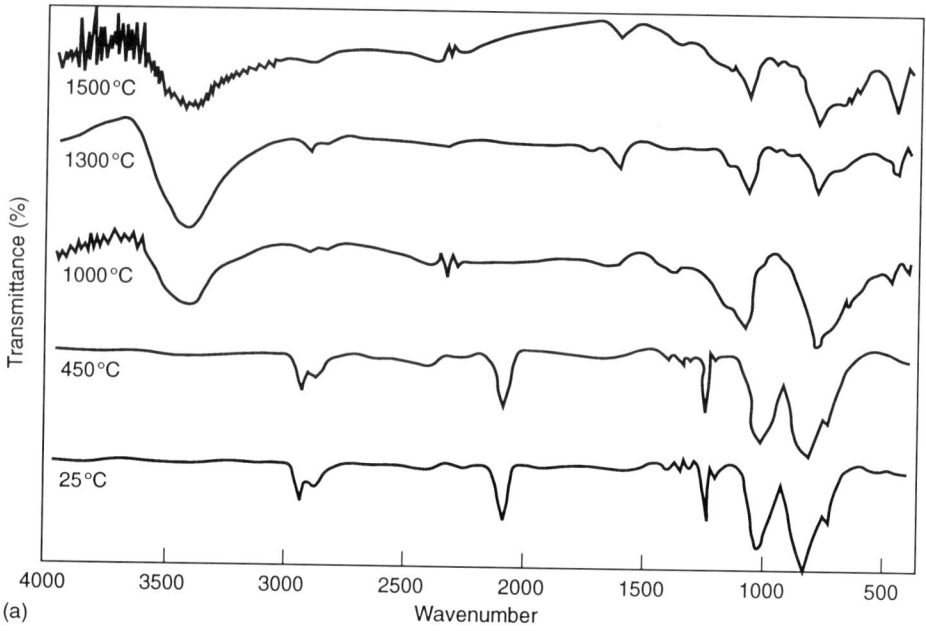

Figure 2. FT-IR transmitance spectra of preceramic polymers after crosslinking and pyrolysis at different temperatures: (a) PCS, (b) PSS, (c) PV.

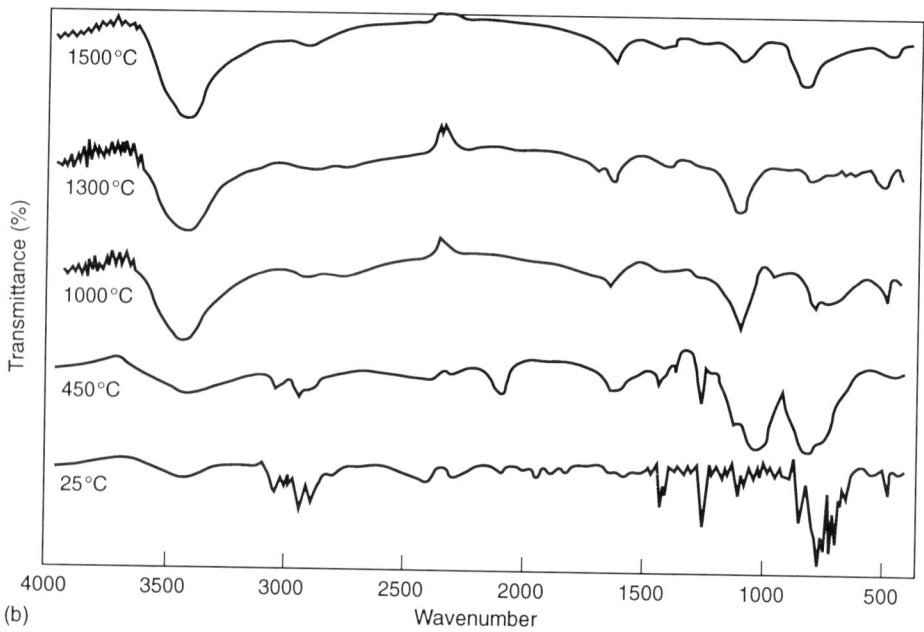

Figure 2(b).

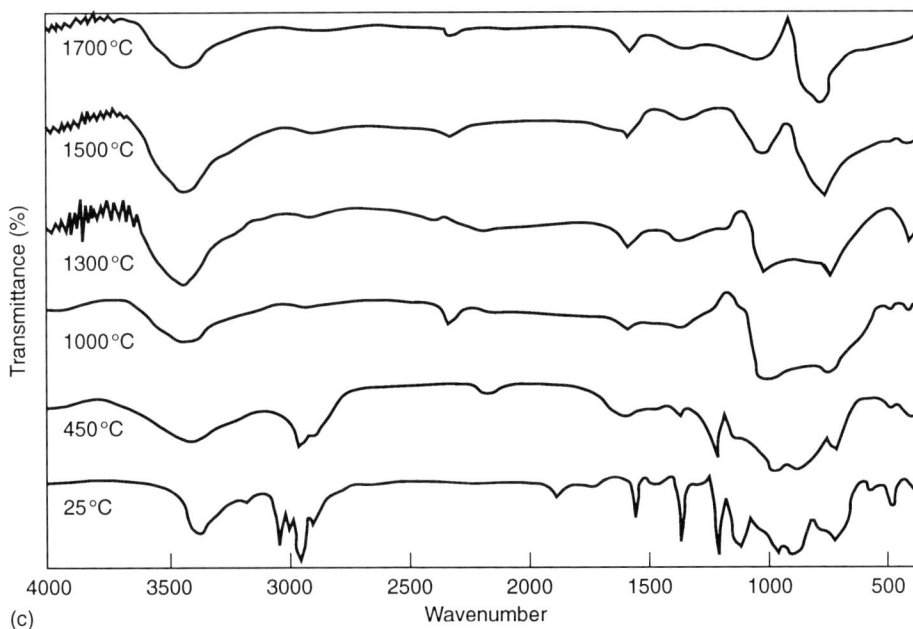

Figure 2(c).

1500°C, a dramatic change in Si–O and Si–C peaks (1080 and 800 cm^{-1}) occurred due to the carboreduction reaction which is less significant for PCS. For PV, all vinylic-bands (3050, 3010, 1600, 1410 cm^{-1}) disappeared but the Si–H band at 2100 cm^{-1} appeared after crosslinking. When the pyrolysis temperature was below 1300°C, a broad peak in the range of 750 to 1100 cm^{-1} is overlapped by Si–C (800 cm^{-1}), Si–O (1080 cm^{-1}) and Si–N–Si (790, 950, 1010 cm^{-1}) bands. When the pyrolysis temperature was above 1300°C, only Si–C and Si–O bands remained, indicating that the Si–C–N–O network underwent a decomposition process with the evolution of nitrogen present in the material,[11] which is also evident from the pyrolysis yield measurement, where PV exhibited a second large weight loss (about 15%) after pyrolysis at 1500°C.

The XRD patterns of the three polymers pyrolysed at different temperatures under H$_2$-Ar are shown in Fig. 3. The results are consistent with the IR spectra. All polymers will produce both α– and β–SiC particles at temperature above 1300°C but the crystallisation processes are different in each case. For PSS, a small cristobalite peak appeared at 1300°C and disappeared at 1500°C. The pyrolysis residue of PCS is a mixture of SiC crystals and free carbon (graphite). While for PV, the crystallisation process was delayed due to the presence of nitrogen. The pyrolysis product of PV is still amorphous when the pyrolysis temperature is 1300°C. The particle size calculated from crystalline β–SiC (111) is less than 20 nm when the pyrolysis temperature is below 1500°C.

It is interesting to note that the FT-IR spectrum is somewhat different when using diffuse-reflectance mode, as illustrated in Fig. 4, A strong Si–O band (1100 cm^{-1}) and

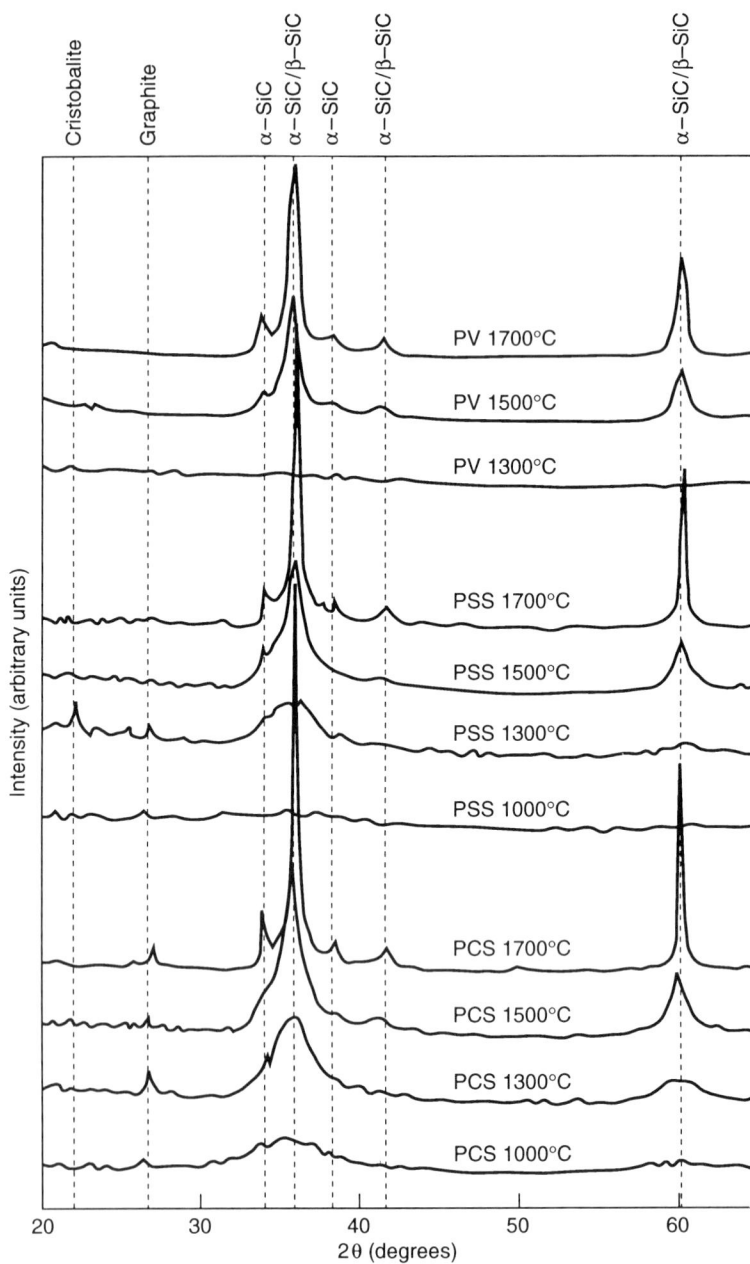

Figure 3.   XRD patterns of the polymers pyrolyed in the temperature range 1000–1700°C
for 2h under 1% H$_2$-Ar atmosphere.

a relatively weak Si–C band (800 cm^{-1}) can be seen in the FT-IR spectra of the pyrolysis product of PCS, in contradiction to the transmittance spectrum (Fig. 2(a)). This indicates that PCS derived SiC particles are oxygen-rich on the surface because the diffuse-reflectance technique is more surface sensitive than the transmittance technique where the IR beams transmit through the KBr pellet. This feature is reported to be a common occurrence in SiC powders especially in ultrafine SiC powders.[12]

We also investigated the influence of different atmospheres (Ar, N$_2$, 1% H$_2$–Ar) during the pyrolysis. XRD and FT-IR analysis show that only the reducing atmosphere (1% H$_2$–Ar) can lower the oxygen content and avoid SiO$_2$ formation. Oxidation was found in the case of both Ar and N$_2$ atmosphere. Correspondingly, mullite was observed in the pyrolysis products of the polymer/alumina powder if cristobalite formed in the polymer derived products. In this case, no SiC particles were observed by TEM investigation. Therefore, it is very important to control the oxygen content in the product to obtain Al$_2$O$_3$/SiC nanocomposite powders.

Figure 5 shows TEM micrographs of the nanocomposite powder using PCS as the SiC precursor at different pyrolysis temperatures under H$_2$–Ar. No crystalline silicon containing particles were found by EDX in the powder pyrolyed at 1000°C. The alumina grains are coated by an amorphous Si–C–O layer. The occurrence of the layer is confirmed by a particular EDX investigation which shows an equal silicon distribution without a specific aggregation. Whiskers formed as the pyrolysis temperature was raised to 1300°C. The whiskers are largely distorted β-SiC as determined by selected area electron diffraction. The SiC whiskers have a width of only 10 to 100 nm and a length up to 1 μm. The nanocomposite powder also contains some

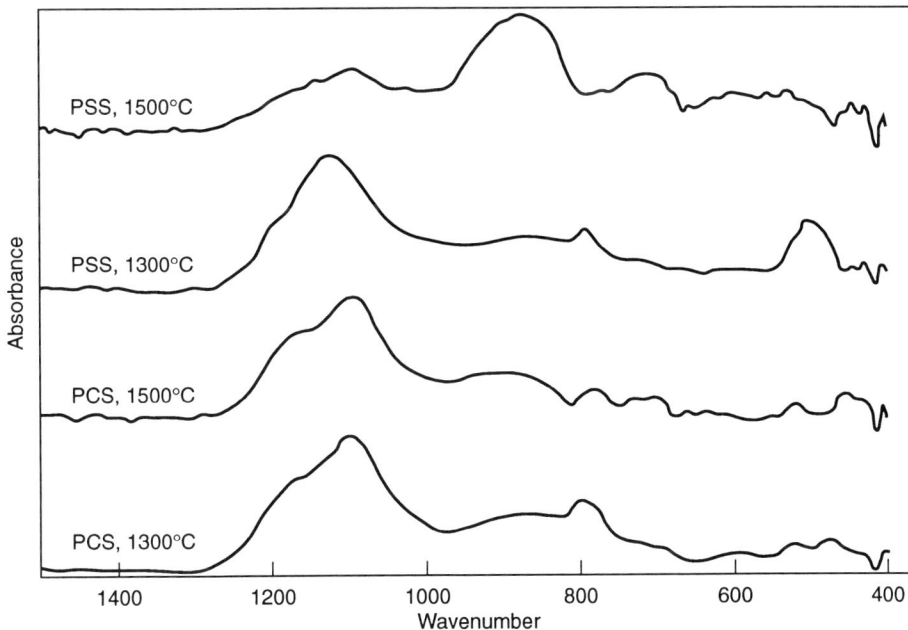

Figure 4.   FT-IR diffus-reflectance spectra of the pyrolysis products of PSS and PCS.

very tiny silicon rich particles corresponding to the start of β–SiC crystallisation. The pyrolysis at 1500°C yields a powder with well distributed, nanosized SiC crystallites as shown in Fig. 5(c). Each alumina grain is surrounded by a number of SiC particles with the particle size less than 20 nm, consistent with XRD results (Fig. 3). No whiskers, however, formed at this stage.

The mechanism of SiC whisker formation is thought to be by vapour-liquid-solid (VLS) process in which materials are transported through the vapour phase. Previous studies on thermal behaviour of Nicalon fibres and PCS-derived Si–C–O ceramics have shown that they will undergo the following reactions with a large decrease in the oxygen and silicon content when the temperature is above 1250°C:[13–15]

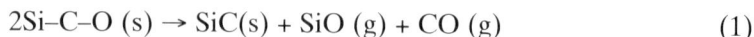

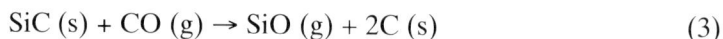

$$2Si–C–O \text{ (s)} \rightarrow SiC(s) + SiO \text{ (g)} + CO \text{ (g)} \tag{1}$$

$$SiO_2 \text{ (s)} + 3C \text{ (s)} \rightarrow SiC \text{ (s)} + 2CO \text{ (g)} \tag{2}$$

$$SiC \text{ (s)} + CO \text{ (g)} \rightarrow SiO \text{ (g)} + 2C \text{ (s)} \tag{3}$$

In the case of hydrogen gas evolved during pyrolysis, free carbon, formed in the PCS derived ceramic as shown in Fig. 3, will react with hydrogen gases to form hydrocarbons:

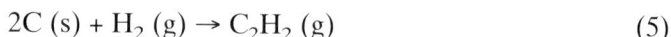

$$C(s) + 2H_2 \text{ (g)} \rightarrow CH_4 \text{ (g)} \tag{4}$$

$$2C \text{ (s)} + H_2 \text{ (g)} \rightarrow C_2H_2 \text{ (g)} \tag{5}$$

Thermodynamically, SiC whiskers can readily grow by the VLS process through the following reactions:[16,17]

$$SiO \text{ (g)} + CH_4 \text{ (g)} \rightarrow SiC \text{ (s)} + H_2 \text{ (g)} + H_2O \text{ (g)} \tag{6}$$

$$SiO \text{ (g)} + 2CH_4 \text{ (g)} \rightarrow SiC \text{ (s)} + CO \text{ (g)} + 4H_2 \text{ (g)} \tag{7}$$

$$SiO \text{ (g)} + C_2H_2 \text{ (g)} \rightarrow SiC \text{ (s)} + CO \text{ (g)} + H_2 \text{ (g)} \tag{8}$$

$$SiO \text{ (g)} + C_2H_2 \text{ (g)} \rightarrow SiC \text{ (s)} + C \text{ (s)} + H_2O \text{ (g)} \tag{9}$$

$$SiO \text{ (g)} + CO \text{ (g)} + 2H_2 \text{ (g)} \rightarrow SiC \text{ (s)} + 2H_2O \text{ (g)} \tag{10}$$

TGA investigation[14] showed that a high weight loss due to reactions (1–3) occurred most rapidly in the temperature range between 1250 and 1500°C. This explains why SiC whiskers can only be observed at pyrolysis temperatures of 1300°C in our experiments. Because the heating rate during pyrolysis is 5°C min⁻¹, the temperature range in which whiskers can grow by the VLS process is passed very rapidly in the case of pyrolysing the nanocomposite powder at 1500°C. Whisker formation is hindered because of kinetic reasons.

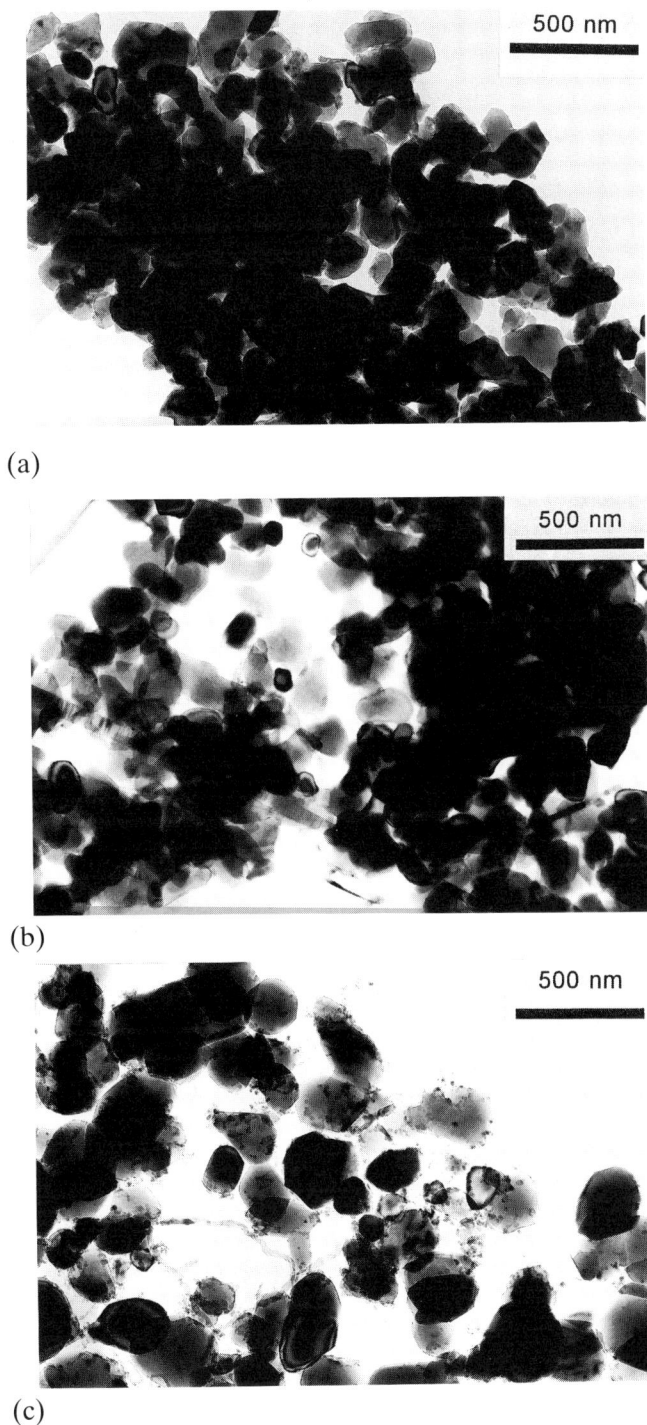

(a)

(b)

(c)

Figure 5.   TEM micrographs of alumina/PCS powders pyrolysed under 1% H$_2$-Ar at different temperatures: (a) 1000°C, (b) 1300°C, and (c) 1500°C.

4. CONCLUSIONS

Al_2O_3/SiC nanocomposite powder can be successfully prepared by using preceramic polymer as the SiC precursor if the oxidation is properly controlled during polymer pyrolysis. Pyrolysis under a reducing atmosphere (e.g. 1%H_2–Ar) is preferred to avoid mullite formation in nanocomposite powders. The preceramic polymers coated on the alumina powder will undergo a structural transformation from amorphous Si–C–O or Si–C–N–O networks to crystalline SiC particles with increase of pyrolysis temperature. β–SiC whiskers formed via the VLS process when samples were pyrolysed at 1300°C, due to the thermal decomposition of the polymers with oxygen contamination. Al_2O_3/SiC nanocomposite powders with SiC particle size less than 20 nm have been obtained by pyrolysis at 1500°C in an H_2–Ar atmosphere.

ACKNOWLEDGMENTS

The polycarbosilane used in this study is generously provided by Changsha Institute of Technology. Thanks are due to Dr. P. Dobson and G. Pethybridge for assistance with FT-IR spectroscopy. One of the authors (Bo Su) would like to thank the Department of Materials at Oxford University for financial support during his stay in Oxford.

REFERENCES

1. K. Niihara: *J. Ceram. Soc. Jpn.,* 1991, **99** (10), 974–982.
2. J. Zhao, L. C. Stearns, M. P. Harmer, H. M. Chan, and G. A. Miller: *J. Am. Ceram. Soc.,* 1993, **76** (2), 503–510.
3. R. S. Haaland, B. I. Lee, and S. Y. Park: *Ceram. Eng. Sci. Proc.,* 1987, **8** (7–8), 879–885.
4. Y. Xu, A. Nakahira, and K. Niihara: *J. Ceram. Soc. Jpn.,* 1994, **102**(3), 312–315.
5. R. Telle, R. J. Brook, and G. Petzow: *J. Hard Mater.,* 1991, **2** (1–2), 79–113.
6. B. Su and M. Sternitzke: *Fourth Euro-Ceramics, Vol. 4,* A. Bellosi ed., Grupps Editoriale Faenza Editrice S.P.A., Italy, 1995, 109–116.
7. S. Yajima, J. Hayashi, M. Omori, and K. Okamura: Nature, 1976, 261, 683–685.
8. R. Riedel, K. Strecker, and G. Petzow: *J. Am. Ceram. Soc.,* 1989, **72**(11), 2071–77.
9. C. Colombier: In Euro-Ceramics, Vol. 1, Edited by G. de With et al., Elsevier Applied Science, London, UK, 1989, 43–52.
10. M. Scarlete, et al.: *Chem. Mater.,* 1994, **6**, 977–982.
11. D. Mocaer, et al.: *J. Mater. Sci.,* 1993, **28**, 2615–2631.
12. R. Vaben, and D. Stover: *ibid.,* 1994, **29**, 3791–3796.
13. T. Shimoo, M. Sugimoto, and K. Okamura: *J. Jpn. Inst. Metals,* 1990, **54**(7), 802–808.
14. D. Bollollon, et. al.: *J. Mater. Sci.,* 1993, **26**, 1517–1530.
15. O. Delverdier et al.: *J. Eur. Ceram. Soc.,* 1993, **12**, 27–41.
16. M. Maeda, T. Funahashi, and R. Uchimura: J. Ceram. Soc. Jpn., 1989, **97** (12), 1505–10.
17. M. Saito, S. Nagashima, and A. Kato: J. Mater, Sci. Lett., 1992, **11**, 373–376.

Design of Microstructures Using the Core-Shell Concept

C. R. BOWEN and R. STEVENS

School of Materials, University of Leeds, Leeds, LS2 9JT, UK

ABSTRACT

Core-shell structures consist of a central core of one phase surrounded by a shell of another phase or further phases and have been observed in a variety of ceramic systems. The core-shell concept presents a novel approach for the design of microstructure to enhance material properties. This paper discusses the production of core-shell zirconia (outer shell cubic and inner core monoclinic), which produces a material with a combination of high thermal shock resistance and high ionic conductivity. Preliminary sintering and thermal shock results are described.

1. INTRODUCTION

'Core-shell' is a term used in the field of electroceramics to describe grains which consist of a central core of one phase surrounded by a shell of another phase or further phases.[1] The most common example is that of $BaTiO_3$ which has been modified by a variety of additives.[2-6] Microstructural observations have revealed that grains consist of a core of pure $BaTiO_3$ into which no additives are diffused and a shell of modified $BaTiO_3$ which contains a high concentration of the additives. It has been proposed that differences in the lattice paramaters between the core and the shell develop a highly stressed state within the grains. For Zr-modified $BaTiO_3$ the core is placed in compression so that the outer shell is tetragonal while the core is pseudo cubic perovskite.[7] This leads to a flat permittivity response with respect to temperature, due to a distribution of Curie points within the structure. In addition to $BaTiO_3$, similar structures have also been observed in yttria tetragonal zirconia polycrystalline ceramics produced using a gas phase route.[8,9]

Although core-shell structures are not new, the design of specific microstructures using the core shell concept is a novel idea for the ceramic scientist. It may be possible to fabricate a precursor powder that produces a core-shell structure which has been designed for a specific purpose, e.g. to enhance mechanical or electronic properties.

The aim of this paper is to discuss how the core-shell concept can be used to fabricate zirconia ceramics with enhanced thermal shock properties. The high melting point of zirconia make it a candidate material for many high temperature applications. However, the monoclinic (*m*) form of zirconia, which is stable at room temperature, undergoes a phase change at ~1150°C to the tetragonal (*t*) form. The ~4% volume expansion associated with the *t*→*m* phase change is usually enough to shatter zirconia bodies which have been sintered to a high density.

One possible solution to this problem is to form solid solutions of cubic c-ZrO_2 with MgO, CaO or Y_2O_3. When the additives are in sufficient quantity they can completely stabilise the c-ZrO_2 structure and avoid the problems associated with the $t{\rightarrow}m$ transformation. The disadvantage of fully stabilised zirconia is that the thermal shock resistance is low due to its high coefficient of expansion ($\sim$7-12$\times$10^{-6}/°C) and low thermal conductivity ($\sim$1.5 Wm^{-1}K^{-1}).

Curtis[10] was able to improve the thermal shock resistance of zirconia by producing bodies consisting of stabilised cubic and pure monoclinic zirconia grains. The effect of the phase transformation on cooling was not enough to crack the body as the monoclinic form was not present in sufficient quantity. The high thermal shock resistance was originally thought to be due to the low thermal expansion of the material. Later work showed that the thermal shock resistance was improved due to the presence of microcracks produced by the $m{\rightarrow}t$ transformation.[11] The high thermal stress resistance of microcracked brittle ceramics is due to the effect of cracks on elastic behaviour, i.e. microcracked materials have elastic moduli lower than crack free materials.[12]

The purpose of this work is to produce a highly microcracked zirconia body with a high monoclinic content, based on the ZrO_2–MgO–SiO_2 (ZMS) system. Using the core-shell concept the ultimate microstructure for thermal shock resistance would be that as shown in Fig. 1. Grains would consist of a core of pure monoclinic zirconia, while the shell consists of a stabilised cubic zirconia. This would produce grains where the inner core is highly microcracked while the outer shell sinters to high density. In addition to a high thermal shock resistance, the structure in Fig. 1 would also have a high ionic conductivity.

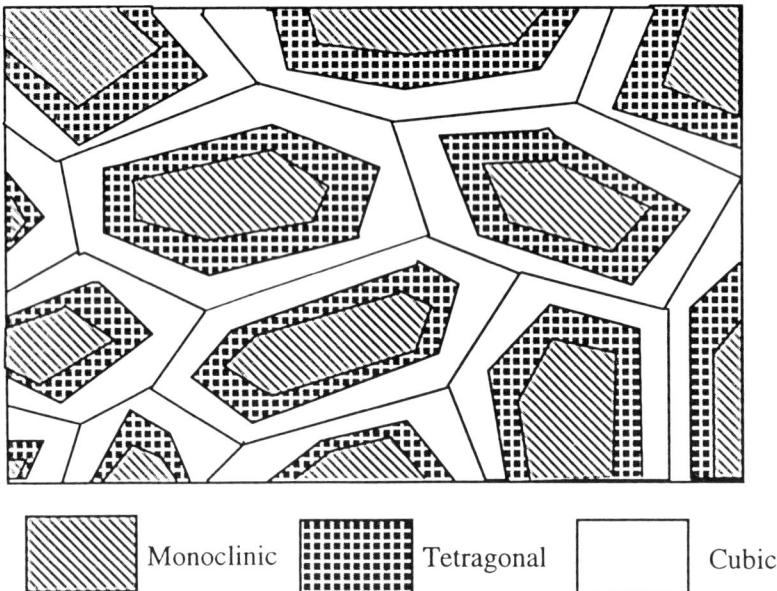

Figure 1.   Core-shell structure design for a ceramic with a combination of ionic conductivity and thermal shock resistance.

Table 1.   Composition of ZMS ceramics

Oxide	Content/wt.%
ZrO_2	96.55
MgO	2.45
SiO_2	1.10

molar ratio MgO/SiO_2 = 3.32

2. EXPERIMENTAL

The mechanical properties of the ceramic are likely to be highly dependent on the core-shell structure (grain size, shell size etc.). Variation of the core-shell structure can be attained by variation of the thermal treatment (such as calcination temperature or sintering temperature), or by modification of the initial particle size. Two zirconia powders were used from MEL Chemicals, Grade S (D_{50} = 14µm) and Grade SC15 (D_{50} = 0.4µm). MgO and SiO_2 were used as additives in the amounts shown in Table 1. It is thought that the additives form a liquid phase sintering aid which surrounds the ZrO_2 grains. During sintering the MgO may stabilise the outer shell of the ZrO_2 grains and thus produce the desired core-shell structure. As more MgO stabilises the ZrO_2 the concentration of MgO in the liquid phase decreases, which leads to an increase in the melting temperature and solidifcation to produce an intergranular glass phase.

ZrO_2–MgO–SiO_2 (ZMS) powder mixtures using the zirconia powders were prepared by ball milling in propan-2-ol for 4 hours. PEG200 binder (2 wt%) was then added and the mixture was ball milled for a further hour. After drying, the powders were sieved at 100µm and uniaxially cold pressed. Pressed green bodies were sintered for 1 hour at various temperatures.

The final density of sintered samples was estimated using the Archimedes principle. Fracture strength was measured using biaxial ball on ring testing with a minimum of 8 specimens for each measurement.[13] Sample diameters were approximately 13mm and 1.5mm thick and test surfaces were polished to 3µm. Thermal shock experiments were carried out by quenching heated samples in water at 25°C and subsequently testing the strength. Fracture surfaces and microcracking was examined using scanning electron microscopy (SEM).

It would be expected that the strength of the ZMS bodies would be sensitive to the final sintered density due the presence of a high volume fraction of monoclinic zirconia. Initial work therefore concentrated on examining the effect of sintering temperature on strength, density and microstructure. Both Grade SC15 ZMS and Grade S ZMS mixtures were sintered to a variety of densities using different sintering temperatures for 1h. In addition, the Grade S ZMS powder was calcined at 1000°C for 2h to investigate the effect of calcination on sintering behaviour.

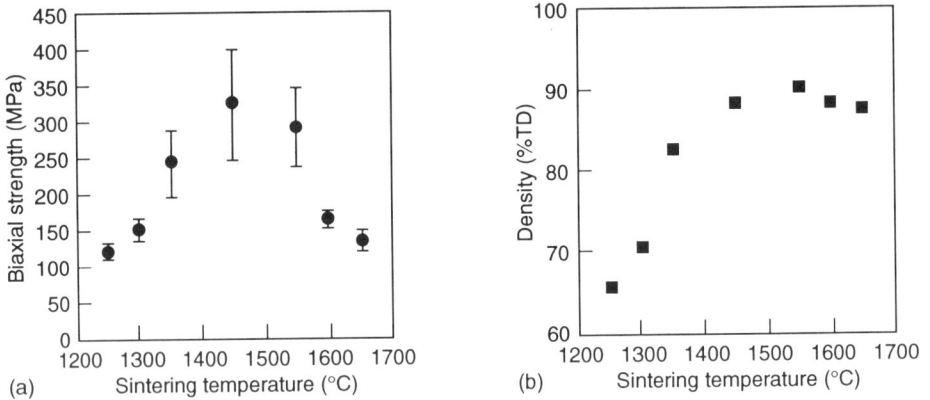

Figure 2.   Variation of strength (a) and density (b) with sintering temperature (1h) for
Grade SC15 ZMS ceramics.

3. RESULTS AND DISCUSSION

3.1 Grade SCIS ZMS Ceramics

Figure 2 shows the effect of sintering temperature on strength and density for the ZMS bodies using Grade SC15 zirconia. Initially the strength and density increase with sintering temperature. However, above a critical sintering temperature of 1450°C h^{-1} (σ=325MPa and ρ=90%TD), the strength begins to decrease with increasing sintering temperature. This can be explained by examining the microstructure of the samples as shown in Fig. 3. On increasing the sintering temperature to 1450°C the strength increase is due to an increase in density (Fig. 3a). Above 1450°C the higher density body is subjected to larger stresses as the ceramic cools through

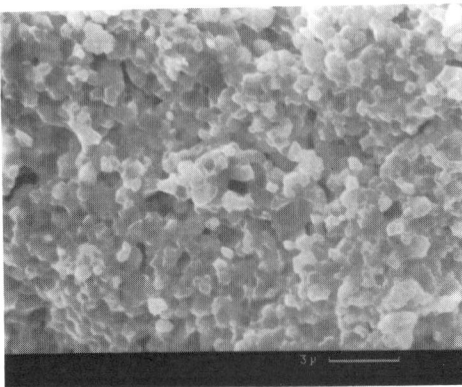

Figure 3(a).   Fracture surface SEM of Grade SC15 ZMS at the optimum sintering
condition (1450°C 1h^{-1}). No microcracks are observed.
(b) Higher sintering temperature (1650°C 1h^{-1}) produces macrocracks which decrease strength.

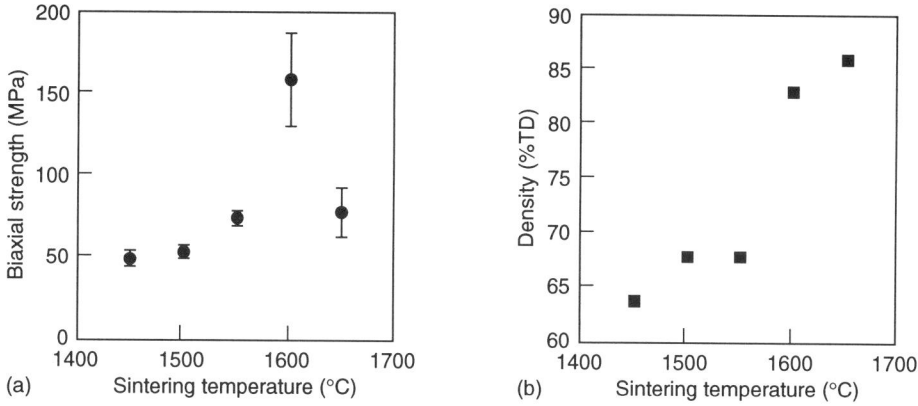

Figure 4.   Variation of strength (a) and density (b) with sintering temperature for Grade S ZMS ceramics.

the $t{\to}m$ temperature. This leads to the formation of macrocracks which can be seen in the samples sintered at 1650°C/1h (Fig. 3b).

3.2 Grade S ZMS Ceramics

The variation of strength and density with sintering temperature for the Grade S ZMS material is shown in Fig. 4. Again the strength begins to decrease after exceeding a critical density. The maximum strength is approximately 150MPa and occurs at a density of 85% theoretical. The lower maximum strength is due to the lower density, larger grain size and microcracking which is observed along grain boundaries (Fig. 5a). Above the critical density the microcracks coalesce to form macrocracks which cause a significant decrease in strength (Fig. 5b).

Figure 5(a).   Fracture surface SEM of Grade S ZMS at the optimum sintering condition (1600°C/1h). Note microcracks along grain boundaries.
  (b) Grade S ZMS sintered at 1650°C 1h^{-1}. Macrocracks are observed.

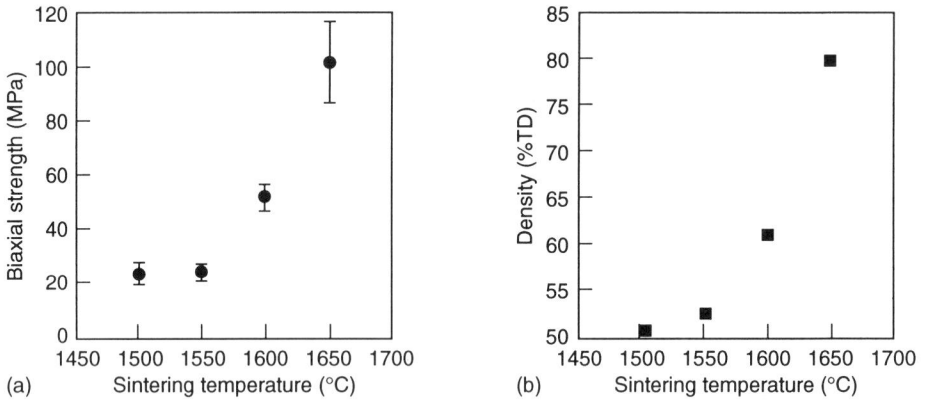

Figure 6.   Variation of strength (a) and density (b) with sintering temperature for Grade S ZMS ceramics (calcined 1000°C 2h^{-1}).

The larger grain size may explain the lower critical density, when compared to Grade SC15 ZMS, as there is a decrease in the surface area in contact with the intragranular MgO/SiO$_2$ phase. This results in a lower volume fraction of stabilised zirconia.

3.3 Calcination of Powder to Sinter at Higher Temperature

In order to increase the thickness of the stabilised shell, the amount of MgO which has diffused into the ZrO$_2$ must increase. This could be achieved by increasing the sintering temperature or time; but this would lead to a decrease in the strength, as observed in Figs 2 and 4. Calcination of the powder prior to sintering decreases the sintering rate as it causes an increase in crystallite size. The effect of calcination (1000°C/2h) on the Grade S ZMS powder prior to sintering is shown in Fig. 6. Comparing Figs 4 and 6 it is observed that the calcined powder requires a higher

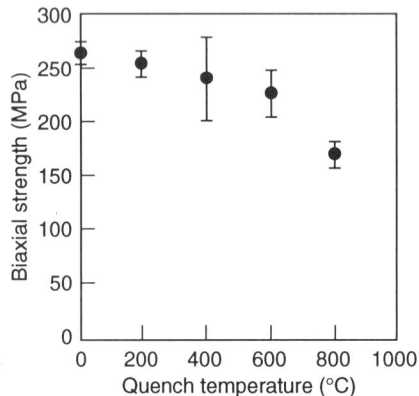

Figure 7.   Thermal shock resistance of Grade ZMS ceramic (1450°C1h^{-1})

sintering temperature to attain a similar density and strength to that of the non-calcined powder. This allows further control over the core-shell structure as sintering at a higher temperature is possible without the formation of macrocracks.

3.4 Thermal Shock Resistance

Thermal shock results for the Grade SC15 ZMS ceramic (1450°C 1h⁻¹) are shown in Fig. 7. There is no evidence of a critical ΔT_c at which a significant reduction in strength occurs, which is observed in many ceramic systems. There seems to be only a slight degradation of strength with increasing severity of quenching. These observations are consistent with stable crack propagation which occurs in highly microcracked materials.[14]

4. CONCLUSIONS

The core-shell approach has been used in an attempt to design zirconia material with enhanced thermal shock resistance. The strengths of the ZMS ceramics are sensitive to the final sintered density, due to the high volume fraction of monoclinic zirconia. Above a critical density, the expansion associated with the $t{\to}m$ transformation on cooling leads to the formation of a high microcrack density and macrocracks which significantly decrease the strength. The critical density is higher for the smaller particle size zirconia, possibly due to the greater volume fraction of stabilised zirconia. Initial thermal shock testing has been promising. The ZMS composition tested has shown no evidence of a critical ΔT_c and the ceramic undergoes stable crack propagation. Further work will concentrate on investigation of the microstructure using transmission electron microscopy and the effect of the variation of core-shell structure on thermal shock resistance.

REFERENCES

1. K. Okazaki and S. Kashiwabara: *J. Jpn. Ceram. Soc.,* 1965, **73**, 834.
2. C. A. Randall, S. F. Wang, D. Laubscher, J. P. Dougherty and W. Huebner: *J. Mater. Res.,* 1993, **8**(4), 871.
3. S. Pathumarak, M. Al-Khafaji and W. E. Lee: *Brit. Ceram. Trans.,* 1994, **93**(3), 114.
4. D. Hennings and G. Rosenstein: *J. Am. Ceram. Soc.,* 1984, **67**, 249.
5. T. R. Armstrong and R. C. Buchanan: *J. Am. Ceram. Soc.,* 1990, **73**(5), 1268.
6. C. Saucy, I. M. Reaney and A. J. Bell: 'Nanoceramics', *Brit. Cer. Proc., 51*, R. Freer, ed., 1994, 31–52.
7. T. R. Armstrong, L. E. Morgens, A. K. Maurice and R. C. Buchanan: *J. Am. Ceram. Soc.,* 1989, **72**, 605.
8. S. Lawson, G. P. Dransfield, A. G. Jones, P. McColgan and W. M. Rainforth: 'Enhanced performance of Y-TZP materials through compositional zoning on a nano-scale', *8th CIMTEC World Ceramic Congress*, Florence, June 1994.
9. G. P. Dransfield et al.: 'Microstructural engineering through gas phase synthesised Y-TZP powders', *8th CIMTEC World Ceramics Congress*, Florence, June 1994.

10. E. C. Curtis: *J. Am. Ceram. Soc.,* 1947, **30**, 180.
11. R. C. Garvie and P. J. Nicholson: *J. Am. Ceram. Soc.,* 1972, **55**, 152.
12. D. P. H. Hasselman, J. Singh: *Ceram. Bull,* 1979, **58**(9), 856.
13. G. de With and H. H. M. Wagemans: *J. Am. Ceram. Soc.,* 1989, **72**(8), 1538.
14. E. H. Lutz, M. V. Swain and N. Claussen: *J. Am. Ceram. Soc.,* 1991, **74**(1), 19.

Laser Drillable, Green-State Dielectrics

D. HOLLAND, M. W. G. LOCKYER, R. PITTSON*,
P. MORILLON** and F. GILBERT**

Physics Department, Warwick University, Coventry, CV4 7AL, UK
*Gwent Electronic Materials, Monmouth House, Mamhilad Park, Pontypool, NP4 0HZ, UK
**Société Anonyme de Télécommunications, BP130, Rue de la Violette, 23103 Dinan, France

ABSTRACT

High density interconnections for microelectronics systems have been produced by laser drilling of vias in layers of dielectric and in-filling these with conductor metal which can then be connected with the conductor pattern printed on each layer. The drilling was carried out in the green state of the dielectric which had been produced by standard thick film techniques. The requirements for a suitable dielectric are that the holes produced by laser drilling should be well defined; the definition should be retained on firing of the dielectric; and the sintering of the dielectric should be sufficient to prevent any compromise of the insulating properties of the layer. Therefore, various combinations of glass and glass-ceramic materials have been subjected to the laser-drilling process and to subsequent sintering. The optimum combination included a medium softening point glass-ceramic and a high softening point glass which underwent crystallisation during the sintering process.

1. INTRODUCTION

There is a continuing drive to reduce the size and increase the speed of electronic circuits, whether they be for industrial, medical, military or domestic use. This enhanced performance relies on the development of a suitable packaging and interconnection system, as well on the integrated circuits to be used. 'Ultimately, progress in packaging is likely to set the limits on how far computers can evolve'.[1] This evolution involves increasing the density of packing of active and passive components and shortening the interconnection between them. This has been addressed to some extent in the various forms of multi-chip module (MCM) and by moving to a multilayer 3D structure. Here, layers of conductors, voltage reference planes, decoupling capacitors, resistors etc. are separated by layers of dielectric.

To accommodate this increased density of components and the reduction in interconnect length, there needs to be a reduction in the spacing between the conductor tracks. In 3D, connection has not only to be made to the components but also between the different layers in the structure. This is done by means of 'vias', holes, filled with conductor, which pass vertically between the layers. These provide connection between the horizontal conductors on each plane (Fig. 1) and thus their separation places a constraint on the minimum separation of the conductors. Several methods are being developed for increasing the density (decreasing the size and

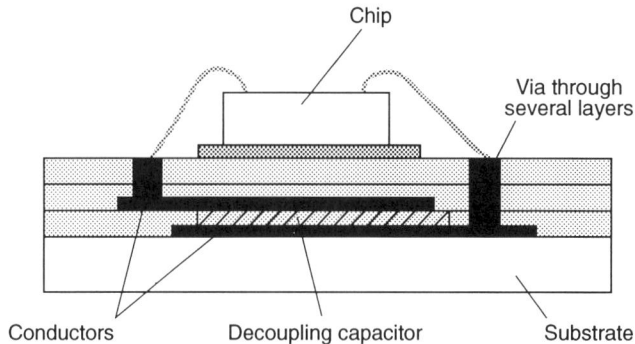

Figure 1.   Diagram showing multilayer structure interconnected with vias.

separation) of these vias. One method, which is readily integrated with the conventional thin or thick film technologies currently practised, is to laser drill the vias.[2,3] This technique has the advantage that small vias (diameter 50–60 microns) can be achieved with a CO_2 laser using only a small number of steps and conventional screen printing technology. Screen printing alone could only produce vias of approximately 250 microns and ink rheology must be closely controlled. Smaller vias (down to 30 microns) can be produced by photolithographic techniques applied to photo-imageable dielectric thick films.[4] The problem with this technique is the need for additional photolithographic stages and clean room conditions. Etching processes also require multistage photolithography and produce vias approximately 150 microns in diameter.

The laser drilling process involves a localised energy input to the dielectric, dependent on the absorption characteristics of the material at the wavelength of the laser. This results in the ablation of the organic binder and the inorganic dielectric powder. The laser drilling process is selective, only the dielectric is drilled, since it is absorbing at the laser wavelength. It is self-limiting because the underlying metal conductor reflects the laser light. However, as we shall see later, there can be problems with fusion peripheral to the ablated region. Laser optics have been developed to drill both thin film organic dielectrics (using excimer lasers) and thick film, inorganic dielectrics (using a CO_2 laser). This paper deals with the requirements of a thick film, inorganic dielectric to be suitable for laser-drilling.

A schematic diagram of the laser drilling process, as applied to a thick film dielectric, is shown in Fig. 2. There are certain properties which are required for the dielectric film:-

(a) as a laser drillable material it is required that it should:
   (i)   be absorbing at the laser wavelength (10.6µm) for CO_2 laser
   (ii) show no distortion of the via profile during the drilling operation
   (iii) show no distortion of the via profile during subsequent thick film firing operation (typically at 850°C). Therefore the dielectric must 'freeze' before flow has destroyed the via shape.

LASER DRILLING PROCESS

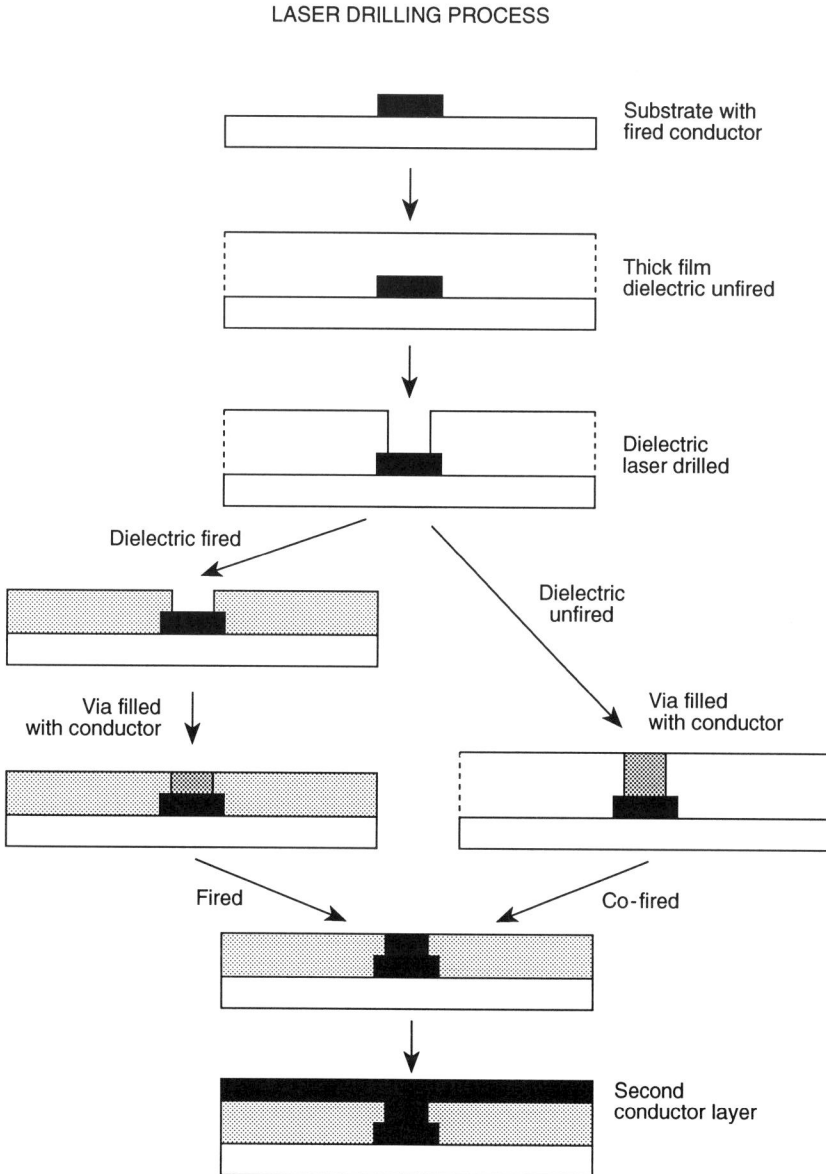

Figure 2. Schematic flow diagram of the sequence of operations to produce a laser-drilled interconnect system.

(b) as a dielectric layer it is required that it should:
  (i) be a dense, pin-hole free film (to provide hermetic coverage of underlying components and tracks). Therefore it must sinter, flow and wet to eliminate any through porosity.
  (ii) support high frequency operation and therefore have low permittivity (<5) and low dielectric loss (<0.001).

(iii) be chemically and thermally inert to any subsequent processing (e.g. plating)

(iv) be compatible with current firing practices

(v) have high thermal conductivity to help to dissipate the extra heat generated as a result of the increased density of components.

Glass-ceramics will densify (in the glassy form) by viscoelestic flow and then crystallise to prevent excess flow. Suitable composition selection will thus give a material which will sinter without distortion of the via, since densification will take place by shrinkage perpendicular to the plane of the film, i.e. the film becomes thinner but there is no change in the diameter of the vias. The resulting glass-ceramic is more refractory than the original glass and will thus withstand subsequent thermal processing without distortion or flow. The aluminosilicates on which most glass-ceramics are based can also be chemically relatively inert, particularly at low pH. The need for low permittivities and low dielectric loss is best met by low mass elements, e.g. Li, Mg. We report here an investigation of lithium and magnesium aluminosilicates as potential thick film dielectrics for laser drilling of vias. A barium aluminosilicate system was also included, since this is used as a thick film dielectric in some applications. Thermal expansion matching to the substrate (alumina in this case) was necessary, to prevent stress generation due to differential contraction. The compositions were thus selected to give glass-ceramics with TCEs compatible with that of alumina. It should be noted however, that the TCE of the final, via-filled system would become modified by the relatively large volume fraction of conductor which would become incorporated in the layer.

2. EXPERIMENTAL

The dielectric powders were prepared by conventional glass-melting and fritting. The glass frit was ball-milled to a particle size of typically 6–7 microns and then dispersed in a screen printing medium using a bead mill. This process reduced the particle size to approximately 3 microns, which is more suitable for laser drilling. The resulting inks were screen printed to a thickness in the range 20–90 microns over a printed and fired layer of silver or platinum conductor. The dried thick films (in the green state) were then subjected to the laser drilling operation to produce 60 micron vias with 60 micron separation. After SEM examination of the quality of the vias produced, the film was fired using various belt furnace profiles and speeds. The quality of the vias and sintering of the film was then assessed by further SEM examination. Some estimate of the extent of densification could be obtained by measuring the fired film thickness to give a measure of the extent of shrinkage.

Further inks were prepared in which the glass powders were used in combination with glass-ceramic powders obtained by precrystallising the glass powders, regrinding, mixing and then dispersing the mixed powders in the screen printing vehicle. The thick films produced by various combinations of composition and glass-ceramic:glass ratios were also tested for laser drilling quality, effectiveness of sinter and via-shape retention.

Table 1.   Some parameters of the base glass-ceramic materials and the fired films of their mixtures.

Sample	Firing T (C)	Dielectric (1 MHZ)	Loss tangent	Crystal phases
MAS bulk		5±0.4	0.001	cordierite
LAS bulk		6.6±0.3		stuffed quartz
BAS bulk		5.8±0.5	0.0005	celsian
5:1 MAS:LAS	850	5.5–5.8	0.01–0.02	cordierite
	950	5.4–6.0	0.01–0.02	cordierite
5:1 LAS:MAS	850	–	–	stuffed quartz
	950	10.1–10.4	0.004–0.01	stuffed quartz
5:1 BAS:LAS	850	8.2–9.7	0.004–0.015	celsian
	950	11.6–12.4	0.006–0.013	celsian

The dielectric properties of the fired films were measured using the platinum conductor layer as the ground electrode and depositing a sputtered gold, top electrode and guard ring arrangement. Capacitances were measured at 1MHz using an HP 4192A LF impedance analyser with an HP 16451B dielectric test fixture.

3. RESULTS AND DISCUSSION

The dielectric properties of the basic glass and glass-ceramic materials are listed in Table 1 together with values from some of the fired films.

3.1 Glass Powders – Single Composition

Laser drilling of glass-only materials produced vias of the required shape but also left debris, in the form of beads of glass dispersed on the base, walls and rim of the via (Fig. 3). These were sometimes as large as 20 microns and cannot be tolerated in the final product since they interfere with the via-filling operation and the printing of the next layer of conductor tracks (Fig. 4). Fired films had permittivities similar to those obtained from pressed powder pellets which had been fired with the same profile. Transformation to the required glass-ceramics was confirmed by X-ray diffraction.

3.2 Glass-Ceramic + Glass – Single Composition

Glass-ceramic:glass ratios from 5:1 to 1:6 were produced and tested. The formation of the glass beads during the laser drilling was found to be dependent on the quantity of glass phase present. Ratios of 5:1 yielded good results. Higher crystal fractions were even better and lower crystal fractions were much poorer. These findings were true for all three systems, LAS, MAS and BAS. However, the three systems did differ when subjected to the thick film firing. After firing at 850°C, there was little sintering in the MAS system (depending on the glass content). No sintering occurred in the

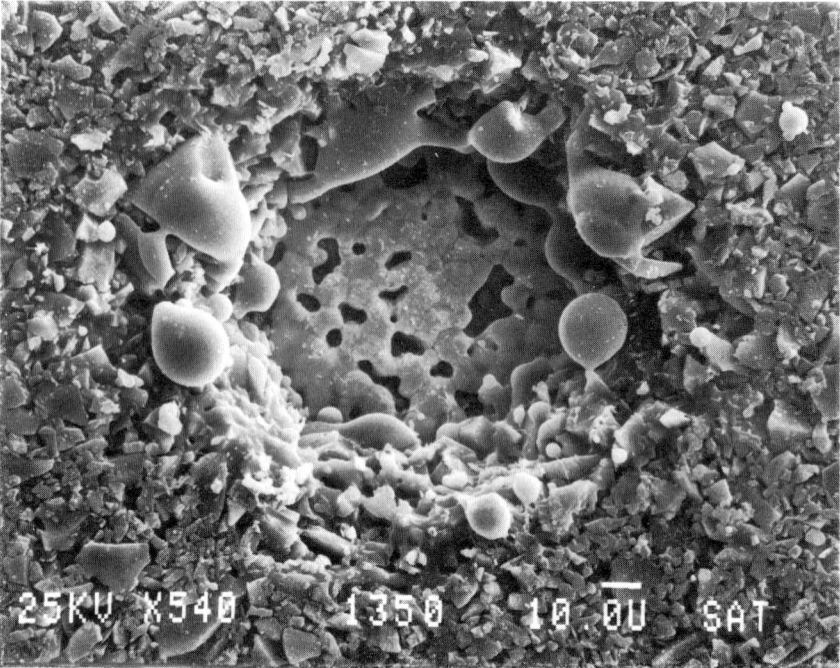

Figure 3.   Scanning electron micrograph of a 120 micron via laser-drilled in a green-state MAS glass film. Large glass beads form debris on the sides and rim of the via.

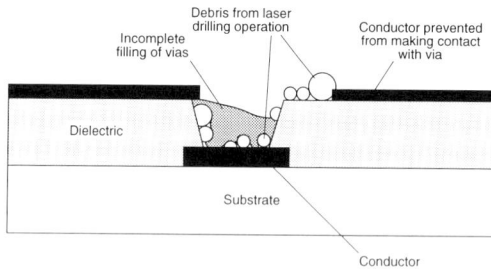

Figure 4.   Effect of laser-drilling debris on subsequent via-filling and conductor printing operations.

BAS system until a ratio of 1:6 glass-ceramic to glass. Conversely, there was complete reflow and hole closure in the LAS system for all ratios. This behaviour correlates with the known softening points of these materials, in both glass and crystallised forms. The dielectric properties were also dependent on the amount of glass phase present. In general, for each composition, the permittivity increased as the proportion of glass increased e.g. giving a value of 5.0 for a 5:1 glass-ceramic:glass ratio but 8.7 for a 1:5 ratio. Some samples gave very low permittivities (3–3.5) but this was a result of porosity. This may appear to be very attractive from a dielectric point of view but

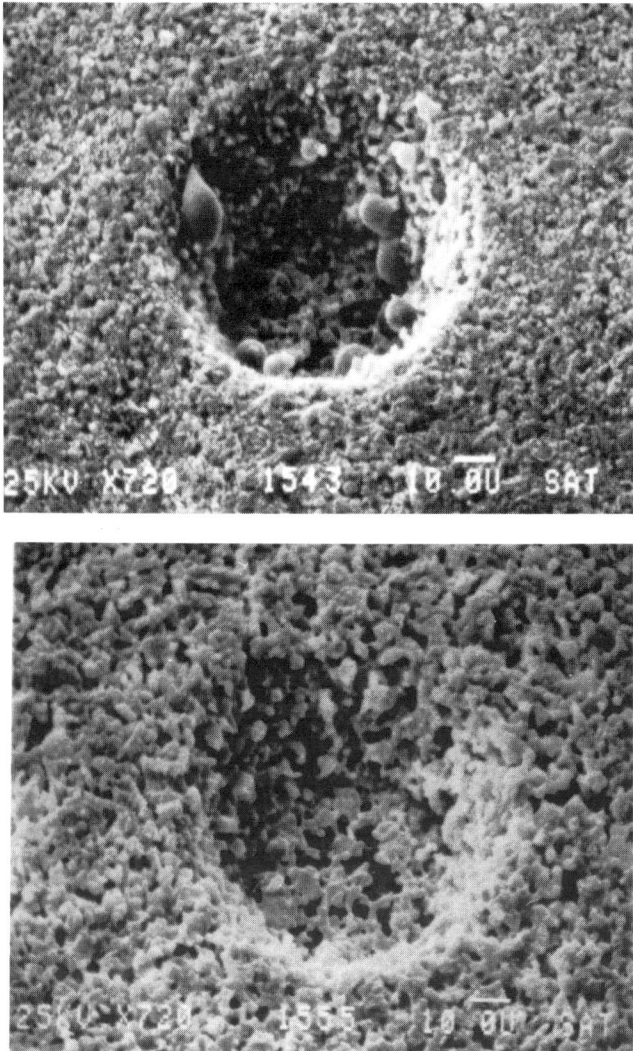

Figure 5. Scanning electron micrographs of 5:1 BAS glass-ceramic:LAS glass films. (a) 60 micron via as-drilled in the green-state dielectric, showing good definition and only a few LAS glass beads. (b) Via after firing, showing insufficient sintering of the BAS and disappearance of the LAS glass.

porosity can give major problems with electrical insulation and track definition. The dielectrics need to be well densified and have a high degree of planarity to be of use.

3.3 Glass + Ceramic – Mixed Composition

These mixed compositions were of the form LAS or BAS glass plus MAS or BAS glass-ceramic. Two firing profiles were also used – with top temperatures of 850°C or 950°C and the ceramic:glass ratios used were 5:1 and 2:1. For the MAS:LAS and the BAS:LAS

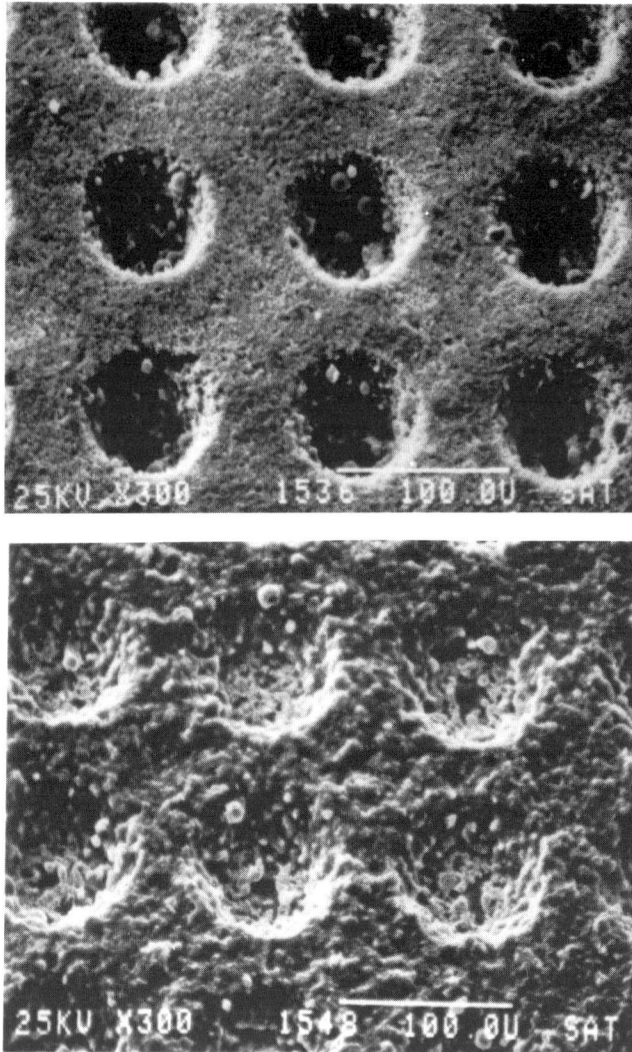

Figure 6.   Scanning electron micrographs of 5:1 LAS glass-ceramic:MAS glass films. (a)
An array of 60 micron diameter, 60 micron separation vias as-drilled in the green-state film,
showing good definition and few beads. (b) The array after firing, showing excessive flow,
loss of via definition and decrease of separation.

combinations, the permittivities of the low firing temperature sample were less than
those for the high firing temperature. In the former case this may reflect the difference in
phase composition developed at the two temperatures but, in the latter, the phase
compositions are identical and thus the lowered permittivities result from porosity arising
from insufficient sintering. Figures 5–7 illustrate some of the extremes of behaviour
observed.

Figure 5 shows a via in a thick film of 5:1 BAS glass-ceramic: LAS glass; a) as laser
drilled into the dried, green-state film and b) after firing at 850°C. In the green state,

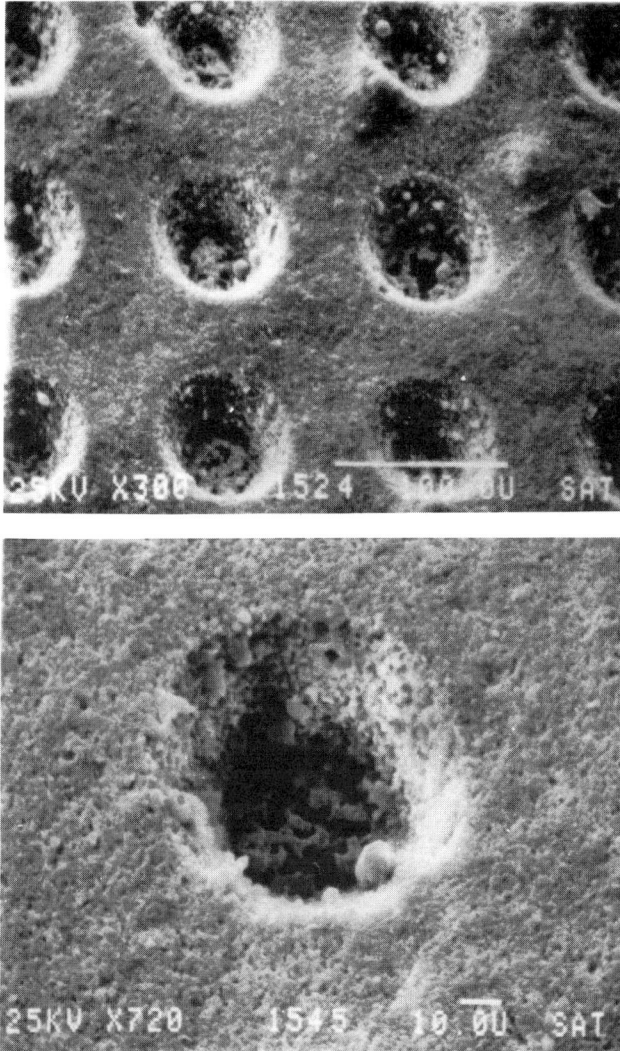

Figure 7.   Scanning electron micrographs of 5:1 MAS glass-ceramic: LAS glass films. (a)
An array of 60 micron diameter, 60 micron separation vias which have been laser drilled in
the green-state film. They show good definition and little bead formation. (b) A single via,
after firing, showing good shape retention and sintering.

it can be seen that the via shape is good, with only a few glass beads formed. These must be formed from the LAS glass present since, after firing, they have disappeared, having melted into the porous structure formed by the refractory BAS glass-ceramic, which has partially sintered but not fully densified. The porosity of such a film leads to an artificially low measurement of permittivity. Such a film cannot be regarded as hermetic as the porosity must be interconnected. Moisture uptake by the porosity could lead to high dielectric loss and also to electromigration corrosion problems.

Figure 6 shows an array of vias drilled in a 5:1 LAS glass-ceramic: MAS glass film. Micrograph (a) shows the array as-drilled in the green-state. The via definition is good and there are few glass beads formed. However, after firing (Fig. 6b) there has been considerable distortion of the vias as a result of excessive flow of the LAS glass-ceramic which has a low softening point.

Figure 7 shows the results for a 5:1 MAS glass-ceramic:LAS glass film. Micrograph (a) shows an array of vias with good definition and few beads. After sintering, this definition is retained. There is still some porosity (Fig. 7b) but much reduced and possibly isolated. An increase in the proportion of glass present may remove most of this remaining porosity.

4. CONCLUSIONS

The properties needed by a thick film dielectric to make it suitable for the laser drilling and subsequent firing of vias, may be achieved by combining glass and glass-ceramic powders, of different compositions and properties, in appropriate ratios. Many commercial thick film dielectrics combine a low melting point glass with a refractory ceramic powder to 'freeze' the flow of the glass during the conventional furnace firing. The glass does not devitrify during the firing. These materials, and particularly their relative quantities (more than 50% glass) are unsuitable for laser drilling since the glass gives rise to bead formation and also gives deformation during firing. Use of a devitrifiable glass can overcome this problem.

ACKNOWLEDGEMENTS

This work was carried out as part of a EUREKA project and the DTI and French government are thanked for the funding provided.

REFERENCES

1. R. E. Gomory: *Microelectronics Packaging Handbook*, Tummala and Rymaszewski, eds, Van Nostrand Reinhold, 1989.
2. P. Kersten, V. Glaw and H. Reichl: *Circuit World*, 1993, **20**, 20–22.
3. D. M. D'Ambra, M. C. A. Needes, C. R. S. Needes and C. B. Wang: *Proc. 42nd Electronic Components and Technology Conference*, 1992, 1072–1081.
4. G. Shorthouse, A. Berzins, J. Smyth and K. Wardell: *Proc. 12th Int. Electronic Manufacturing Tech. Symp.*, 1992.

Joining Ceramic Materials

J. A. FERNIE

TWI, Abington Hall, Abington, Cambridge, CB1 6AL, UK

ABSTRACT

Engineering ceramics offer designers a new range of materials from which the next generation of, for example, engines and heat exchangers will be produced. However, ceramics are infrequently used in isolation and a means is required for joining them both to themselves and to metallic alloys.

Joining of ceramics is not simple, due to their high chemical stability and low coefficient of thermal expansion (CTE). Therefore a means of bridging the mismatch in thermal expansion between ceramics and particularly metals is required. One way of achieving this is to use an interlayer of intermediate expansion which is also designed to be flexible, so that during thermal cycling, the interlayer can flex and thus accommodate some of the stress within the joint. These interlayers are produced from ductile metals and are joined via processes such as brazing. A second way of combatting this problem is to 'design' the joint such that effects, such as stress concentrators/ sharp edges are removed from the joint area.

This paper describes the design of such interlayers and routes to producing ceramic-metal joints with high CTE mismatch, and also how the use of design can benefit the joining process.

1. INTRODUCTION

Although the interest in monolithic ceramic materials, such as silicon nitride (Si_3N_4) and silicon carbide (SiC) is high, their use has been inhibited by the cost and difficulty of manufacturing complex shaped components. Limitations in size and geometry can potentially be overcome by joining small and simple shaped parts together to form a complex component. However, improved joining techniques are required so that the joint is not the performance limiting 'weak link' of the component.

A further problem to be overcome is that ceramics are rarely used in isolation, and frequently it is necessary to produce a metal-ceramic bond. For this, two complications arise: in general metals and ceramics have very different coefficients of thermal expansion (CTE), and this reveals itself as thermal stress in the joint. Secondly, ceramics do not readily wet (or react) with metals due to the highly stable nature of the atomic bonding, (ceramics generally exhibit strong covalent or ionic bonding).

In some instances due to the large CTE mismatch, a 'direct' bond is not feasible. In such instances stress-relieving interlayers can be introduced between the two components to be joined. These are usually made from materials of intermediate CTE, or soft, ductile metals which can yield and hence absorb stress at the joint region. Typical designs of such interlayers are given in Fig. 1, which shows how single or multiple layers may be used. A natural extension of the multiple layers is to make the interlayer with a continuous gradient; a functionally gradient material. Also shown is

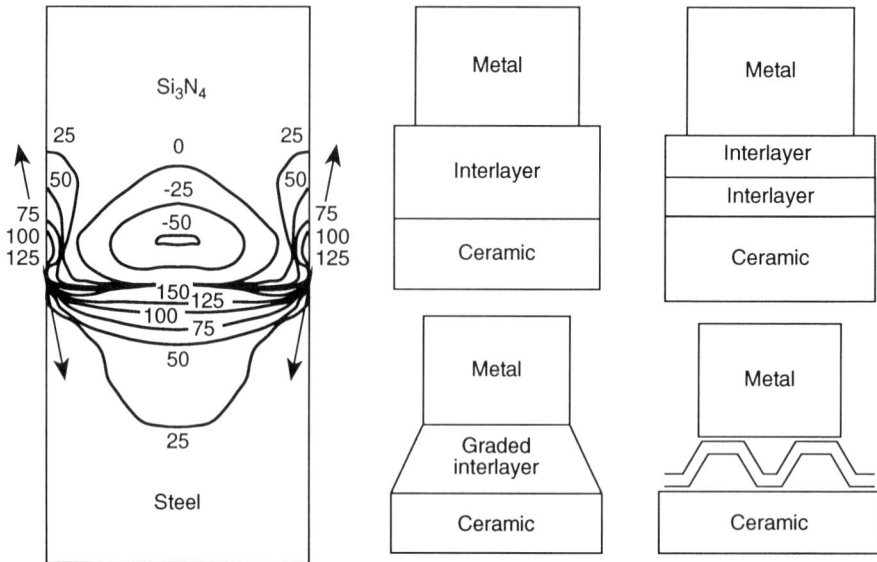

Figure 1.   Designs of stress-releaving interlayers.

an interlayer based on mechanically flexible materials. Such designs are potentially useful for a number of applications.

2. ACTIVE METAL BRAZING

One of the simplest joining mechanisms for metal-ceramic bonding is the use of active metal brazing. Brazing is a liquid phase joining process where a braze (a metal whose liquidus temperature is lower than the components to be joined) is placed between the two components, the entire assembly is then heated up, usually in controlled atmosphere or vacuum, allowing the braze alloy to melt and flow between the components and thus produce a bond. In active metal brazing a metal (usually Ti) is deliberately added to the braze alloy to promote reaction with the ceramic and render it wettable by changing its surface composition. There are a number of commercially available active braze alloys used in the joining of engineering ceramics such as alumina, zirconia and silicon nitride. Active metal brazing has found commercial use in Japan where it is used to produce ceramic turbocharger rotors,[1] using interlayer geometries similar to those shown in Fig. 1.

The following sections describe three contrasting ways in which a metal-ceramic bond has been produced using active metal brazing and appropriate joint design.

2.1 Brazing Copper to Graphite

Copper–graphite assemblies can be used as targets for pulsed proton beams in the production of sub-atomic particles (in this case muons), for instance, in nuclear research facilities. The graphite used is pyrolytic grade and is engineered to have a lamellar structure, perpendicular to the proton beam, such that the heat generated within the target during operation flows vertically through the graphite to the copper base. Copper is used because it has a high thermal conductivity and it is subsequently water cooled to keep the heat removal as efficient as possible.

The graphite targets, produced previously for a number of years, had been brazed with a commercial Ni–Cr–P alloy. This was selected since it was known to wet onto graphite. The required brazing temperature is a minimum of 940°C and the inter-metallics formed between the braze and the graphite, which enable wetting, are based on chromium carbide. Although, a relatively brittle compound the process is well established and generally accepted.

To alter the yield of muons, a design modification was specified and a change made to produce a range of graphite thicknesses, ie. greater bond area, where a greater thickness gives a greater muon yield. The generic shape of the copper heat-sinking disk was to remain relatively unchanged, with the graphite rhombohedral section required in three thicknesses, which would then offer a range of operational parameters.

Successful shapes had previously been produced using Ni–Cr–P braze described earlier; however, when the brazing of the 'new' shapes was attempted, the graphite sections all failed as a result of cracking at the sharp edged corners, Fig. 2. Examination of the failed bonds produced showed classical residual stress cracks emanating from each corner of the graphite. This was due to both the large difference in coefficient of thermal expansion (CTE) between graphite and copper (0.1×10^{-6} °C^{-1} and 18×10^{-6} °C^{-1} respectively) and also the presence of sharp corners which act as stress concentrators.

Figure 2.   Failed graphite-copper bond.

The ways used to resolve this problem were:
(i)   a change in braze alloy – to reduce the brazing temperature and hence reduce the thermal stresses induced during the heat treatment a lower temperature active metal braze based on Ag–Cu–Ti was used. Ag–Cu–Ti alloys are brazed at temperatures typically 100°C lower than Ni based alloys.
(ii)   modification to the copper heat sink – either by changing the composition to reduce the CTE (by alloying with a low CTE metal such as tungsten), or a change of dimensions to reduce the cross sectional thickness. This reduced thickness allows the copper to distort during cooling and so absorb stress more easily.
(iii) modification to the design of the graphite – removal of the sharp edges which act as local stress concentrators. This was achieved by rounding the edges of the graphite – and polishing the surface to be bonded to reduce surface defects.

Figure 3 shows a successful bond produced using all three of the above techniques.[2] Solid, stress relieving, interlayers were not considered appropriate as they would have increased the thermal impedance of the joint.

Figure 3.   Successful graphite-copper bond.

2.2 Brazing Ceramics to Metals Using Interlayers

In some cases it is not practicable to produce a direct bond and an interlayer must be used. Selection of interlayer type is made by considering a number of joint requirements. One such example is a radiant burner nozzle. Required to withstand temperatures in excess of 1400°C at the hot end, they are manufactured from SiC based materials. A typical component is shown in Fig. 4. Relatively cool gas enters at one end and leaves very hot at the nozzle end. When considering how to introduce the

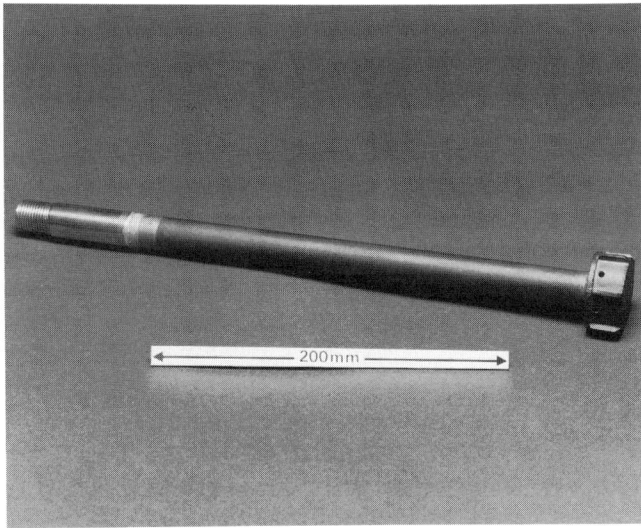

Figure 4. Radiant burner nozzle (Courtesy of British Gas plc).

component into the whole assembly a number of factors relating to performance should be considered. Firstly, the entire assembly does not see high operational temperature and therefore the whole component need not be made from ceramic. Secondly, the ceramic tube must be fitted into the furnace, the simplest means of which is a screw thread. To produce a screw thread in a ceramic would be expensive and also prone to brittle fracture; therefore the preferred route is to make the screw thread from steel and produce a SiC–steel joint elsewhere, at the cold end. However, there is a large CTE mismatch between SiC and stainless steel of ~5 to $15 \times 10^{-6} \, °C^{-1}$. A direct butt-joint was not possible and therefore a cylindrical insert was introduced. In this case a low expansion alloy with high fracture strength has been incorporated. The material has intermediate expansion and was readily wettable by the braze alloy used. Such a joint is relatively easy to produce and lends itself well to mass production if required. The technique of using solid interlayers is also the one used to produce the ceramic turbo charger assembly.[1]

2.3 Mechanically Flexible Interlayers

Where weight (or volume) is critical then a heavy, solid interlayer may not be acceptable. In such cases a non-solid, flexible, interlayer geometry may be used. A number of such flexible interlayer designs are being investigated. Two typical designs are shown in Fig. 5, which shows one interlayer of dimpled formation, and another produced as a half-honeycomb. These interlayers are produced from soft, ductile metals and alloys such as iron or stainless steel and can also be manufactured in higher temperature materials such as nickel alloys. They are then brazed into position. The important property of the interlayer is that it is ductile and able to absorb strain caused by differing CTE's during cooling from the joining temperature. This

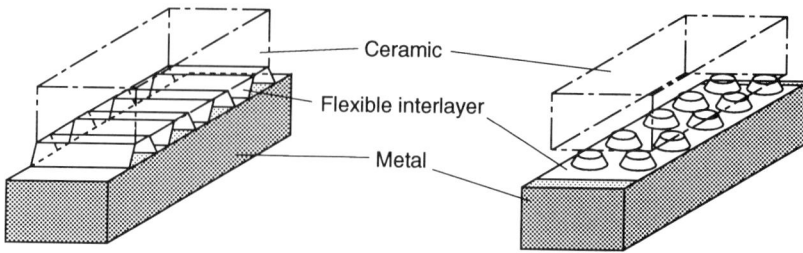

Figure 5.   Dimpled and honeycombed flexible metallic interlayers.

interlayer design is particularly versatile and has been demonstrated in a number of applications, such as ceramic faced tappets and gas turbine components.[3,4]

Flexible (or compliant) interlayers have been used to produce prototype ceramic faced tappets for the automotive industry. Silicon nitride discs were active metal brazed to a stainless steel housing. The flexible interlayer was produced in iron, in the form of a convoluted washer. This had the advantage that it was able to overcome the significant thermal expansion difference in a much smaller volume than any solid interlayer, with a complimentary saving in weight. It was also required to withstand the shear forces imposed in service and be sufficiently strong to withstand the vertical service stresses.[3]

The concept has also been developed for higher temperature applications and a recent programme has shown how the concept could be applicable to gas turbine technology.[4] In this case, the flexible interlayer allows the joint to be made to withstand lower temperatures than the working temperature of the ceramic because cooling air can be forced through the interlayer.

Again the major problem to be overcome is the CTE mismatch between the metal and ceramic components. Since the engine assembly cannot be made from entirely ceramic (due to cost and engineering constraints), a metal-ceramic bond, in this case nickel alloy and silicon nitride, is required. Typical values of CTE for a nickel alloy substrate are ~11×10^{-6} °C^{-1} and for a ceramic, such as silicon nitride are ~3×10^{-6} °C^{-1}. To overcome this, a combination of good design and the use of interlayers is required. The design problem can be addressed using Finite Element Analysis to model the stresses at the joint region and to optimise the design of the interlayer.[5]

An SEM micrograph of a dimpled iron interlayer showing the wetting of Ag–Cu–Ti braze onto both metallic and ceramic substrates is given in Fig. 6. Results were also obtained using Nimonic 75 honeycomb brazed using Ag–Cu–Ti. Examples of these types of sample are given in Fig. 7. The honeycombed design showed an orientation effect during mechanical testing, such that the strength was better parallel to the applied force and poorer perpendicular to the applied force. Therefore, for strength applications, the dimpled design has more potential since it shows no orientation affect. The general trend observed is that a small number of large dimples is preferable to a large number of small dimples for the given sample surface area.

The identification of a high temperature capability interlayer, Haynes 230, looks likely as a possible material for use in the future. It showed comparable shear

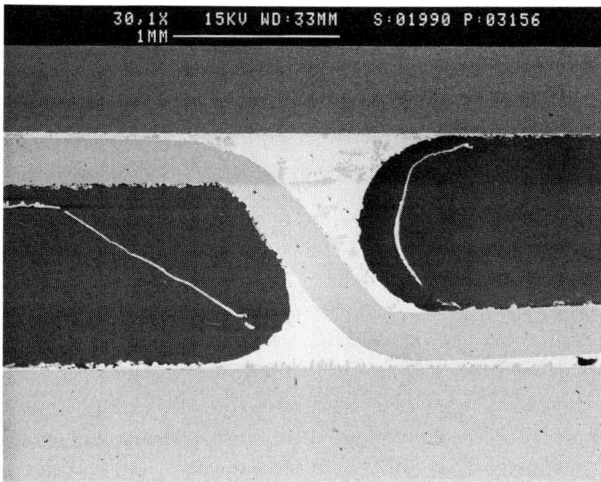

Figure 6.   Wetting of iron/ceramic interface by Ag-Cu-Ti alloy braze.

strength values to those achieved through the low temperature iron interlayers. Full implementation into commercial applications could require the production of the alloy in the corrugated design (since Haynes 230 is difficult to produce in dimple configuration, due to tearing of the interlayer on formation). Other alloys which may also be considered for such applications are stainless steel and titanium alloys.

However, before widespread implication is achieved, a high temperature active metal braze alloy is required, or some form of other joining media.

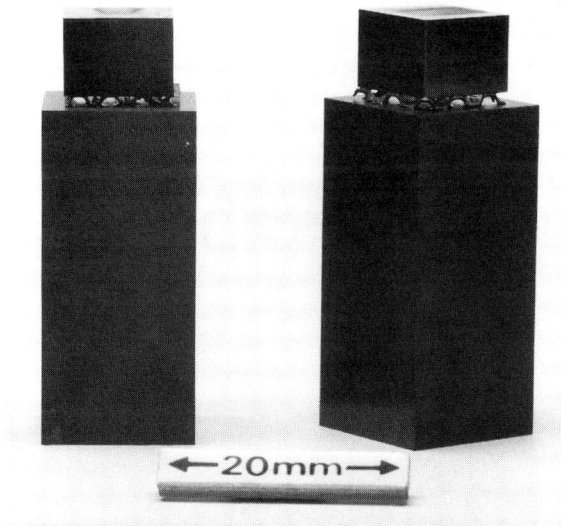

Figure 7.   Samples brazed using honeycombed interlayer design.

3. EMERGING TECHNOLOGIES

Brazing is only one of numerous techniques which can be used to produce ceramic–ceramic and ceramic–metal bonds. New techniques are constantly under development and among the leading contenders likely to play an increasingly strong role in joining are: friction welding, microwave bonding and glass–ceramic bonding.

3.1 Friction Welding

In the friction welding process, the two surfaces to be joined are aligned and made to rub together. The most common arrangement used in friction welding is where one of the components is held stationary while the other is rotated. The rotation provides friction between the two components when they are brought into contact, under uniaxial load. This not only generates heat at the interface, but also allows the break up of surface contaminants. Friction welding of aluminium to ceramics forms a soft/hard material combination such that the only material to be macroscopically deformed and extruded is the aluminium.

The majority of work with ceramics has investigated the potential for friction welding in the electronics industry. Heat sinking metals such as aluminium, have been joined to substrate materials such as alumina, aluminium nitride and silicon carbide. An example of a friction weld between SiC and aluminium is given in Fig. 8.[6]

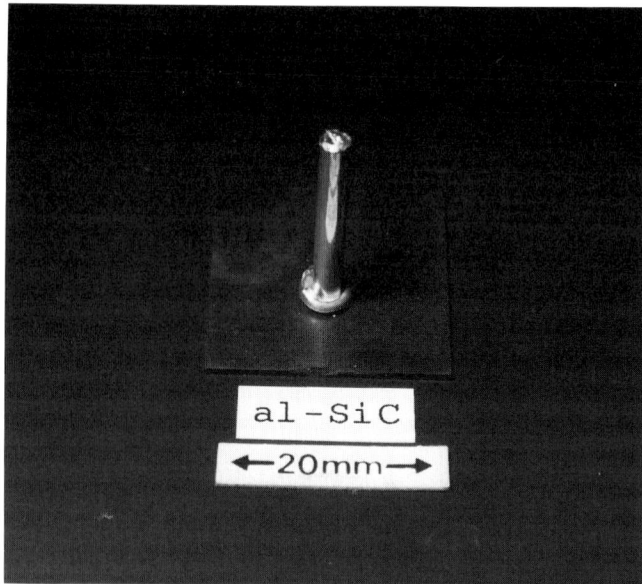

al–SiC
←20mm→

Figure 8.   Friction weld between SiC and Al.

The process is quick and provides good contact between the components with little thermal impedance. However initial work has shown that bond sizes must be restricted to relatively low values (<100mm^2) to avoid damage to the ceramic.

3.2 Microwave Bonding

For ceramic joining applications, microwave processing offers the possibilities of: rapid heating with modest power levels compared to conventional techniques; localised heating of joints due to furnace configuration and sample design; selected heating of materials and on-line process control. The application of microwave processing to joining appears to be a viable approach for ceramic-ceramic bonding. The technique has already been demonstrated for alumina and work is underway with zirconia.[7] An example of a bond produced between alumina is given in Fig. 9, with the corresponding SEM micrograph in Fig. 10. The bond itself has been produced at 1600°C in under 10 minutes.

Such high quality bonds produced in short processing times makes this process highly economic, with potential for great savings in energy consumption. Among the potential markets for the technology is the production of high temperature heat exchangers, where a number of ceramic–ceramic joints must be produced simultaneously.

3.3 Glass–Ceramics

The microstructure of a glass–ceramic can be tailored to meet a range of properties– most notably the coefficient of thermal expansion. They can also be designed to

Figure 9.   Microwave diffusion bonding of alumina

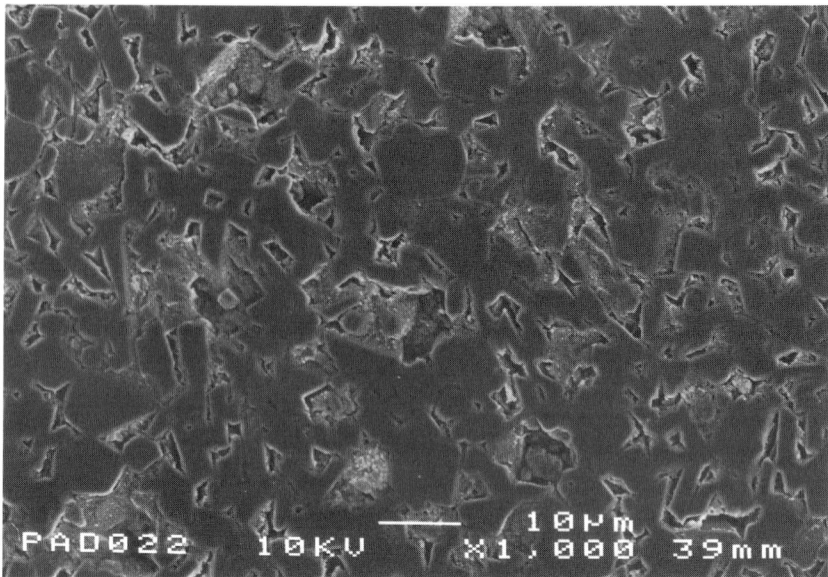

Figure 10.   SEM micrograph of microwave diffusion bonded alumina.

provide excellent thermal shock resistance, dielectric properties, chemical durability, high strength and toughness.

Among the most common of the glass–ceramic systems are Lithium–Alumino–Silicate (LAS), Zinc–Alumino–Silicate (ZAS) and Magnesium–Alumino–Silicate (MAS). These systems have been studied in detail and many formulations are commercially available.

Probably the most versatile feature of glass–ceramics is their wide range of thermal expansions. This is shown in Fig. 11, where the thermal coefficients of four glass–ceramic systems are given. Taking the LAS system as an example, it is observed that there are two CTE ranges. At the low (near zero) end, the glass–ceramic CTE can be matched to that of tungsten and nilo. However, the real versatility of this particular system is then exemplified. When the composition is altered by decreasing the percentage of alumina, the CTE is raised and can be matched to that of nickel alloys and stainless steel. Similar dexterity is also shown by other glass–ceramic systems.

As with all glass containing ceramic materials, the final usage temperatures are dictated by the amount and composition of the residual glass phase. The low residual glass content of glass–ceramics allows them to withstand extremes of temperature making them ideally suited as sealing and bonding media in high temperature and/or corrosive environments.[8,9] Also the nature of the system and its ability to have a tailored microstructure lend it to applications where the CTE between two substrates needs to be accommodated through the joint or interlayer.

4. FUTURE REQUIREMENTS FOR JOINING TECHNOLOGIES

Three important requirements for the joining technologist are:
– higher temperature active metal braze alloys
– better definition of ceramic material properties
– more emphasis in the design of ceramic components and joint geometries

4.1 High Temperature Braze Alloys

Current commercial active braze alloys have a maximum temperature of operation of under 700°C, or are prohibitively expensive. This means that although suitable brazed joints may be produced for the automotive industries, aerospace applications (with a temperature requirement of >1000°C) are not yet feasible. There is a market need to produce high temperature active braze alloys. To do so will require new fabrication techniques, probably based on powder metallurgy.

4.2 Ceramic Property Information

The generation of a standard set of mechanical, thermal and electrical property information on ceramics and ceramic matrix composites is essential if these materials are going to be of future use. Currently the literature quotes large discrepancies in figures, making it extremely difficult for the designer/engineer to have confidence in the design tolerances (such as strength etc). In addition the properties of the reaction products at ceramic–metal interfaces must be established. These products are frequently the weak areas of a joined component. Both the materials engineer and the designer would benefit from knowing more about the performance of the interface. Such data would also make numerical analysis of these systems much more reliable.

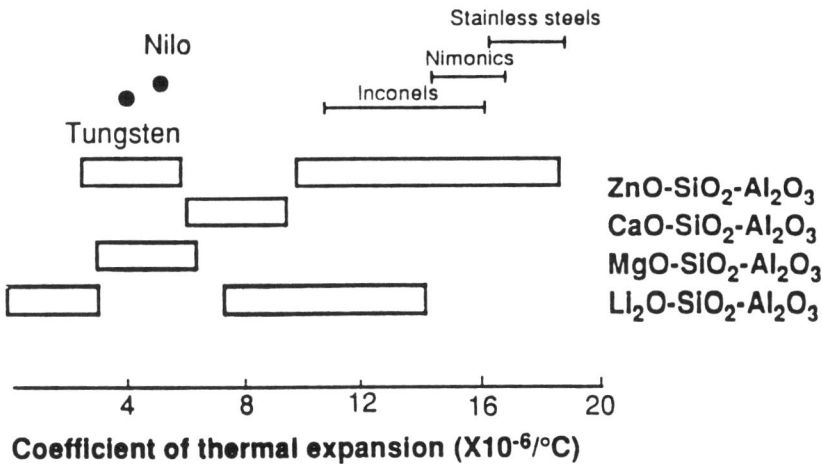

Figure 11.   Comparison of coefficients of thermal expansions for a range of glass-ceramic systems, metals and alloys.

4.3 Emphasis on Design

One of the greatest challenges faced by the joining technologist is the re-education of designers to think in ceramic, rather than metallic, terms. All too frequently the component is far advanced into the production stage (or before the material engineer is consulted) before it is realised that the ceramic will not 'stick' in position the way in which a metal would. More often than not, it is design rather than chemistry which precludes the production of a sound joint. Ceramics are only 'strong' when held in compression and designers have to find a means of designing the joint to avoid all sources of tensile stress and also the removal of sharp edged corners which act as stress-raisers.

5. CONCLUSIONS

Whilst the use of a wide range of engineering, chemical and design solutions can overcome the difficulties in joining of metals to ceramics and ceramics to ceramics, we must also re-educate the designers to treat the ceramic as a stand-alone material rather than a 'high temperature metal'. This will only be achieved by the widespread knowledge and acceptance of the properties of ceramic materials and the advancement in joining technologies.

REFERENCES

1. K. Katayama, T. Watanabe, M. Matoba and N. Katoh: SAE Tech. Pap. Ser., 1986, 861138.
2. W. B. Hanson: *Joining of graphite and copper for nuclear applications*. TWI Bulletin, May/June 1995, 60–61.
3. I. A. Bucklow, S. B. Dunkerton, W. G. Hall and B. Chardon: 'A Brazed Ceramic to Metal Joint in a Car Engine Tappet', *4th International Symposium on Ceramic Materials and Components for Engines*, Ed. Carlsson, Pub. Elsevier, 1991, 324–331.
4. W. B. Hanson, J. A. Fernie, P. A. Singleton and P. H. Bond: 'Joining of Metals and Ceramics for Turbine Applications'. *Second International Conference on Ceramics in Energy Applications*, Pub. Inst of Energy, 1994.
5. P. H. Bond, T. H. Hyde, A. A. Becker and J. A. Fernie: 'The Analysis of Flexible Interlayers for Ceramic to Metal Joining using the Finite Element Method'. To be published in Proc. 8th CIMTEC World Ceramic Congress, Florence, Italy, 1994.
6. J. A. Fernie: 'Friction Welding of Aluminium to Ceramics'. *Mater. Manuf. Process.*, 1994, **9**, 379–394.
7. J. G. P. Binner, P. A. Davis, T. E. Cross and J. A. Fernie: 'Microwave Joining of Ceramics'. (Keynote Lecture). Presented at American Ceramic Society Annual Meeting, Cincinnati, USA, 1995.
8. G. Partidge: 'Joining Glass–Ceramics to Metals', *Joining of Ceramics*, Ed. Nicholas, Pub. Chapman and Hall, 1990, 31–55.
9. P. A. Walls and M. Ueki: 'Joining SiAlON Ceramics using Composite β-SiAlON–Glass Adhesives'. *J. Am. Ceram. Soc.*, 1992, **75**(9), 2491–2497.

Joining of Sialon to Stainless Steel

P. bin HUSSAIN, A. ABED and A. HENDRY

Metallurgy and Engineering Materials Group, University of Strathclyde, Glasgow, G1 1XN, UK

ABSTRACT

The problems of ceramic to metal joining are well known and result primarily from the difference in the coefficients of expansion of the ceramic and metal. The strain caused by cooling such a bond from the joining temperature results in thermally-induced stresses which cause failure in the interface or in the ceramic. A common approach to solution of this problem is to use a ductile barrier between the ceramic and the metal in order to minimise the strain in the interface and give a gradual transition between the two materials. The disadvantages of these methods are that either the ductile interlayer has a relatively low melting point (such as braze alloys) of that reactive and potentially embrittled metals of higher melting point are used (for example, titanium).

The use of a sialon ceramic impeller to assist circulation of hot, aggressive gases in an enclosed furnace depends on the ability to couple the rotating ceramic head within the furnace to an external drive train and motor. It was proposed that the drive shaft be of stainless steel which therefore must be joined to the sialon impeller adjacent to the furnace wall. Mechanical joints proved unreliable due to the thermal cycles experienced by the components and brazed joints were insufficiently strong. It was necessary therefore to provide a direct diffusion bond between the sialon and steel and the present work describes the methods employed.

Two approaches were taken to provide a suitable join. First, a method of direct joining sialon to treated steel by diffusion bonding was devised. The join was examined by microscopy and X-ray diffraction in order to characterise the microstructural nature of the bond. A second approach was to use a ceramic-metal composite interlayer to provide a graded join and this method was also investigated. A limited mechnical characterisation was carried out which showed that both methods gave joins which were suitable for the application.

1. INTRODUCTION

Joining of ceramics to metals is in principle a relatively simple diffusion bonding process but in practice the large difference in coefficients of thermal expansion between the two materials results in the generation of unacceptably high stresses on cooling from the bonding temperature. In many cases this stress exceeds the fracture stress of the interface or of the ceramic as the principal concentration of stress will be in these parts of the composite structure. The metal, as a ductile material, can deform plastically to accommodate the stresses which are therefore concentrated (in tension) on the brittle parts of the composite. Many methods have been devised to overcome these problems where the need for direct ceramic/metal bonding cannot be overcome by alternative joining methods. The principal approach among these is the use of ductile metal interlayers between the ceramic and metal.[1-4] The aims of this are twofold; first, the interlayer(s) provides a gradual transition between the relatively high coefficient of thermal expansion of the metal and the lower value for the ceramic, and secondly, allows plastic deformation to accommodate the stresses generated on cooling. To satisfy the first of these criteria the interlayer must have a coefficient of thermal expansion intermediate to those of the parent materials while remaining sufficiently ductile to fulfil the second condition. In the specific case of silicon nitride-based materials bonded to stainless steel, several methods have been successful in achieving sound bonds. The use of nickel has been successful in the first category,[3] while braze alloys have been used to provide the interlayer ductility.[4] Indeed combinations of these materials have also been used in an individual join.[5]

The principal of selecting an interlayer of property characteristics intermediate between those of the two parent materials suggests a combination of the parent materials themselves, provided that the 'average' of those properties can be sensibly achieved in a composite. There is an additional advantage in that there is no new material involved which may have an influence on the overall behaviour of the composite structure; for example, in the case of sialon-steel joining, operational parameters such as maximum temperature or corrosion resistance which can be satisfied by the parent steel will also be tolerated by the composite.

The present study is concerned with joining of sialon ceramics to austenitic stainless steel in applications in combustion engineering. There is a requirement therefore, not only to provide a sound joint but also to obtain corrosion resistance in combustion gases combined with mechanical properties appropriate to extensive thermal cycling of the bonded structure. Previous work at Strathclyde[6] has shown that ceramic matrix composites of sialon with stainless steel particles are attractive materials in such applications and studies of the ceramic-metal interfaces which have been reported elsewhere,[7] have shown that the chemical reactions can be controlled to give a sound material by pressureless sintering. These latter results are of particular importance as they show that the chemical characteristics can be controlled whether the interface is on a microscopic level between matrix and particle, or on a macroscopic level where the interface is that at which the two parts of the structure are bonded.

The first part of the present work shows that direct bonding of sialon to stainless steel can be achieved when the steel is nitrided before the joining operation. Thus two conditions are achieved. First, the steel is saturated with nitrogen, preventing the decomposition of the sialon matrix at the interface with the metal,[7] and secondly the coefficient of thermal expansion of the metal close to the interface is decreased by the presence of compounds in the nitrided layer.

The results presented here show that direct bonding of nitrided steel to sialon can be attained and that, alternatively, sialon-steel composites provide properties intermediate between those of the parent ceramic and metal and have the ability to accommodate the stresses created by the bonding process to give a sound join. Either of these methods, based on the same materials science principles, can be used in the present application.

2. EXPERIMENTAL METHODS

The sialon used in these experiments is Cookson 201 material available from Cookson Zyalon Ltd. This is a β'-sialon of Z-value 0.75 and was supplied as a precursor mix of mean particle size 10μm with 8% Y_2O_3 as a sintering aid. The ceramic powder was used in the monolithic, fully densified form by hot-pressing at 1650°C to provide the ceramic parent for the joining experiments. The same powder was also pressureless sintered at 1550°C with 20% stainless steel (25Ni–20Cr–1.8Ti) particles of 35μm mean particle size which had previously been nitrided.[6] The parent steel in the experiments involving the composite sialon-steel interlayer was AISI 316 (18Cr 12Ni Mo austenitic grade).

The parent steel (10 Cr, Mo V martensitic grade) used in the direct bonding experiments and the steel powder for the composites were nitrided in ammonia: nitrogen mixtures at temperatures between 700 and 1200°C depending on the conditions required.

In all experiments the parent steel and sialon parts of the join were used in the form of 10mm diameter rods and the composite parts for the interlayers were cut to appropriate thickness of the same diameter. The joining process is carried out in a graphite holder in which the parts to be joined are held in firm contact with each other with a minimal applied force (the force is applied simply by the expansion of the steel component on heating to the joining temperature and can be calculated approximately from knowledge of the expansion coefficients of the parts and the holder). The surfaces to be joined are prepared by polishing to a 6μm diamond finish and rinsed in an ultrasonic bath immediately before the experiment.

The joining process for the direct bonding of sialon to nitrided steel is carried out at 1250°C, but the process involving the composite interlayer is carried out in two stages. First the sialon parent is bonded to the composite interlayer by diffusion at 1550°C for one hour in nitrogen gas giving a strong bond between the two β'-sialon phases across the interface and with no evidence of metal transfer from the composite to the parent ceramic. Second, the bonded pair is in turn joined to the steel parent by reaction at 1220°C for one hour, also in nitrogen. Several other combinations of ceramic/steel/composite couples were also prepared to examine specific features of the process and these are described in the sections below.

The samples after joining were sectioned and examined by standard metallographic techniques and the joins were tested by four-point bending. Finite-element methods (FEM) of stress analysis were also applied to calculate the stresses in a number of joint combinations.

3. RESULTS AND DISCUSSION

The initial experiments carried out involved the direct-joining of sialon to austenitic stainless steel in order to characterise the interface both by optical microscopy methods and by FEM. This gave a clear picture of the nature of the joining problems for this specific combination of materials rather than relying on published data. An example of such an interface formed at 1300°C is shown in Fig. 1 for a sample in which fracture did not occur and it can be seen that there is considerable chemical reaction across the interface. In the ceramic side of the couple there has been considerable penetration of metal, while the steel shows extensive precipitation of prismatic crystals adjacent to the

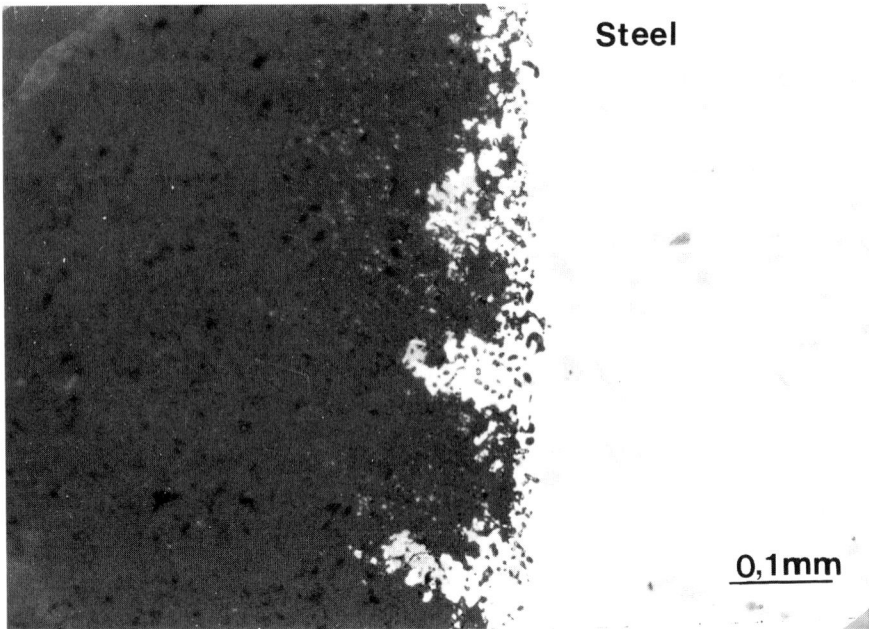

Figure 1.   Optical micrograph of the interface between sialon and steel directly bonded
at 1300°C.

interface (these are grey in colour in Fig. 1) and which are not present in the original
steel. These two effects are due to the chemical reaction which takes place, at the
joining temperature, between silicon nitride and steel.[6–8] The net effect is that silicon
nitride decomposes to give silicon dissolved in the steel and releases nitrogen gas, some
of which may also dissolve in the steel. The solidus temperature of such alloys is below
the joining temperature and so a liquid metal phase is formed at the interface which
penetrates the ceramic as the nitride is attacked. Similarly, on the steel side of the
interface and immediately adjacent to it, there is dissolution of silicon and nitrogen in
the metal which on cooling reprecipitates as silicon nitride in the metal matrix. This is
the grey phase in Fig. 1 and has been identified by SEM analysis and by X-ray
diffraction as α-silicon nitride which is precipitated in the steel.

Interfaces formed by the reactions described above are coherent and strong but
are, of course, subjected to stresses on cooling after the reaction. In general this leads
to failure of the join by fracture in the ceramic half of the couple. By analysis and
experiment it was shown[9] that successful joins were only obtained where the stresses
were symmetrical and the strain was less than a critical amount. These experiments
also showed clearly the nature of the chemical reactions at the microscopic interface
and have been discussed elsewhere[7] in terms of the microscopic interfaces in composites. The regions of maximum tensile stress generation for the geometry used in
this work lie in the ceramic side of the couple and have been analysed in previous
work.[7,9]

Following these initial studies, it was clear that for directly bonded joins which were to be free of the residual thermal stresses which produce cracking on cooling from the joining temperature, it was necessary to fulfil two conditions. First the interfacial reaction to produce silicon in the steel at the interface should be minimised, and also the coefficient of thermal expansion of the metal in the immediate interface region should be decreased. The first part of this has been explained previously,[6–8] and work on the metallurgy of nitrided steels at Strathclyde had shown that the formation of nitrides and other high-nitrogen steel phases had the effect of decreasing the mean coefficient of expansion of the 'composite' metal surface. It was therefore decided to investigate the joining of a nitrided steel surface to sialon where the optimum steel characteristics were obtained in a nitrided martensitic steel grade. The surface in this example has a thin layer of approximately 30μm of very high nitrogen content and a sub-surface layer of about 1 mm of a coarse acicular ferrite structure (Fig. 2a). The structure of the un-nitrided part of the bar is shown in Fig. 2b and retains some evidence of the bar texture. After joining to sialon at 1250°C the structure is as shown in Fig. 3 where a clear interfacial layer of precipitation can be seen in this un-etched sample. This is the high-nitrogen region of the nitrided surface and the presence of this layer has resulted in a much reduced penetration of silicon into the metal from decomposition of the silicon nitride solid solution (compare Fig. 1 for a typical un-nitrided case). There is no evidence of re-precipitation of silicon nitride crystals in the steel in Fig. 3. This is the result of the extremely high nitrogen content which inhibits the decomposition of the sialon solid solution[8] and which

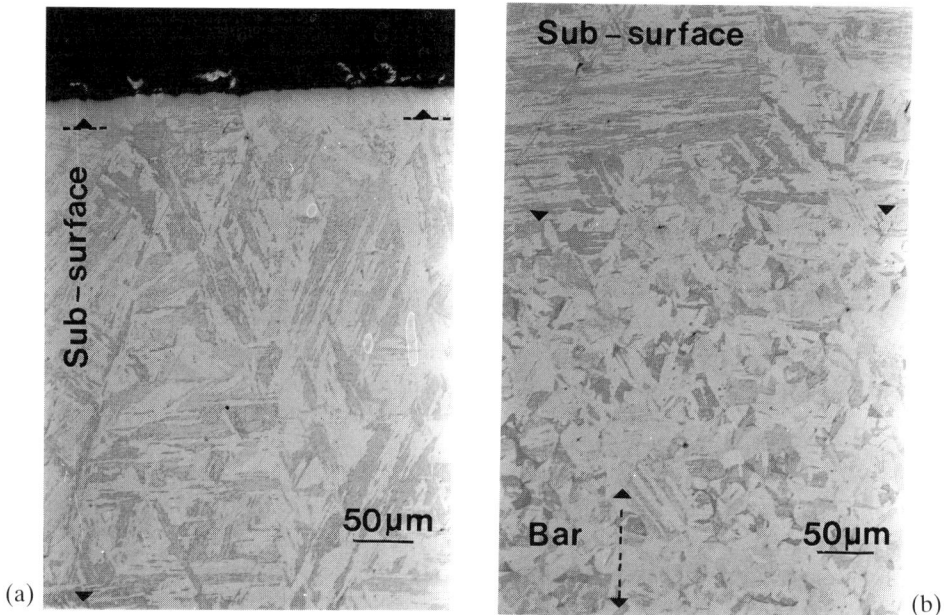

Figure 2(a).   Optical micrograph of the surface layer of 10Cr Mo V stainless steel nitrided at 1200°C.
(b) Optical micrograph of the structure of the untreated steel bar as in Figure 2(a).

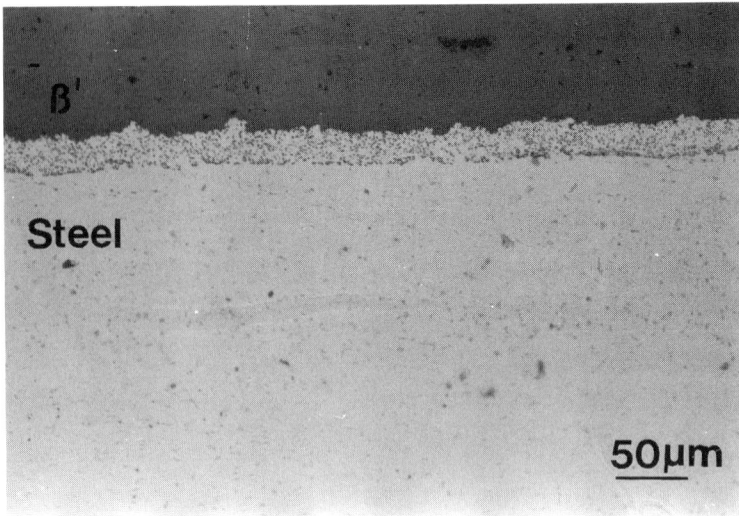

Figure 3.   Optical micrograph of the interface between sialon and nitrided martensitic
stainless steel joined at 1250°C (unetched).

reacts immediately adjacent to the interface with any small amount of silicon which
diffuses to the metal.

It is also evident from Fig. 3 that the interface is well bonded and coherent and it
was observed in this series of experiments that there was a complete absence of
cracking in the joins, in contrast to the system shown in Fig. 1 where a coherent
interface and no cracking was the exception.

Following these studies of direct bonding, it was clear that a martensitic grade of
stainless steel could be used in this application. However the long-term high-
temperature strength of this grade is considered to be insufficient for such applications and so a method of joining utilising an austenitic grade was sought. Preliminary
experiments using nitrided surfaces of AISI 316 were less successful than those with
the 10Cr steel described above and although this is believed to be simply a case of
finding a suitable nitriding treatment, another (proven) method was used. Previous
work[6,9] had shown that sialon-metal composites could be manufactured and were
suitable for use as an intermediate layer in the joining of sialon to austenitic stainless
steel. Thus samples of sialon were joined to stainless steel by a sialon/metal composite layer.

The interfaces between sialon and composite, and between steel and composite are
shown in Fig. 4. The nature of the bonding in the first step of the process – sintering
of sialon to composite at 1550°C for one hour – is equivalent to a liquid-phase
sintering process. The densification addition in Cookson 201 is 8% yttria which has a
solidus temperature in the Y_2O_3–Al_2O_3–SiO_2 system of approximately 1350°C but
that temperature is decreased further by the effect of dissolved nitrogen. It therefore
follows that during joining at 1550°C there is diffusion of the species in the ceramic
structure to give a coherent bond across the interface. This has been confirmed by

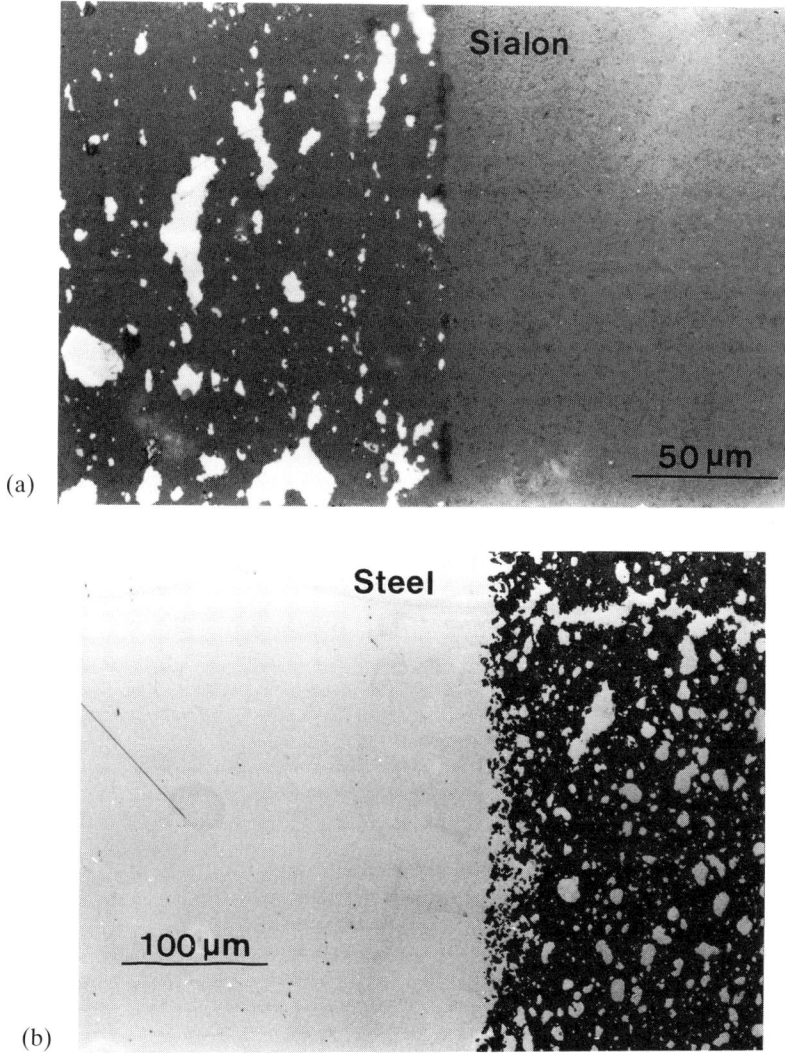

Figure 4(a). Optical micrograph of the interface between sialon and composite joined
at 1550°C.
(b) Optical micrograph of the interface between composite and austenitic stainless steel
joined at 1220°C.

SEM and X-ray analysis which show that the sialons on either side of the interface
are indistinguishable. The same analyses show that there is no apparent effect of the
metal particles in the composite extending into the ceramic parent, and from previous work[6–8] it is clear that this is to be expected as the composite preparation at the
same temperature has already brought the steel into chemical equilibrium with the
sialon matrix. The second stage of the bonding process is to join the metal parent to
the composite at 1220°C in a nitrogen gas atmosphere. This gives the join illustrated
in Fig. 4b. Again the join is coherent and with no evidence of fracture on either this

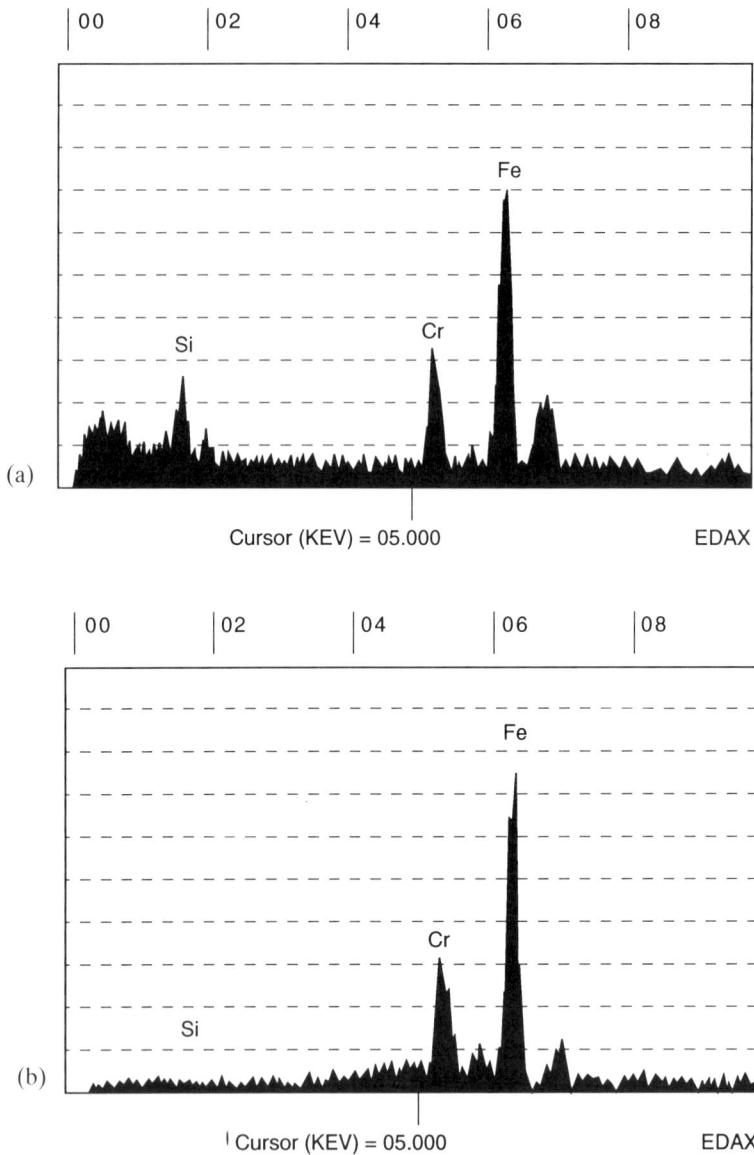

Figure 5(a).   SEM/EDAX analysis of a steel particle in the interface in Figure 3(b).
(b) SEM/EDAX analysis of a steel particle in the composite away from the interface.

half of the join or on the previously bonded ceramic/composite side. There is, however, clear evidence in Fig. 4b of an interfacial layer in which the ceramic matrix in
the composite has reacted with the stainless steel parent. This is due to the reaction at
the bonding temperature discussed above for the case of an un-nitrided steel/sialon
direct bonding, in which silicon transfers across the interface by decomposition of the
sialon. This results in a decreased solidus temperaure for the alloy and incipient
melting at the interface, which gives good bonding but penetration of liquid metal

into the composite (as in the case in Fig. 1). This can be confirmed by SEM micro-analysis as shown in Fig. 5 where 5a shows the spectrum for a metal particle in the interface and 5b shows the steel analysis at some distance from the interface. When compared with Fig. 1, there is no evidence of silicon nitride precipitation in the steel parent in Fig. 4b as the temperature of reaction is lower in the latter case and the particles in the interface set up a barrier to silicon transfer.

The integrity of the bonds formed by the interlayer method has been tested in four-point bending at room temperature. The modulus of rupture of the monolithic sialon sintered at 1650°C in these experiments is 650MPa and the bonded joints were found to have a mean failure strength of 370MPa. The failure always occurred in the sialon and no interfacial cracking was observed in samples tested in this work. Although the fracture occurred in the sialon, the failure stress is considerably less than that of the monolithic material. It must be concluded therefore, that there is a residual tensile stress of approximately 650–370 = 280MPa in the ceramic inherited from cooling and arising from the difference in coefficient of thermal expansion between the sialon and the composite. The strength of the interface itself must be greater than the applied stress of 370MPa at which the bond bar failed.

In an attempt to model the stresses in the macro-composite bonded bar, measurements were made of the coefficient of thermal expansion of the sialon-20% steel composite in a dilatometer and an estimate of the value was taken as $8.5 \times 10^{-6} C^{-1}$. Using this value, FEM calculation was carried out to model the residual stresses in the bonded composite bar and compare them with the values for the direct sialon to steel bond as described above. The results of the computed profiles show that the maximum residual stress in the ceramic is 1.9 times greater in the case of sialon/steel than for sialon/composite/steel. Thus, if the maximum thermal stress in the case of the direct join is 640MPa, then the composite bond gives a residual stress of 336MPa which is to be compared with the experimentally derived value of 280MPa. It should be noted however, that the FEM calculation is particularly sensitive to the values of coefficient of expansion, Young's modulus and Poisson's ratio used and that the agreement is therefore reasonable.

4. CONCLUSIONS

The experiments reported here confirm that acceptable bonding can be achieved between sialon and stainless steel by two different methods. By pre-nitriding the surface of the steel to be joined when using a martensitic steel grade it is possible to obtain a strong direct diffusion bond between the ceramic and steel. It is also considered that it is possible to devise a similar route using austenitic steel grades.

In the case of austenitic stainless steel bonding to sialon, the value of steel/sialon composites as an effective means of joining without interface cracking or failure in the ceramic due to cooling stresses has been demonstrated. Microscopic analysis provides a model for the process which highlights the importance of the interfacial reaction between sialon and steel in both microscopic and macroscopic interfaces.

FEM calculations confirm that there is a residual thermal stress in the sialon when a composite layer is used in bonding but that the stress is considerably lower than the fracture stress of the sialon or the failure stress of the interface.

The method of fabrication of these joins is simple and can be performed in nitrogen gas by pressureless sintering. The performance of the join in the application is then dictated by the load on the ceramic parent and the temperature capability of the steel parent.

REFERENCES

1. P. Batfalski, R. Godzimemba-Maliszewski and R. Lison: 'Difference between diffusion welded and brazed metal to ceramic joints', *Joining of Ceramic, Glass & Metal,* W. Kraft, ed., DGM, Oberursel, FRG, 1989, 81.
2. V. A. Greenhat: 'Joining of cermic-metal systems', *J Am. Cer. Cos.,*1990, **73**, 2463.
3. T. Ishikawa, M. E. Brito, Y. Inoue, Y. Hirotsu & A. Miyamoto: 'Interfacial structure and mechanical strength of β'-sialon/Ni bonded systems', *I.S.I. Jap. Int.,* 1990, **30**, 1071.
4. S. M. Johnson: 'Mechanical behaviour of brazed silicon nitride', *Ceram. Eng. Sci. Proc.,* 1989, **10**, 1846.
5. K. Suganuma, T. Okamoto, M. Koizumi & M. Shimada: 'Joning Si_3N_4 to Type 405 steel with soft metal interlayers', *Mater. Sci. & Tech.,* 1986, **2**, 1156.
6. A. C. Smith, A. Abed, H. J. Edrees & A. Hendry: 'Ceramic matrix composites of silicon nitride with conducting particles', *Silicon Nitride, '93,* M. Hoffman, P. Becher & G. Petzow, eds, Trans Tech. Publications, Aedermansdorf, Switzerland, 1993, 423–428.
7. A. Abed, H. J. Edrees & A. Hendry: 'Interfacial behaviour in ceramic-metal composites', Silicates Ind., 1995, in the press.
8. A. Abed, A. C. Smith & A. Hendry: 'The effect of nitrogen alloying on the densification behaviour of a sialon ceramic matrix composite containing stainless steel', *Materials Science & Engineering,* 1995, in the press.
9. A. Abed: 'Joining of sialon-stainless steel and sialon-sialon', PhD. Thesis, University of Strathclyde, 1994.

Hydroxyapatite Filters for the Removal of Heavy Metal Ions from Aqueous Solutions

J. G. P. BINNER and J. REICHERT*

Department of Materials Engineering and Materials Design, University of Nottingham, Nottingham, UK

ABSTRACT

A novel method of producing highly porous hydroxyapatite ceramics with densities $\geq 15\%$ and an open cell structure has been developed via the foaming of ceramic slips. Three grades of hydroxyapatite with varying purity were investigated with respect to their potential for removing heavy metal ions from aqueous solutions. The materials were evaluated as both loose powders and as ceramic foams. The results showed that all three grades of HA were capable of removing a number of different ionic species although the more impure grades generally yielded the best performance. It is believed that the increased impurity levels resulted in increased numbers of lattice defects which were ideal adsorption/exchange sites. 100% removal could be achieved for some ions under the correct experimental conditions.

For the ceramic foam filters the optimum filtration parameters were found to be: a high surface area; long filtration times; a low pH and a high filtrate temperature. Ion adsorption was positively detected as a mechanism of ion removal. Ion exchange was not observed but could not be completely ruled out.

1. INTRODUCTION

In recent years waste water treatment has been raised in importance due to increased concern about the environment and tighter international regulations on water pollution.[1] Toxic heavy metal ions are generally removed from industrial waste water by one of three mechanisms:[2] coagulative precipitation, reverse osmosis or ion exchange, with ion exchange resins being favoured in many applications. Although these can provide adequate performance, it is always desirable to investigate alternatives.

Although hydroxyapatite (HA) receives considerable attention due to its potential for use as a bioactive bone substitute,[3] it has also recently begun receiving attention as a potential water filter due to its capacity for removing heavy metal ions.[4] Currently, such filters suffer from two disadvantages which make them too expensive for commercial applications. Firstly, only high purity grades of HA, which are intended for biomedical applications, have been investigated. Secondly, all of the techniques for producing porous HA ceramics examined to date (see reference 5 for a review) are felt to be too complicated or too expensive for commercial production.

* Now with Schott Glaswerke, Germany

Kingery et al.[6] have shown that a non-dispersed slip produces clusters during sintering which lead to increased surface area due to pore formation. This effect would be quite beneficial for the production of ceramic foams for applications as filters since a high surface area would increase the effectiveness, as long as the strength of the filter remained adequate. Hence this paper describes both a novel method of producing hydroxyapatite-based ceramic foams via the foaming of ceramic slips which are not fully dispersed and the results of experiments aimed at determining the potential for using cheaper grades of precursor powder.

2. EXPERIMENTAL

2.1 Production of HA Ceramic Foams

Ceramic foam bodies were produced using three grades of HA-based powder* of varying purity; these are identified as H (high), M (medium) and L (low purity). The Ca/P ratio was 1.66 or 1.67 for all three grades. Further information on the powders may be found in Tables 1a and b.

Although a number of routes for producing foamed slips were investigated,[7] only the most successful is described here. Known amounts of HA powder and agar powder† were added to a specific volume of deionized water and a known amount of Decon 75‡ added under constant stirring with a high speed stirrer at 1200 rpm for one minute. After homogenisation, the mixture was slowly heated up to 80°C on a hotplate under constant stirring with a laboratory mixer. Meanwhile more deionised water and Decon 75 were mixed in a predetermined ratio at 1200 rpm for one minute to produce a foam. The hot slip and foam were then blended together under further stirring at 1200 rpm. After approximately 2 minutes the foam structure became uniform and the foamed slip was then poured into filter paper moulds. These containers were transferred to a refrigerator to cool the green body down to 6°C and set the gel before drying. The pH of the slips was monitored throughout production and the viscosity and zeta potential of various slip compositions was measured to determine their rheological properties.

Drying time varied from 1 to 2 days depending on room temperature and humidity. The use of filter paper boxes as moulds promoted even drying from all sides since water from the slip could penetrate the filter paper and evaporate. Some specimens were dried in an environmental cabinet in order to work under more controlled conditions. Various temperatures and humidities were used within the ranges of 45°C

---

* All powders were supplied by Jesse Shirley and Son, Stoke-on-Trent, UK.
† Agar is a linear polymer which is extracted from seaweed and is thus a heterogeneous material. It does not dissolve in water at room temperature, but gelling can be initiated at temperatures above about 70°C. A stable, heat resistant gel is then formed on subsequent cooling to below 20°C. Although gel strength is primarily determined by the agar concentration and its molecular weight, different batches can contain impurities such as polysaccharides, salts and proteins and a range of additives. The latter are generally used to produce special grades of low gelling-temperature agar and hence the temperatures quoted above can vary slightly.
‡ A general purpose surfactant added to deagglomerate the powder and encourage the foaming of the slip. Decon Laboratories Ltd, Sussex, UK.

Table 1. (a) Physical and (b) chemical analysis of the three grades of HA-based powders.

Material	Grade H	Grade M	Grade L
Particle size/μm	42% <1	40% <1	36% <1
Crystal size/μm		1	1
Moisture content	dry	dry	12.5% ± 1%
Density/g cm^{-3}	3.11	3.11	3.11

Chemical analysis/ weight %	Grade H	Grade M	Grade L
SiO_2	0.05	<0.2	1.81 ± 0.23
TiO_2		<0.01	<0.01 ± 0.002
Al_2O_3	<0.01	<0.2	0.10 ± 0.03
Fe_2O_3	<0.01	0.03	0.04 ± 0.016
CaO	54.4	54.8	53.18 ± 0.24
MgO	0.95	1.12	1.17 ± 0.03
Na_2O	0.75	0.95	0.60 ± 0.04
K_2O	0.02	0.01	0.02 ± 0.008
P_2O_5	40.9	41.0	39.91 ± 0.19
BaO		0.03	0.02 ± 0.005
SrO		0.09	0.10 ± 0.015

and 70°C and 50% to 95% relative humidity. Once dry the green bodies were carefully removed from the filter paper boxes and placed on to alumina plates in a furnace for sintering at temperatures ranging from 1000°C to 1350°C. Unless otherwise stated a heating rate of 1°C min^{-1} and a hold time of 90 minutes were used. The cooling rate averaged at 4–5°C min^{-1}.

The sintered samples were characterised by optical and scanning electron microscopy and an image analysis system* which allowed the percentage of porosity of the specimens to be determined. Pore size distributions were calculated using a purpose-written computer program which allowed for the fact that when a porous body is cut in any particular plane the statistical probability that the pores are cut at the maximum diameter is very low.

2.2 Filtration Experiments

33 mm diameter disc-shaped filters were produced by machining the sintered foams with a diamond tipped core drill. The filters were then rinsed with distilled water to remove any loose powder and other impurities and then dried. Unsintered powder was also used as a filter material to compare effectiveness. In these cases the same mass of powder as ceramic filter material was used in each individual experiment. All three filter grades had an approximate surface area of 1 m^{2}g^{-1}, whilst the powders had a surface area of approximately 9 m^{2}g^{-1}.

---

* Foster Findlay, Newcastle-upon-Tyne, UK.

Two types of heavy metal ion-containing solutions were used to determine the filtration characteristics of the ceramic and powder filters. These were as follows:

- Standards used for atomic absorption spectroscopy (AAS). Each of the solutions contained 1000 mg 1^{-1} ($\pm$ 0.5 %) of the metal ion. The metal was dissolved in approximately 1 M nitric acid, with the exception of Ti which was dissolved in 2 M hydrochloric acid and La which was supplied as 10% $LaCl_3$ solution. The pH of these solutions was generally extremely acidic. With the exception of the Si (pH = 12.3) and La (pH = 1.6) solutions, the pH generally ranged from 0.6 to 0.9. This was a disadvantage since the acid could dissolve the filter if the solution was used in very high concentrations. The standard solutions were diluted to the concentrations required for individual experiments.
- Less acidic solutions were obtained when nitrates of the metal ions were used which were dissolved in deionised water to make up standard solutions of various concentrations.

Two different filtration methods were used. The first, a column method involving only the ceramic filters, involved simply pouring the filtrate through the filter contained in a pyrex filter holder with O-ring seals at both top and bottom. The residence time depended on the structure of the ceramic foams and its thickness, but was typically about 10 seconds.

The second approach, a batch method, enabled a direct comparison of powder and ceramic filter effectiveness. The filters were completely immersed individually in the filtrate for a fixed time period after which the ceramic filters were removed and dried whilst the powder filters were retrieved using filter paper and dried.

Initial experiments focused on determining which ions could be removed from aqueous solution by HA filters. Subsequently, a matrix of experiments was performed to evaluate the influence of a number of parameters on filter effectiveness. A distinction was drawn between:

- characteristics of the filter material, such as surface area, grade and sintering temperature, and
- characteristics of the filtrate, such as ionic species present, ion concentration, filtrate pH and temperature. Filtration time was also examined in this section.

Filter effectiveness was determined by measuring the concentrations of the heavy metal ions in solution before and after filtration using atomic absorption spectrometry.

3. RESULTS AND DISCUSSION

3.1 Production of HA Ceramic Foams

Production ranges of the three different slip compositions are shown in Figs 1–3 for two stabilising additive levels. For slips with a lower solid content a drying temperature of 60°C was favourable because this meant that the time to immobilise the particles was shorter and less drainage would occur. For higher solid content slips a lower temperature, 40°C, was preferred because a low moisture gradient was desired to prevent cracking. Drainage was not the main problem in these cases. With agar

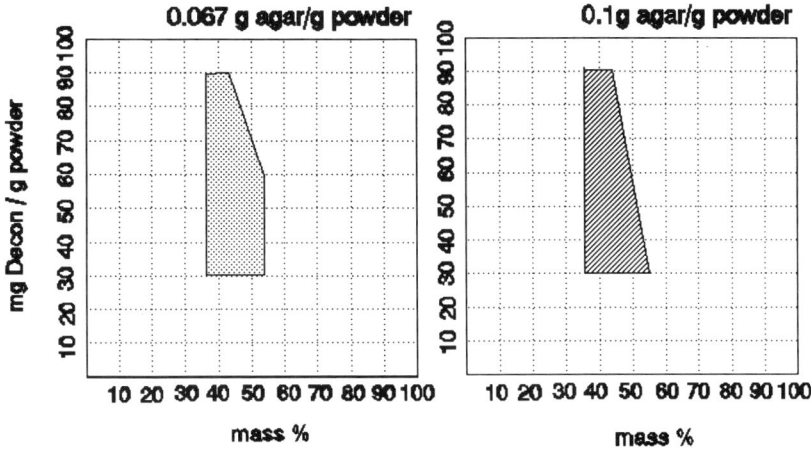

Figure 1. Production range of foamed H-grade slips with agar.

used to stabilise the slips, it was found unnecessary to dry specimens in the humidity chamber because the gel gave the specimens sufficient stability. In all cases, the shrinkage of the foam volume during drying was typically approximately 30%. The optimum sintering temperature was found to be 1350°C with a heating rate of 4–5°C min⁻¹. Specimens sintered at lower temperatures were not sufficiently strong. In these experiments it was found necessary to maintain the sintering temperature for 90 minutes and then allow the furnace to cool at an average rate of approximately 1°C min⁻¹. In this way thermal stresses in the sintered foams were reduced. The average shrinkage in volume of the porous specimens during sintering was approximately 60%.

The H-grade ceramics did not show well defined pores (see Fig. 4a) due to the thin walls between the pores rupturing during processing as a result of drainage. During sintering the edges of these ruptured walls were rounded off producing a

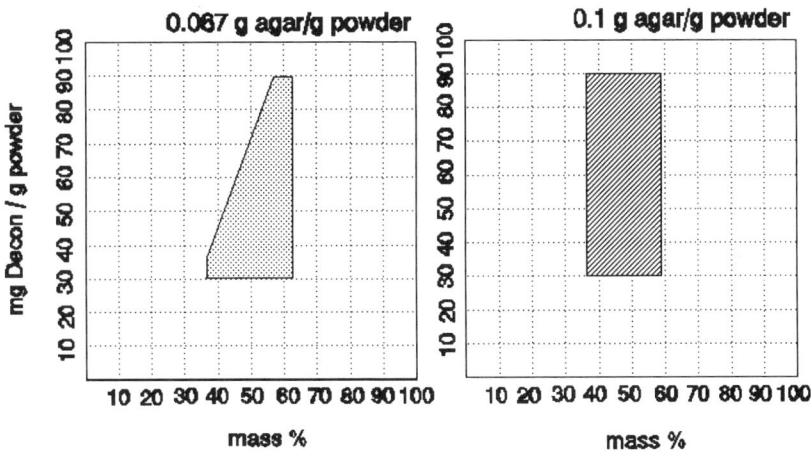

Figure 2. Production range of foamed M-grade slips with agar.

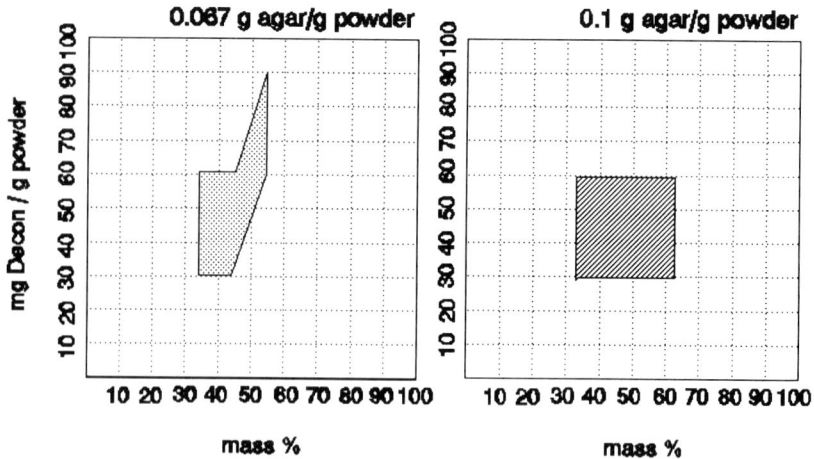

Figure 3.   Production range of foamed L-grade slips with agar.

network of open channels which made it impossible to determine pore size distributions. The bridges showed 'microporosity' in the range of <1 to 2 µm at high magnifications (see Fig. 4b). These micropores arose from the non-fully dispersed nature of the ceramic slip and resulted in a higher surface area. The grain size was fairly uniform at about 3 µm.

The M-grade ceramics exhibited more structured pores (see Fig. 5a) compared to the purer H-grade filters but a less uniform microstructure. Again the interpore walls ruptured at their thinnest parts during production however the sharp edges remained during subsequent processing. Higher magnifications (see Fig. 5b) revealed that the average grain size was larger than in the H-grade filters, individual grains varying between 2 and 12 µm in size, and the majority of the grains were not spherical. The bridges showed microporosity, but were denser than the H-grade filters.

For the L-grade ceramics, well separated pores were clearly detected at low magnifications (see Fig. 6). Again the pore walls ruptured during production at the thinnest part. The entire network appeared less structured than the M-grade ceramics but

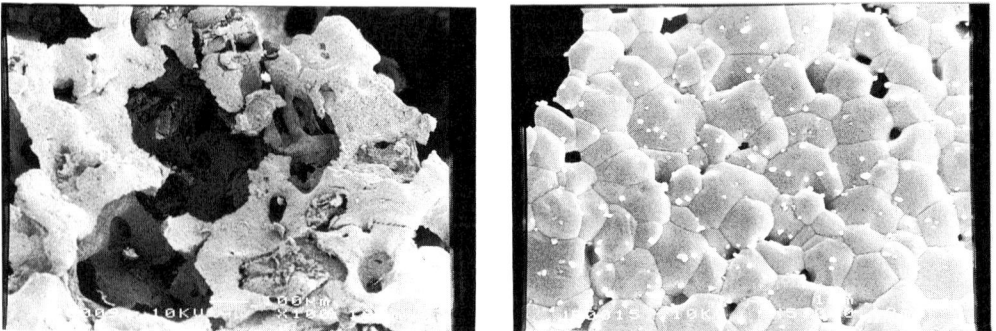

Figure 4.   Micrograph of an H-grade filter at a magnification of (a) × 100 and (b) × 5000.

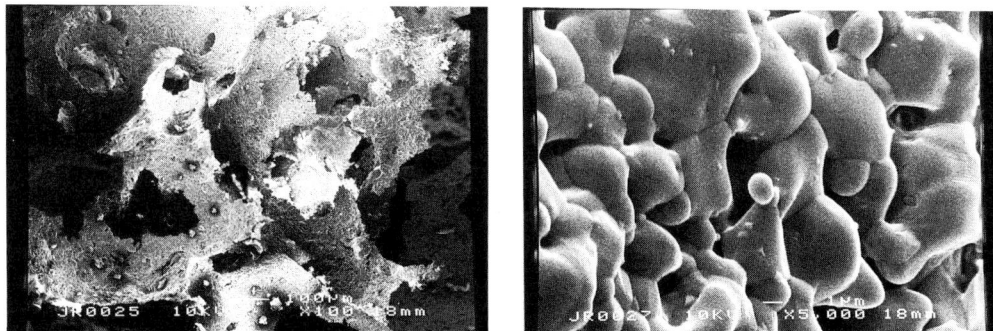

Figure 5.   Micrograph of an M-grade filter at a magnification of (a) × 100 and (b) × 5000.

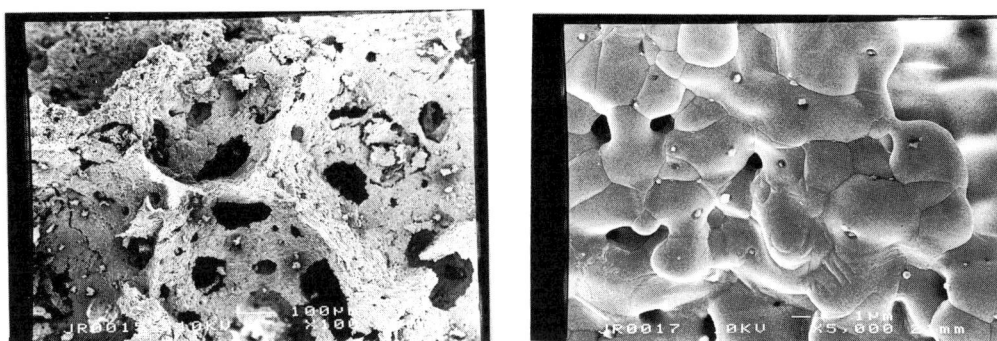

Figure 6.   Micrograph of an L-grade filter at a magnification of (a) × 100 and (b) × 5000.

more structured than for the H-grade material. The pore bridges exhibited a very small amount of microporosity and showed a grain size of 2 to 5 μm. The grains seemed well bonded and the struts denser than those of the other filters.

3.2 Filtration Experiments

Table 2 divides the full range of ions examined into two groupings, those that could be at least partially removed from solution and those that were not affected by any of the filtration processes described above.

Figure 7 shows that Pb^{2+} removal for all grades decreased with increasing sintering temperature and an increasing purity of the filter material. This situation was also found with the Cr^{3+}, Al^{3+} and Cu^{2+} ions, that is the purest material removed the least ions from the filtrate. In general, the lower the purity of the HA the higher the fraction of lattice defects which will be present due to the presence of impurity ions in the lattice. These defects will form ideal sites for ion exchange or adsorption due to the lower energy required for the processes to occur compared with the ideal crystal

Table 2. Classification of heavy metal ions in terms of success for removal by HA filters.

Ions removed by filtration	Ions not removed by filtration
Cr^{3+}	Ca^{2+}
Co^{2+}	K^+
Ni^{2+}	Na^+
Al^{3+}	Ti^{2+}
Cu^{2+}	V^{5+}
Pb^{2+}	La^{3+}
Fe^{3+}	Mg^{2+}
	Zn^{2+}
	Si^{4+}

structure. This theory would account for the superior ability of the M- and, particularly, L-grades in removing ions from solution when compared to grade H. It is also a very satisfying result from an economic point of view since the cost of the M- and L-grades was much lower than that for the purer H-grade. However, this cannot be the complete story since, as will be seen later in Figs 10 and 11, the H-grade filters were the most effective of the three in removing Co^{2+}, Ni^{2+} and Fe^{3+} ions. Further work is required to understand the complete relationship between grade and ion removal capability.

Figures 8 and 9 show the ion removal properties of equal weights of H-grade powder and ceramic filters for the full range of removable ions. As expected, due to

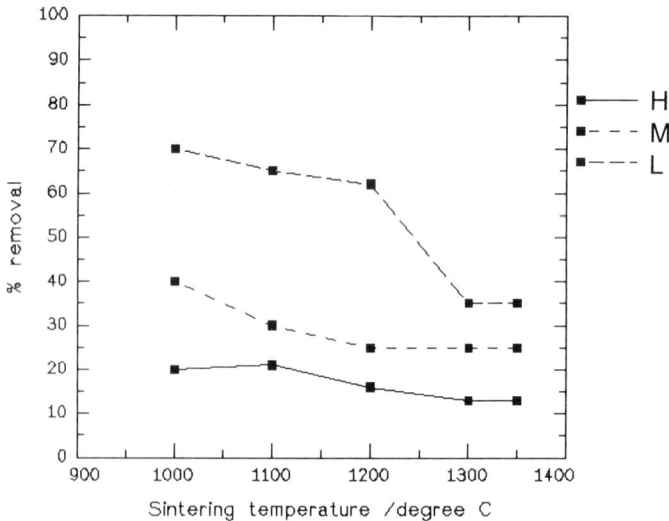

Figure 7.   Effectiveness of 1 g ceramic filters, sintered at different temperatures, in removing Pb^{2+} ions from 30 ml of a 10 mg/l Pb^{2+} filtrate after 1 h.

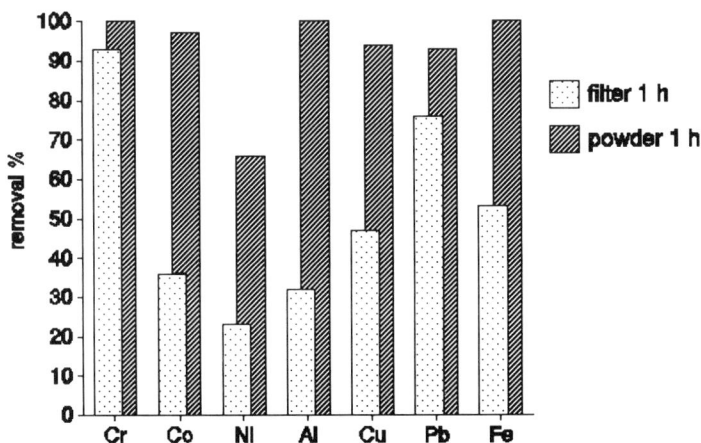

Figure 8.   Comparison of removal effectiveness of equal masses of H-grade powder and ceramic filters (filtrate volume 30 ml, filtration time 1 h).

its higher surface area the powder was found to be much more effective at removing heavy metal ions from solution, particularly at the shorter filtration time. The results shown in Fig. 10 show the effect of increasing filter mass (and hence surface area) on the removal of Pb^{2+} ions. Only 3.5 g of powder was sufficient to remove all the ions within the limits of measurement error, whilst 6 g of ceramic filter only removed approximately 40%.

Figures 11 and 12 show the removal efficiency of all 3 filter grades for all the ions removable from solution over 1 and 24 hours respectively. More detailed results for the removal of Pb^{2+} ions by L-grade ceramic filters are presented in Fig. 13. As expected ion removal is initially very rapid but decreases as a function of time as the concentration of the heavy metal ions in the filtrate was reduced and hence the

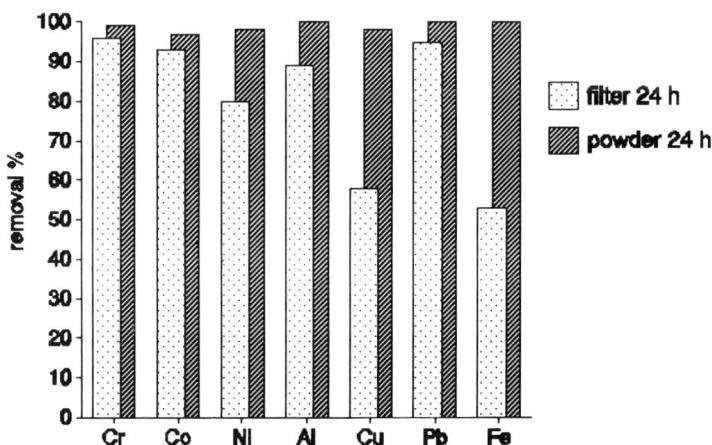

Figure 9.   Comparison of removal effectiveness of equal masses of H-grade powder and ceramic filters (filtrate volume 30 ml, filtration time 24 h).

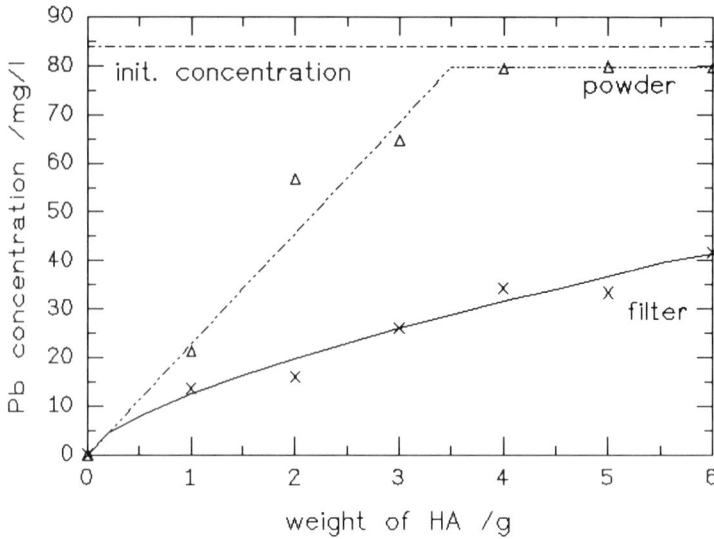

Figure 10. Ion removal from a 30 ml solution of 84 mg/l Pb^{2+} by various masses of H-grade powder and ceramic filters after a filtration time of 10 min.

average diffusion distances of the ions in the bulk of the solution towards the surface of HA became longer. In addition, as the surface exchange or adsorption sites became exhausted the ions had to move deeper into the filter. Evidence was found for Fe^{3+} saturation with all filter grades and Cu^{2+} with the H-grade.

Evidence for the effect of initial concentration has already been seen in Fig. 13. Figure 14 shows that the same behaviour was found for all filter grades. An increase in the filtrate concentration decreases the removal rate although it should be remembered that the absolute number of ions removed is greater.

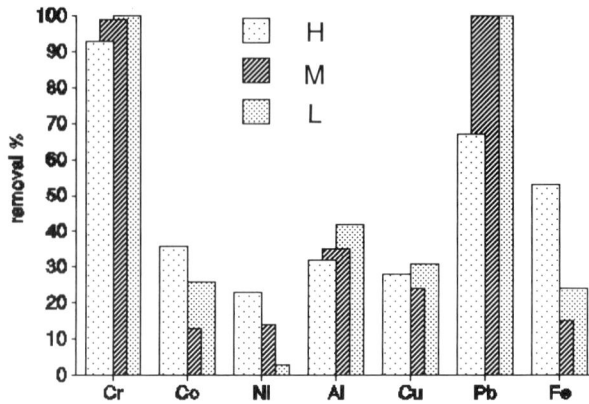

Figure 11. % ion removal results for ceramic filters of all grades, 1 h filtration time, 30 ml filtrate volume.

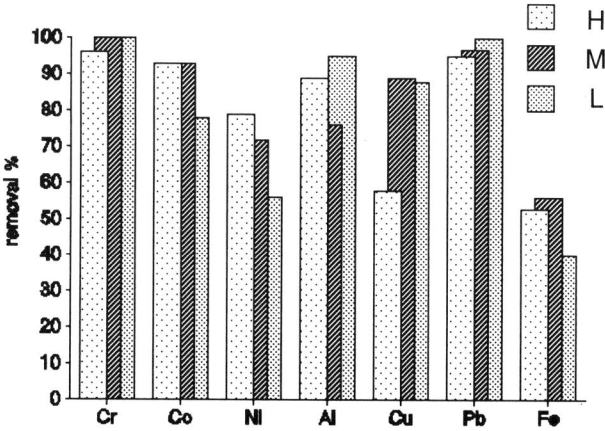

Figure 12.   % ion removal results for ceramic filters of all grades, 24 h filtration time, 30 ml filtrate volume.

It should be noted that in all cases the quantity of Ca^{2+} ions increased after filtration, even when no heavy metal ion removal had occurred. This was due to the acidity of the filtrates which chemically attacked the HA. Table 3 shows the increasing Ca^{2+} ion release values in moles per litre for each grade of ceramic filter as a function of pH. To ensure that the results corresponded to the ion filtration tests presented in Fig. 11, the same filter masses were used in each individual experiment. It can be observed that with increasing solution pH the Ca^{2+} ion release per gram of filter decreased.

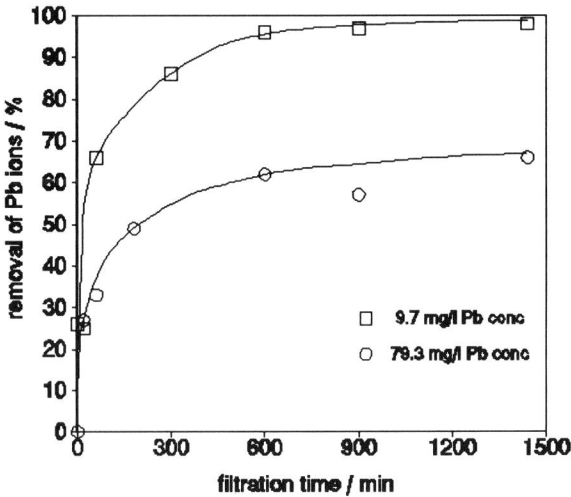

Figure 13. Pb^{2+} ion removal versus filtration time for L-grade filters using 30 ml of filtrate with an initial Pb^{2+} concentration of i) 9.7 mg l^{-1} (5.7 g filters), and ii) 79.3 mg l^{-1} (1 g filters).

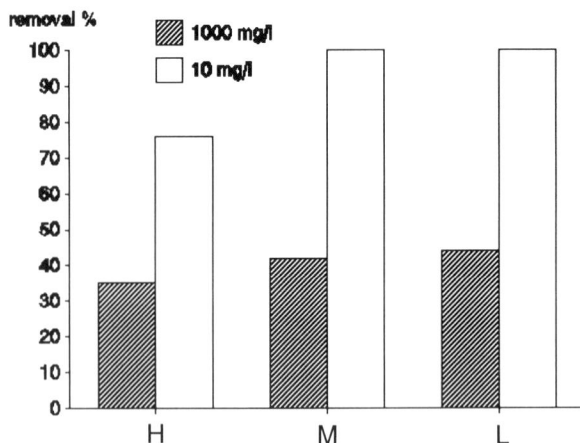

Figure 14.   Comparison of filtration effectiveness for all filter grades at different initial
Pb^{2+} concentrations; 5 g filters, 30 ml filtrate, 1 h filtration time.

Figure 15 shows the influence of temperature on the removal efficiency of L-grade filters. The removal level increased for all ions when the filtrate temperature was increased from 20°C to 40°C. However increasing the temperature to 60°C only significantly increased the percentage removal further in the case of the Cr^{3+}, Co^{2+} and Ni^{2+} filtrates.

3.3 Ion Removal Mechanism

Initially it was assumed that ion exchange was the only operational mechanism for HA since it features strongly in all the literature (see for example references 12 and 8) and ion removal by adsorption remains unaddressed. During experimentation however, it became increasingly apparent that ion adsorption was responsible for at

Table 3.   Ca^{2+} ion release of the filters as a functon of filtrate pH. Results correspond to
the 1 hour ion filtration tests (see Figure 11). Ions in brackets refer to the equivalent
filtration test.

pH	Filter mass/g	Ca^{2+} ion release after 1 h/mol 1^{-1}		
		Grade H	Grade M	Grade L
0.7	3.4 (Cu^{2+})	0.02545	0.03393	0.02375
0.7	3.2 (Al^{3+})	0.02396	0.03194	0.02375
1.0	5.0	0.03493	0.04590	0.03293
1.5	3.5 (Ni^{2+})	0.01477	0.01188	0.00849
1.7	3.9 (Pb^{2+})	0.01168	0.00681	0.00487
1.8	3.9 (Co^{2+})	0.01128	0.00642	0.00389
1.8	3.3 (Fe^{3+})	0.00897	0.00560	0.00339
1.9	5.0 (Cr^{3+})	0.01317	0.00689	0.00494
2.0	5.0	0.01347	0.00399	0.00299
3.0	5.0	0.00699	0.00070	0.00040

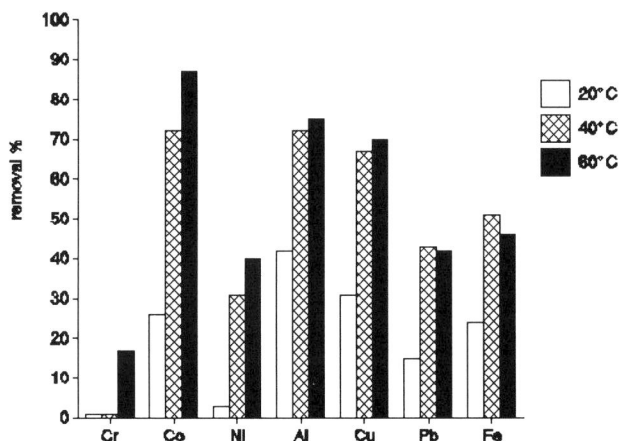

Figure 15.   % ion removal for various ions from 30 ml filtrate at various temperatures after 1 h (Cr^{3+} and Pb^{2+} 10 min) by L-grade filters.

least part of the filtration capability of HA. From a practical point of view it matters little whether the uptake of ions is achieved by ion exchange or adsorption, but an attempt was made to find out which process took place or whether it was a combination of both.

Ion exchange is a stoichiometric process. Every ion that is removed from the filtrate is replaced by another ion from the filter. In adsorption on the other hand the ions are taken up by the filter without being replaced by another species. This distinction seems clear cut. However, it is difficult to apply it in practice since most ion exchange processes are accompanied by adsorption.

If the heavy metal ions were exchanged with the Ca^{2+} ions in the HA lattice the spacings between the lattice planes would have to expand to make room for a bigger ion or would have to contract if a smaller ion than Ca^{2+} was incorporated. Although XRD is a very sensitive means of determining cell parameters, no significant change in the a and c lattice parameters were noted after ion filtration. This might have been due to insufficient ion removal from solution. A similar lack of proof was forthcoming from the EDX analyses. Again, possibly due to insufficient ion removal from solution and the resolution limits of the technique, EDX failed to detect the presence of heavy metal ions in the structure of the HA after filtration. Hence ion exchange could not be ruled out by these results, but no positive indication of the process could be provided either.

Zeta potential measurements were more informative. If the positively charged heavy metal ions were adsorbed on to the surface of the HA then, for all ions, the surface charge would be expected to increase (or become less negative). If the ions were being exchanged into the lattice then the zeta potential would increase for exchange of Ca^{2+} with trivalent cations but would remain unchanged for the divalent cations. The results of the zeta potential measurements are shown in Fig. 16. They show that ion filtration increased the zeta potential significantly in the case of all ions

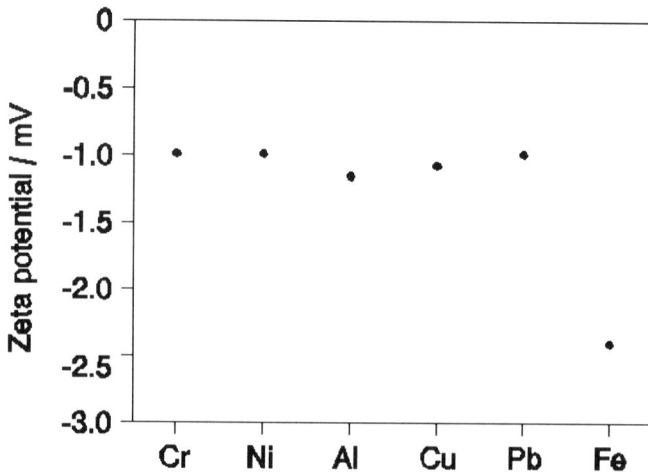

Figure 16.   Zeta potential of H-grade powder samples after use as filters for 30 ml of various ion containing solutions for 1 h. (Original zeta potential: −10.2 mV).

from −10.2 to ≤−2.5 mV (with a measuring inaccuracy of about 20%). Hence the results of Fig. 16 can be taken as a clear indication of ion adsorption having been present with no evidence for ion exchange.

The amount of Ca^{2+} released during filtration ($[Ca]_{measured}$) can be taken as a further indicator for ion exchange or ion adsorption. If ion exchange was the exclusive mechanism of ion removal, every mole of heavy metal ion taken up by the filter must have caused the release of 1 mol of Ca^{2+} ions into the filtrate. This fraction of the Ca^{2+} release is named $[Ca]_{ion\ exchange}$. However, Ca^{2+} ions will also be released due to the dissolution of the HA in the highly acidic solutions used, this quantity is named $[Ca]_{pH}$. Thus if ion exchange was the only mechanism for ion removal from the filtrate then:

$$[Ca]_{measured} = [Ca]_{ion\ exchange} + [Ca]_{pH}$$

or:

$$\frac{[Ca]_{measured}}{[Ca]_{pH} + [Ca]_{ion\ exchange}} = 1 \qquad (1)$$

The lower the ratio below 1, then the greater the amount of adsorption that must have taken place. A ratio of > 1 should not occur in theory.

The Ca^{2+} concentration after filtration ($[Ca]_{measured}$) and the Ca^{2+} release due to the high acidity of the filtrate ($[Ca]_{pH}$) were determined from Table 3, and the $[Ca]_{ion\ exchange}$ values determined from the number of moles of heavy metal ions removed from solution assuming that the only mechanism was ion exchange. From these values the ratio in eqn 1 was calculated for each grade of filter; these are

Table 4. $[Ca]_{measured}/([Ca]_{ion\ exchange} + [Ca]_{pH})$ **ratios determined from equation 1 for a range of ion solutions and all three filter grades.**

Ion solns Filter grade	Co^{2+}	Cu^{2+}	Ni^{2+}	Pb^{2+}	Al^{3+}	Cr^{3+}	Fe^{3+}
H	0.89	2.33	1.08	0.65	0.02	0.73	1.32
M	0.27	1.31	0.54	0.58	0.01	0.32	0.45
L	0.22	1.55	0.35	0.52	0.02	0.12	0.77

shown in Table 4. A number of observations result. First, that for all ionic species except Cu^{2+} the ratio of the Ca^{2+} concentration in the filtrates after filtration, $[Ca]_{measured}$, to the Ca^{2+} release measured in the reference system where the pH of the solution was the same but heavy metal ions were absent, $[Ca]_{pH}$, is significantly less than 1. This is a clear indication of ion adsorption having taken place. Second, it may be seen that there is an increasing trend towards adsorption and away from exchange as the purity of the hydroxyapatite filters decreases, except for Al^{3+} for which apparently negligible exchange occurs for any of the filters. It is believed that the variation in nature of the removal mechanism between the different ions as a function of hydroxyapatite grade is probably linked to the different removal tendencies discussed earlier. No clear trends are readily identifiable however and further work would be required to determine the precise nature of any relationship.

In the case of the Cu^{2+} filtration there are several possible explanations of the high ratios. Firstly, it is known that the presence of Cu^{2+} in the filtrate interferes with the measurement of Ca^{2+} by AAS. Secondly a (Cu/Ca) complex might have been formed in the filtrate. The free Ca^{2+} ion concentration would therefore be reduced encouraging further dissolution of Ca^{2+}. Thirdly, a Cu-complex might have formed on the surface of the HA as reported by Shimabayashi et al.[9]

4. CONCLUSIONS

Hydroxyapatite-based ceramic foams with densities > 15% and an open cell structure have been produced by the foaming of stabilised hydroxyapatite-based ceramic slips. These foams can be used to remove a number of different heavy metal ions from aqueous solutions. In particular, the following specific conclusions can be drawn:

- All grades of HA were found to be capable of removing Al^{3+}, Co^{2+}, Cr^{3+}, Cu^{2+}, Fe^{3+}, Ni^{2+} and Pb^{2+} ions from solution. The ion removal level depended on experimental conditions, but virtually 100% ion removal could be achieved in most cases under the correct conditions.
- In order to achieve maximum ion removal levels the filtration parameters required were found to be:
  - use of impure rather than pure grades of HA. It is believed that increased impurity levels resulted in increased numbers of lattice defects which are ideal adsorption/exchange sites.

- a high surface area. In this respect the powders were superior to the ceramic filters and for the latter a low sintering temperature was preferable.
- long filtration times. Since removal rate decreases with increasing filtration time, a suitable compromise was found to be a period of one hour.
- a low pH. This encouraged initial dissolution of the HA which resulted in an increased removal level. It is believed that the HA was protected from continued acid attack by the adsorption of the heavy metal ions on the material's surface.
- a high filtrate temperature. With increasing temperature the diffusion rate of the ions increases, leading to higher ion removal levels.
- Ion adsorption was positively detected as a mechanism of ion removal. Ion exchange was not observed but could not be completely ruled out.

ACKNOWLEDGEMENTS

The authors would like to thank Dytech Corporation Ltd, Sheffield, and Jesse Shirley and Son Ltd, Stoke on Trent, for financial support.

REFERENCES

1. R. D. Letterman: *Management and Operations Journal AWWA,* Dec 1987, 26–32.
2. Maruzen: *Kagaku-Kougaku-benran*, 5th Edition, Tokyo, 1988, 1315.
3. J. C. Heughebaert and G. Bonel: *Biological and Biomechanical Performance of Biomaterials*, P. Christel, A. Meunier and A. J. C. Lee eds, Elsevier Science Publishers, Amsterdam, 1986, 9–14.
4. J. Reichert and J. G. P. Binner: Accepted by *J. Mater. Sci.*
5. A. Slósarczyk and J. Parzuch: *Sprechsaal*, 1989, **122**(8), 745–746.
6. W. D. Kingery, H. K. Bowen and D. R. Uhlmann: 'Introduction to Ceramics', 2nd edition (John Wiley & Sons, New York, 1976), 484–485.
7. J. G. P. Binner and J. Reichert: Accepted by *J. Mater. Sci.*
8. S. Suzuki, T. Fuzita, T. Maruyama, M. Takahashi and Y. Hikichi: *J. Am. Ceram. Soc.*, 1993, **76**, 1638–1640.
9. S. Shimabayashi, C. Tamura and M. Nakagaki: *Chem. Pharm. Bull.*, 1981, **29**(8), 2116–2122.

Rheological Behaviour of Aqueous Injection Moulding Mixtures

R. J. HUZZARD and S. BLACKBURN*

*Interdisciplinary Research Centre in Materials for High Performance Applications,
*School of Chemical Engineering, University of Birmingham, Edgbaston,
Birmingham, B15 2TT, UK*

ABSTRACT

The rheology of a water-soluble system for the formation of alumina and other components was evaluated. Capillary rheology measurements were carried out on a paste based on a hydroxy-propylmethylcellulose (HPM5000DS) at six temperatures between 18°C and the gel temperature. The viscosity depends exponentially upon the inverse of absolute temperature up to 40°C where the influence of hydrophobic interaction between cellulose molecules produces higher viscosities than predicted. The optimum temperature for moulding was found to be close to 50°C, just below the gel point of the binder. A cylindrical mould was filled with no significant deformation on release from the die under the optimised conditions.

1. INTRODUCTION

Injection moulding of complex shapes is well established in the plastics industry. This technique has also gained favour as a route for the production of ceramics where the binder system, usually a wax or thermoplastic, is removed and the powder sintered to confer the final properties required.[1,2] The technique requires the control of defects in the compact.

Aqueous binder systems which use low moulding temperatures are available.[3,4,5] In these systems, moulding temperature and die temperature can be similar, reducing shrinkage and consequently producing fewer moulding defects. Most of the binder system can be driven off as water vapour, cutting process time, reducing burn-out problems and leaving a rigid structure with better shape retention. Solutions of cellulose derivatives have low viscosities below the gel point, and gels form on heating. Injection moulding mixes based on cellulose derivatives therefore require a barrel temperature below the gel point and a die temperature above the gel point.

The rheological properties of the paste have important consequences for mould filling and the subject has received considerable attention in the literature.[1,2,6] Rheology has generally been modelled by considering the variation of shear stress with shear rate in a capillary rheometer. The usual method is to fit the resultant line or curve to a standard rheological model. Account is taken of capillary entry effects using the Bagley correction and any apparent viscosity dependence of shear rate must be considered, for example, using the Rabinowitsch correction.[7] Newtonian flow produces a straight line passing through the origin with a gradient defined as the viscosity of the material. Where the line cuts the shear stress axis at the yield stress the flow is termed 'Bingham'. The more general Herschel-Bulkley model may be used to describe a power law relationship with a yield value. This model closely describes the behaviour of many pastes, but often breaks down at low rates of shear. Although the capillary rheometer mirrors the injection of a paste into a mould and gives some insight into flow in the mould, it is not suitable for the low shear rate required to predict the yield stress. Prediction of yield stress by extrapolation is unreliable, but difficult to improve upon. Yield stress has been measured using alternative methods such as plastometers or cone and plate rheometers but it is difficult to take account of departures from Herschel-Bulkley behaviour and the methods are based on different assumptions. Many references use the term 'apparent viscosity' to describe the relationship between shear stress and shear rate in non-Newtonian systems. A review of ceramic injection moulding[6] suggested that an apparent viscosity of less than 1000 Pa.s was required for successful moulding over the usual shear rate range for moulding of 100 to 1000 s^{-1}.

Most of the examples found in the literature relate to injection moulding formulations based on conventional binder systems, but the rheology of extrusion mixes based on hydroxypropylmethylcellulose has been studied in detail[8] and Zhang[9] considered formulations based on agar for injection moulding. Hydroxypropylmethylcellulose (HPM5000DS), a cellulose derivative with high gel strength, is studied here to test suitability for use in ceramic injection moulding.

2. EXPERIMENTAL PROCEDURES

A reactive, calcined alumina, RA107LS from BA Chemicals Ltd, was used for the moulding formulations and had an average particle size of 0.5 μm. An 8 wt% solution of a commercially available cellulose derivative, hydroxypropylmethylcellulose (HPM5000DS), was used as the binder.

Premixing was carried out in a Kenwood planetary mixer prior to 30 minutes cooled mixing in a Werner and Pfleiderer double lobe mixer. As a final mixing stage, the paste was extruded through a 6 mm die prior to evaluation.[10] The paste composition is given in Table 1.

The system was investigated by capillary rheometry. A barrel and piston arrangement (Fig. 1a) was used to force the paste through dies of 2.03 mm diameter and 12.5, 25 and 37.5 mm in length. A water jacket with pumped circulation was placed around the barrel to facilitate controlled heating. The dies were mounted inside the barrel to

Table 1.   Paste composition

Constituent	Weight (%)
RA107LS	79.4
De-ionised Water	19.0
HPM5000DS	1.6

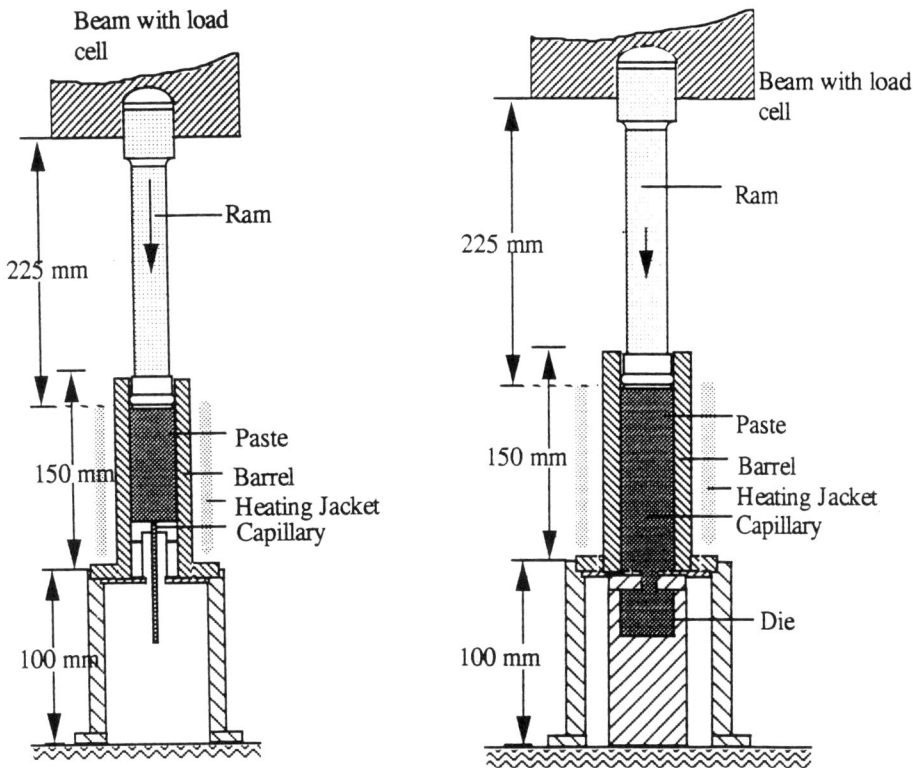

Figure 1(a).   Capillary rheometer measurement equipment
(b) Laboratory injection moulding equipment.

avoid heat loss. An instrumented Avery-Denison load frame provided the force to the ram to give the required extrudate velocities of 0.0026, 0.0052, 0.013, 0.026, 0.052, 0.13 and 0.26 ms^{-1}. The results were obtained at 18°C by first refrigerating the paste and rheometer and at 25, 40, 49, 59 and 69°C as follows. Both the capillary rheometer and paste were preheated to temperature before the barrel was charged and the water jacket attached. The test was carried out when the temperature of the paste in the barrel stabilised. Pressure on the ram rose to a maximum and the material was forced through the die. The pressure was recorded when it stabilised just below the maximum value.

Injection moulding was carried out using the arrangement shown in Fig. 1b with a mould attachment. An injection speed of 150 mm min^{-1} was used. This gave a volumetric flow rate of 1250 mm^{3}s^{-1} into a cylindrical mould of 27.5 mm diameter by 18 mm length. The barrel and die temperatures were 50°C and 80°C respectively. Residence times in the mould of 2 and 5 minutes were used. Distortion of the cylinders was measured by placing the domed end surface on a flat glass plate and measuirng the gap between the plate and the edge of the moulding with feeler gauges.

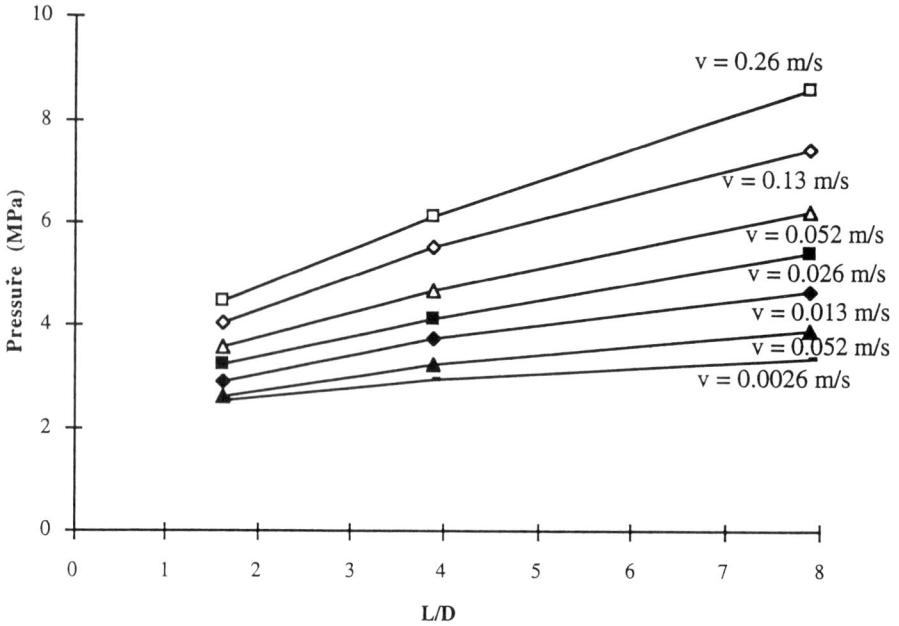

Figure 2.   Pressure versus die length divided by diameter at 40°C.

3. RESULTS AND DISCUSSION

The plot of extrusion pressure against die length (L) divided by die diameter (D) approximated to a straight line at each temperature and velocity, as illustrated in Fig. 2. The intercept at L/D = 0 gave the pressure due to die entry and exit effects and is used to apply the Bagley correction.[7] Shear stress and shear rate were calculated with a preliminary assumption of Newtonian behaviour using.

$$\tau = (P - P_0)D/4L \tag{1}$$

$$\mathring{\gamma} = 8V/D \tag{2}$$

where τ is the shear stress, $\mathring{\gamma}$ is the shear rate, P_0 is pressure at $L/D = 0$, P is pressure, V is the extrudate velocity

The log shear stress against log shear rate gave a straight line indicative of power law behaviour. The shear rate was recalculated using the Rabinowitsch correction

$$\mathring{\gamma} = \frac{8V}{D}\left(\frac{3n + 1}{4n}\right) \tag{3}$$

$$\text{where } n = \frac{\partial \log \tau}{\partial \log \mathring{\gamma}} \tag{4}$$

to take account of the power law behaviour.

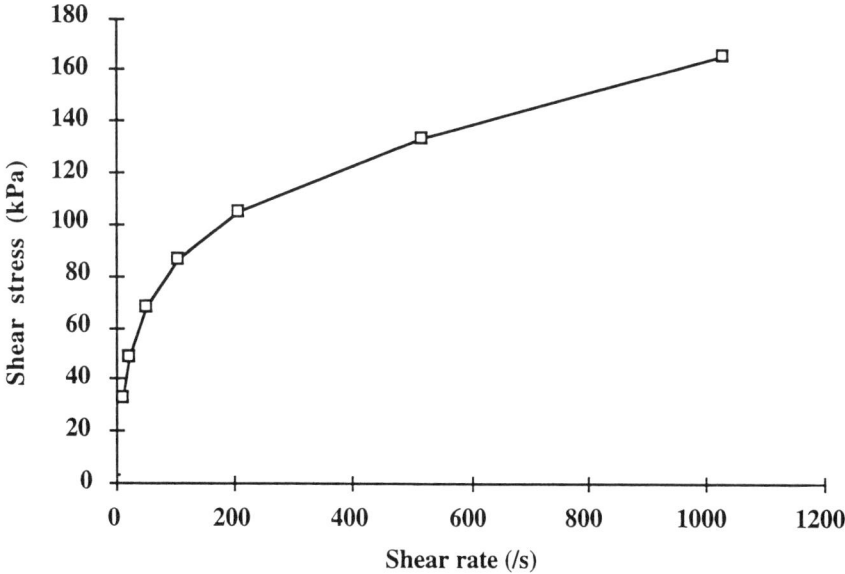

Figure 3. Corrected shear stress against corrected shear rate at 40°C.

The corrected shear stress against shear rate curve is shown in Fig. 3. By extrapolation of this curve the yield stress τ_0 was estimated to be approximately 4 kPa. Power law behaviour with a yield value suggested that the relationship is governed by

$$\tau = \tau_0 + k\dot{\gamma}^n \tag{5}$$

where the consistency, k is a power law parameter of the paste.

Figure 4 shows a log/log plot of the corrected shear stress against shear rate. Approximate yield stress values were obtained at each test temperature by this method. In addition, shear stress values were read from all of the shear stress-shear rate curves at shear rates of 20, 100 and 1000 s^{-1} (Fig. 5). For each shear rate, the shear stress was relatively high at low temperatures, dropping to a minimum at around 49°C. At 59°C the shear stress rose slightly suggesting a greater degree of cross linking of the binder with the onset of gellation. Significantly higher shear stresses were observed at 69°C, suggesting that the gel point lies between 59 and 69°C.

Apparent viscosity, η_{app} was calculated from shear stress and shear rate according to

$$\eta_{app} = \tau/\dot{\gamma} \tag{6}$$

The viscosity shear rate relationship shown in Fig. 6 followed similar trends to the variation of shear stress with temperature.

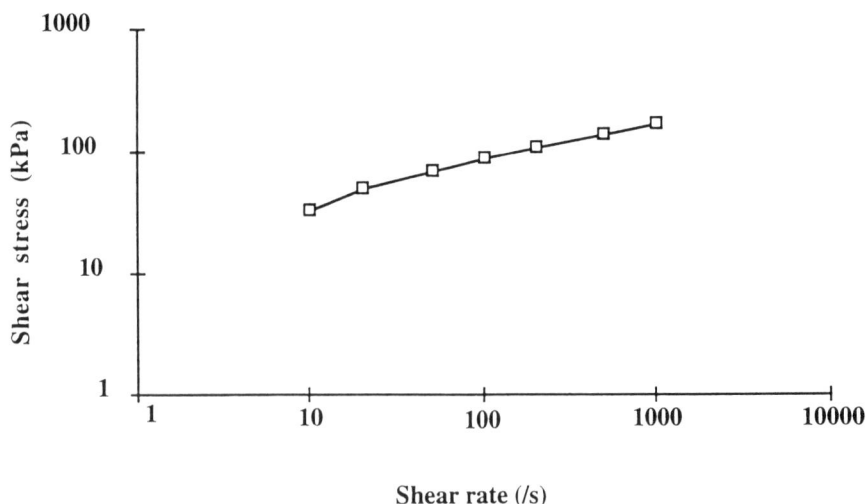

Figure 4.   log log scale plot of corrected shear stress against corrected shear rate at 40°C.

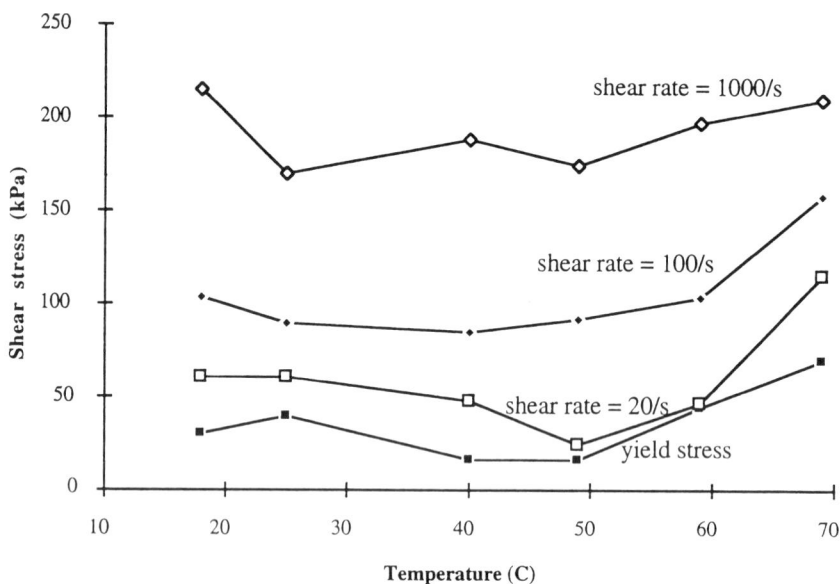

Figure 5.   Corrected shear stress versus temperature.

Figure 7 shows log inverse viscosity against the inverse of the absolute temperature. The lower temperature region followed the Arrhenian behaviour expected from a thermally activated process. Above 40°C, the apparent viscosity was much higher than expected from this process. This was probably caused by the hydrophobic interactions responsible for the formation of cellulose gels. The mechanism may have been accelerated by the excess of the alumina removing water and leaving the cellulose solution more concentrated than the formulation suggested.

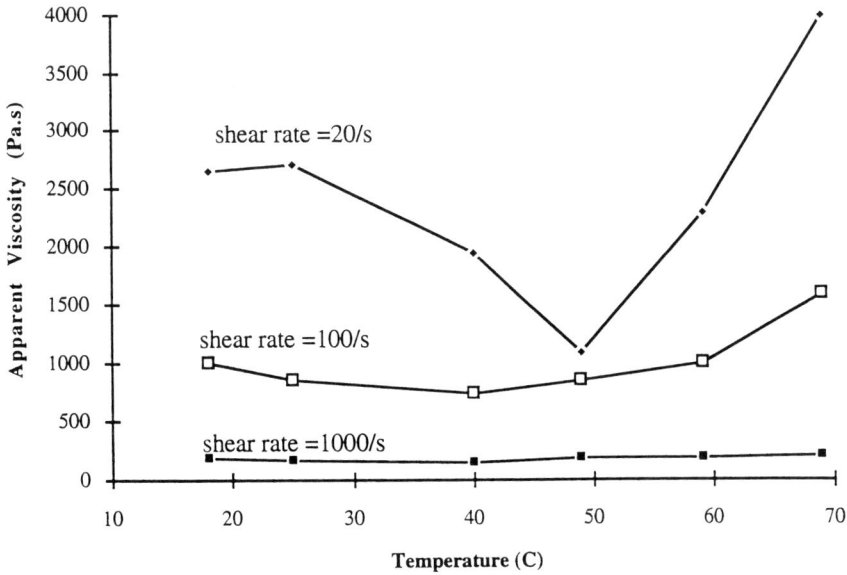

Figure 6.   Apparent viscosity versus temperature.

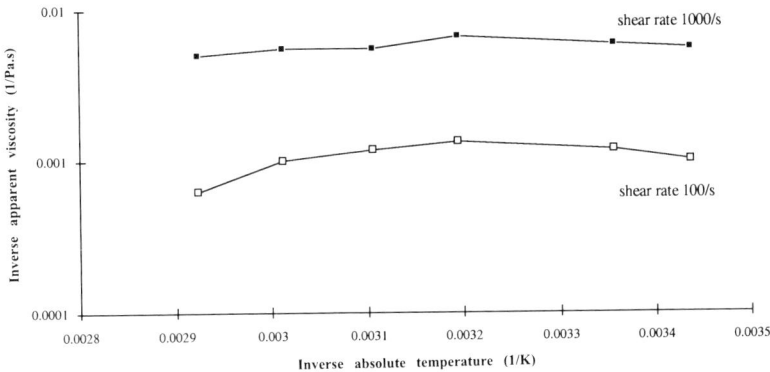

Figure 7.   Log scale of apparent viscosity versus inverse absolute temperature.

During injection moulding of the cylinders, there was an initial pressure rise to overcome the yield value of the paste in the barrel and force the paste through the gate into the empty die. The pressure then rose gradually as the mould filled. A rapid increase in pressure occurred as the air was forced out of the mould.

The paste showed unacceptable levels of distortion (0.4mm) when ejected after 2 minutes in the die with severe cracking resulting. A rigid cylinder was formed by increasing the residence time in the mould prior to ejection to five minutes. The cylinder was removed without significant distortion (0.1mm) or cracking.

4. CONCLUSIONS

An injection moulding formulation based on HPM5000DS was successfully moulded at 50°C. The rheological properties satisfy the requirements for injection moulding with a viscosity of less than 1000 Pa.s over a shear rate range of 100–1000 s^{-1} and the inherent high strength of the binder solution selected allows release from the mould without damage. The fluidity $(1/\eta)$ follows the behaviour expected from a thermally activated process up to 40°C. Hydrophobic interactions within the gel cause an increase in viscosity as the gel point is approached.

ACKNOWLEDGEMENT

Financial support for this work has been provided by EPSRC. Many thanks are due to Dr. R. Oliver, school of Chemical Engineering, University of Birmingham, for his guidance on the rheological aspects of the work.

REFERENCES

1. R. M. German: *Powder injection moulding*, Metal Powder Industries Federation, 1990.
2. B. C. Mutsuddy: Ceramic Injection Moulding, Chapman & Hall, 1995.
3. R. D. Rivers: U.S. Pat No. 4 113 480, 1978.
4. A. J. J. Fanelli: *Am. Ceram. Soc.*, 1979, **72**(4), 1073.
5. J. E. Schuetz: *Ceram. Bull.*, 1986, **65**(12), 1556.
6. M. J. Edirisinghe: *J of Mater*. Sci, 1987, **22**, 269.
7. H. A. Barnes, J. F. Hutton and K. Walters: *An introduction to rheology*, Elsevier, 1989.
8. V. F. Janas: 'Flow and microstructure of dense suspensions', *MRS symposium proceedings*, 1992, **289**, 123.
9. T. Zhang: *British Ceramic Transactions*, 1994, **93**(6), 229.
10. H. Böhm and S. Blackburn: *J. Mater. Sci.*, 1994, **29**, 5779.

The Effect of Absolute Pressure on the Processing Parameters of Ceramic Pastes

D. R. OLIVER and M. WHISKENS

School of Chemical Engineering, Birmingham University, Edgbaston, Birmingham, B15 2TT, UK

ABSTRACT

A new form of apparatus is described which allows the absolute pressure to be changed whilst paste is forced to flow at a fixed rate into an orifice. It is also possible to carry out a similar test in which paste is forced along the cylinder barrel (with no orifice present) at a series of different absolute pressures.

It is shown that the quantity σ_o which controls flow *into* the orifice[1] is little dependent on absolute pressure ($2 \times 10^5 - 3 \times 10^6$ N/m²) and in some cases may even decrease with increased pressure. The quantity τ_0, which is the shear stress at the cylinder wall[1] however, always increases with increasing absolute pressure in a way which has been noted previously.[2,4,6] Possible mechanisms for these effects and their industrial implications are discussed.

The pastes are mainly alumina-based, with different liquid binders, though one paste is hydrated aluminium acetate. A constant flowrate of paste is used in these tests.

1. INTRODUCTION

Paste forming processes such as extrusion and injection moulding[3] involve the application of considerable pressure to the paste. Since pure liquids and polymer solutions undergo only minor viscosity changes within similar ranges of pressure, it is tempting to assume correlations for paste flow which are independent of absolute pressure. In the Benbow-Bridgwater equations* for paste flow through a die[1]

$$P = 2\ln\left(\frac{D}{D_o}\right)(\sigma_o + \alpha V_o) + \frac{4L_o}{D_o}(\tau_o + \beta V_o) \tag{1}$$

and

$$P = 2\ln\left(\frac{D}{D_o}\right)(\sigma_o + \alpha^I V_o{}^n) + \frac{4L_o}{D_o}(\tau_o + \beta^I V_o{}^m) \tag{2}$$

the absolute pressure is not included. The first term in each equation is the pressure required to force the paste *into* the die and the second term the pressure required to

---

* See 'Nomenclature' on p. 96

force the paste *along* the die land. Some pastes show little effect of extrusion velocity *(V$_o$)* on driving pressure *(P)*, in which case $\alpha = \beta = 0$. Equation (1) shows a linear dependence of *P* on *V$_o$* and eqn (2) a 'power law' dependence of *P* on *V$_o$*, which superficially resembles the behaviour of a pseudoplastic liquid.

Benbow and Bridgwater[4] used two forms of apparatus in which the influence of absolute pressure on τ_o could be determined. One was based on an annular ring of ceramic paste which could be loaded normally, whilst the shear force on the loaded surface could be measured. In the other, a plug of ceramic paste was forced along a tube against a piston and spring, the effect of which was to progressively raise the mean pressure in the paste. Both devices, using at least two different pastes, showed τ_0 increasing by a factor of two or more as the system pressure was raised from 0 to 600 kPa. Nevertheless, on the basis of measurements of the linear pressure gradient along dies, it was concluded that, for the majority of α-alumina pastes, the flow parameters were *independent* of pressure. The time scale of the die-flow process is relatively short.

In another type of test used by Oliver and Whiskens,[6] a rotating spindle was inserted into the side of a paste-filled cylinder, which could be pressurised from one end. The experiments were carried out using mullite precursor gels, some of which contained up to thirty per cent by volume of fibre. Tests of this type showed τ_0 to increase by factors of the order three as the system pressure was raised from 60 to 700 kPa.

Each of these tests measures τ_0 (wall shear stress) and not σ_0 (die entry parameter) as a function of pressure. For measurements of σ_0, it is important to retain the same die geometry and paste flowrate in a typical flow test as the absolute pressure is changed. In the device described in this paper, paste is driven upwards through a short die (orifice); most flow resistance is caused by die-entry effects but some is caused by the bulk movement of the paste up the cylinder, a wall shear stress effect. The variation of τ_0 with pressure is found by the simple process of removing the die and using a longer paste plug; further details are given below.

2. APPARATUS

A brass cylinder A, of internal diameter 16.0 mm, wall thickness 2.6 mm and length (total) 142 mm has flanges at B which permit the insertion of orifice plate E and also assist cleaning (see Fig. 1). The dimensions of the orifices are listed in Table 1; they

Table 1.   Orifice dimensions

Orifice	Diameter (mm)	Length (mm)
1	6.5	1.0
2	4.5	1.0
3	3.5	1.0
4	2.5	0.7
5	2.0	0.7

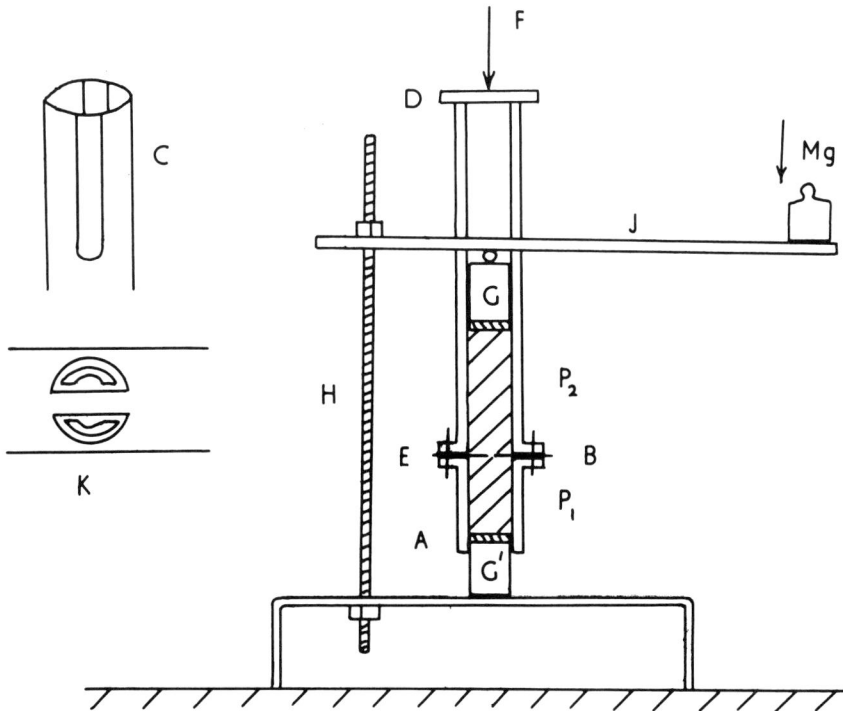

Figure 1.   Apparatus.

are made from stainless steel (1, 2 and 3) and aluminium (4 and 5). The paste is contained between brass pistons G and G[1] fitted with polypropylene seals which are a push fit inside the main cylinder. Piston G has a small steel sphere attached to its upper face; a downward force may be exerted on this sphere by placing weights on the end of aluminium beam J, of width 36 mm and thickness 3.5 mm. Weights of 0.468, 0.944 and 1.30 kg may be placed 184 mm from the fulcrum, which is in line with threaded steel bar H, giving a weight magnification of eight at the steel sphere (23 mm from the fulcrum). Plate J has 'D' shaped holes K which allow an extension (C) of cylinder A to pass through. As illustrated (C), slots are machined in cylinder A to allow this to occur. A downward force F is exerted by an Instron machine on square plate D, which then moves the cylinder A downwards and forces paste upwards through the orifice. The downward speed of the cylinder is 5.1 mm/min in these tests. Force F produces a pressure difference P_1-P_2 between each side of the orifice and the loads placed on beam J increase the value of P_2 (and thereby P_1). Each fixed-speed run is performed by making several changes in the beam loading and observing any changes in F (see method of calculation). All surfaces of the equipment are immediately washed after each run and a very light smear of vaseline placed on the piston seals; the temperature for all tests is 22.0 ± 1.0°C. In some runs, the value of F for zero beam loading was found to rise slowly with time, probably due to paste becoming trapped under the piston seals. If this rise exceeded 10 per cent, the run was discarded.

Table 2.   Pastes used

Paste	Water (gm)	Alumina (gm)	Others (gm)	Comments
A	126	600	Celacol B2/15 (11) Stearic Acid (3.72)	Alumina size of order 1 μm
B	123.5	600	Celacol B2/15 (10.8) Stearic Acid (2.57)	Alumina size of order 1 μm
C	200	0	Aluminium Acetate (50)	Digested 1hr at 200°C, 20 bar
D	48.8 (6.29% w/w)	240 (4.6μm) 240 (8.4μm) 60 (13.5μm) 60 (28μm)	Clay (35) Glucose Syrup (140 ml)	Glucose not fully defined
E	63.3 (8.08% w/w)	240 (4.6μm) 240 (8.4μm) 60 (13.5μm) 60 (28μm)	Clay (35) Glucose Syrup (140 ml)	Like paste D but extra water

3. PASTES USED

The constituents of the five pastes used are listed in Table 2.

The pastes are of three main types, A and B being essentially alumina/polymer/water and D and E essentially alumina/clay/glucose/water. Paste C consists of minute crystals of hydrated aluminium acetate (aluminium hydroxide, or boehmite) dispersed in water. The crystal length is 0.35 μm and diameter 0.03 μm.

4. METHOD OF CALCULATION

For each paste, the apparatus is first set up with the orifice *omitted* and with a paste plug of length about 30 mm. The force F then drives the paste plug along the tube with a driving pressure $\Delta P = F/A$, where A is the cross-sectional area of the tube. The shear stress τ_0 at the tube wall is then

$$\tau_0 = \frac{D\Delta P}{4L} \tag{3}$$

where L is the length of the plug. When a mass M is placed on the lever arm, the overall pressure in the paste is increased by

$$\frac{8Mg}{A}$$

and the average pressure in the paste, P_A, is given by the equation

$$P_A = \frac{\tfrac{1}{2}F + 8Mg}{A} \tag{4}$$

Thus τ_0 is obtained in terms of P_A, which is important in its own right. Additionally, in tests using an orifice, a correction may be applied to the pressure driving the paste through the contraction by subtracting the component of pressure caused by wall friction (at the mean pressure).

An orifice is then inserted and a *shorter* paste plug (10–14 mm) is used, in order to reduce the contribution of wall friction. The process of measuring the value of F for four different values of the load on the arm (including zero) is then repeated. The pressure P_2 downstream of the orifice is given by the equation

$$P_2 = \frac{8Mg}{A}$$
(5)

and the upstream pressure P_1 is then

$$P_1 = \frac{8Mg}{A} + \frac{F}{A}$$
(6)

The mean pressure is

$$\frac{P_1 + P_2}{2}$$

The corrected pressure causing flow through the orifice is

$$(P_1 - P_2)_{CORR} = \frac{F}{A} - \frac{4L\tau_o}{D}$$
(7)

where L is the length of the paste plug in this test and τ_0 is evaluated at the mean pressure.

The value of the constant σ_o relating to die-entry flow is given by the equation

$$\sigma_o = (P_1 - P_2)_{CORR} \left(2\ln \frac{D}{D_o}\right)^{-1}$$
(8)

and this is plotted against the *upstream* pressure P_1, since this represents the region in which the deformation is occurring. It will be noted that no allowance has been made for the changes in orifice velocity V_0, which will occur because the cylinder speed is constant (5.1 mm/min) and the orifice size changes. Values of V_0 are quoted in Table 3. Likewise, it has been assumed that the short die-land makes a negligible contribution to pressure drop.

5. RESULTS

Figures 2, 3, 4, 5 and 6 show values of τ_0 and σ_o plotted against an appropriate value of pressure for the five pastes used. Figure 2 includes some results obtained by the

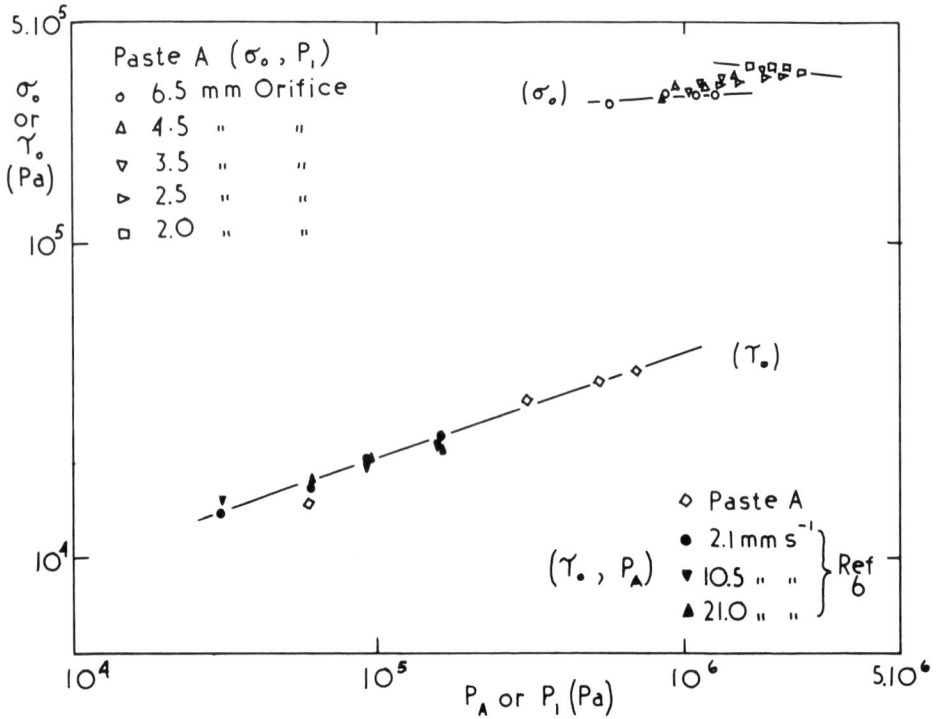

Figure 2.   Values of σ_0, τ_0 versus pressure (Paste A).

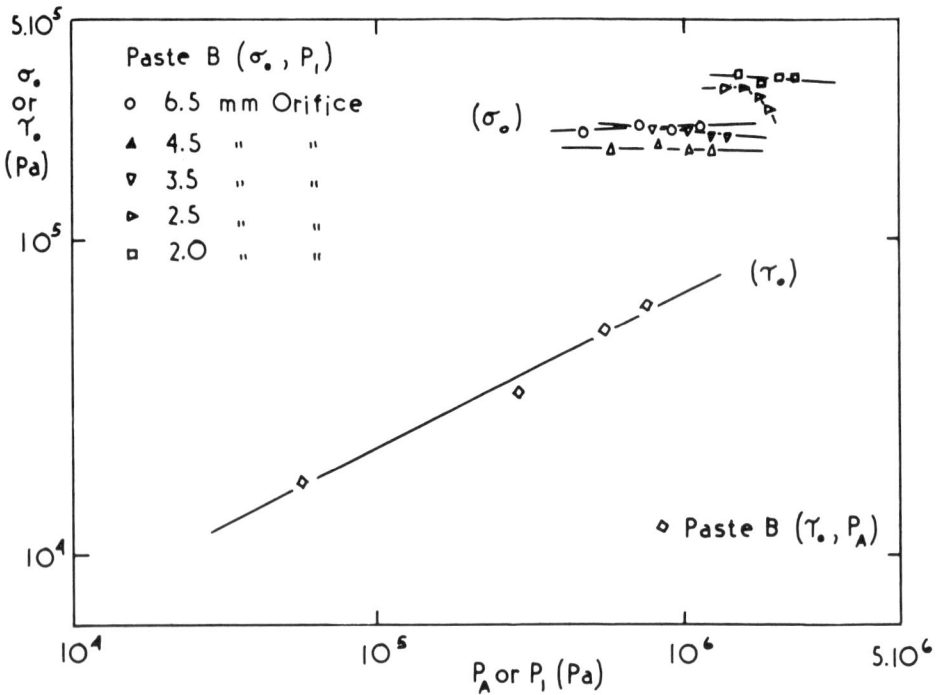

Figure 3.   Values of σ_0, τ_0 versus pressure (Paste B).

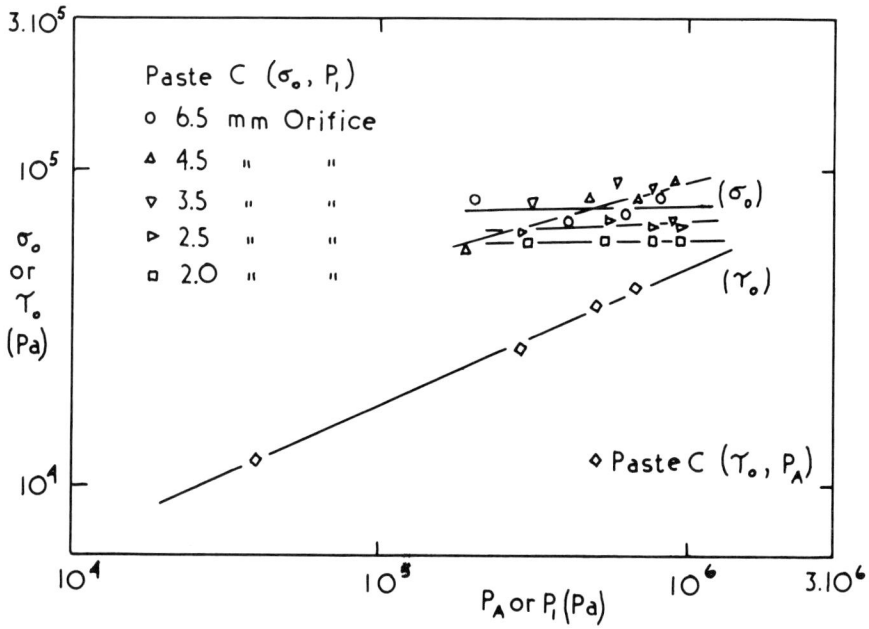

Figure 4.   Values of σ_0, τ_0 versus pressure (Paste C).

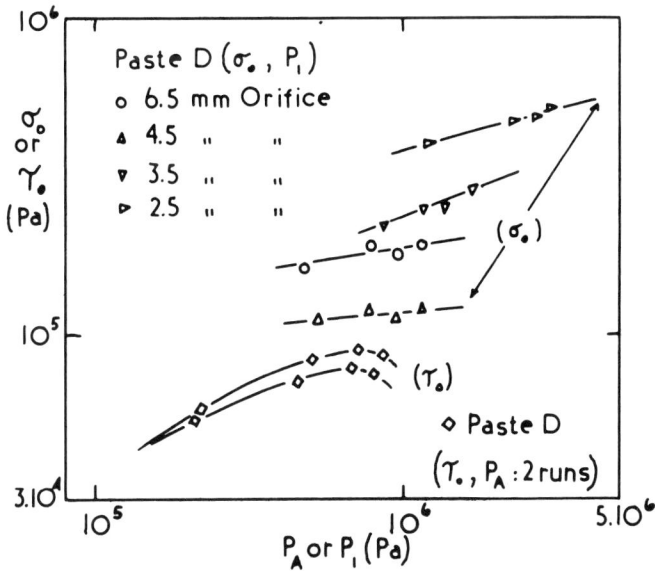

Figure 5.   Values of σ_0, τ_0 versus pressure (Paste D).

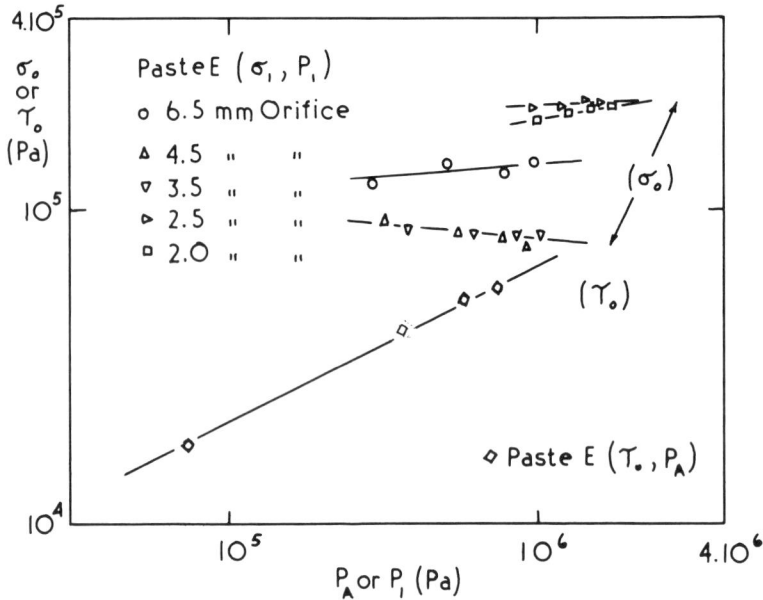

Figure 6.   Values of σ_0, τ_0 versus pressure (Paste E).

rotating spindle method;[6] the values of τ_0 relate to a mullite precursor gel (silica/alumina/water) with little influence of surface velocity on τ_0. The mean pressure P_A in the paste is given by eqn (4) and the upstream pressure P_1 (used in plotting σ_0) is given by eqn 6. Generally, σ_0 is between 3 and 10 times higher than τ_0, but shows much less dependence on pressure. It will be noted that both the orifice size and the load on the beam affect the pressure P_1.

6. DISCUSSION OF RESULTS

The equipment was easy to use and to keep clean; any surface corrosion of the brass was avoided by the removal of all paste after each run. The force F on the cylinder was normally between 1 and 35 Kg (10 and 350 N) and the load caused by the lever arm between 0 and 11 Kg (0 and 110 N). When eqn 7 was used to allow for surface shear stress effects on the pressure causing flow through an orifice, the correction (τ_0 term) was less than 15 per cent when using small orifices, but rose to 25–30 per cent for large orifices.

Figure 2 shows clearly that the wall shear stress τ_0 rises with pressure (gradient of line 0.34), whilst the earlier data points for the mullite precursor paste have a slightly lower slope. The approximate agreement between the magnitudes of τ_0 for these pastes is coincidental. Values of σ_0, however, relating to convergent flow into an orifice, are about eight times higher and show remarkably little variation with pressure. Indeed, sets of data points relating to the smaller orifices show a tendency for σ_0 to *fall* with increasing pressure.

Figure 3 shows broadly similar results, although the gradient of the line relating τ_0 and pressure has increased to 0.47 and the data points for σ_0 are less well grouped, with values for the two smaller orifices lying higher than the other sets of points. This indicates that different pastes may show deviation from the simple behaviour suggested by the logarithmic term in equations 1 and 2, as discussed later.

Results shown in Fig. 4 for the boehmite paste show values of σ_0 only 2–3 times higher than the values of τ_0, at equivalent pressures. The line relating τ_0 and pressure has a gradient of 0.44. Glucose-based pastes are unusual in that the values of σ_0 for the two smallest orifices used are considerably higher than those for the larger orifices (Figs 5 and 6). The gradient of the line relating τ_0 and pressure is 0.50 (Fig. 6). The stiffer of the two pastes (D), in particular, showed effects which could be described as 'strain hardening' in orifice flow. The extensional strain rate of a fluid following a conical path into an orifice is [5]

$$d_{11} = 2 \frac{V_o}{R_o} \sin \phi$$

where terms are defined under 'Nomenclature'. At a constant paste flowrate of 0.0170 ml sec^{-1}, the extensional strain rate may be calculated on the assumption that the cone semi-angle ϕ is 20°. The results are given in Table 3.

It should be borne in mind that the angle of convergence is estimated and may not be the same for each orifice; nevertheless, the values of d_{11} rise quite rapidly as hole size is reduced. Other evidence of strain hardening (or dilatant) behaviour lay in the sharkskinned surface of pastes D and E when they were pushed manually from the tube after an experiment. For paste D, increasing pressure had a larger effect on the values of σ_0 than for the other pastes.

The main feature of the test series is that increasing pressure has much less effect on the value of σ_0 than on the value of τ_0; in Figs 2, 3 and 6 there are runs in which σ_0 actually *decreased* with increasing pressure. The paste is changing shape and increasing speed as it moves towards the orifice, with slip planes (often conical in shape) where the liquid proportion is slightly higher than elsewhere in the sample. It may be suggested that increased pressure might cause a small quantity of liquid to be forced out of the plastic mass and into the slip planes, thus enhancing lubrication and ease of movement. When a paste is sliding along a wall, however, there is probably both a lubrication effect *and* a solid contact contribution to resistance, from asperities in the paste. If this were the only effective friction, τ_0 would be directly proportional

Table 3.   Values of Hole Velocity (V_0) and Extensional Strain Rate (d_{11}) for Flow into Orifices ($\phi = 20°$)

Hole Diameter (mm)	V_O (mm s^{-1})	d_{11} (sec^{-1})
6.5	0.512	0.108
4.5	1.07	0.325
3.5	1.77	0.692
2.0	5.41	3.70

to pressure (related by the coefficient of friction) instead of pressure to the power 0.3–0.5.

These results are relevant to paste processing. The 'worst' cases will be high-pressure, slow flows adjacent to flat surfaces: in the screw of an extruder, along the sprue and in narrow parts of the mould. Other cases are reaction bonding, cold pressing and hot isostatic pressing. If a paste shows any signs of strain hardening in shear then the added effect of increasing absolute pressure on the shear stress τ_0 would cause the process pressure to build up rapidly. The 'best' cases will be those in which the paste is being changed in shape, often by movement into or out of a constriction. Examples are flow into a die, into different parts of a mould and the initial stages of pressing a lump into a sheet. It should be pointed out that these conclusions may not apply to all pastes and that they apply specifically to the effect of *pressure* on processing parameters. In many cases it may be sufficient to be aware that a problem could exist; minor changes in equipment design or paste properties may be sufficient to improve the process. The main conclusions are listed in the Abstract.

7. NOMENCLATURE

A	cross-sectional area of cylinder (internal)
d_{11}	extensional strain rate in flowing paste
D	diameter of barrel or cylinder
D_O	diameter of die or orifice
F	force causing downward motion of cylinder
L	length of paste plug in barrel or cylinder
L_O	length of die land
M	mass placed on lever arm
m	index in equation 2
n	index in equation 2
P	pressure causing paste flow into and along die
P_1	pressure upstream of orifice (below it)
P_2	pressure downstream of orifice (above it)
$(P_1 - P_2)_{CORR}$	value of $(P_1 - P_2)$ corrected for shear stress at wall of cylinder
ΔP	pressure causing paste plug to slide along cylinder
P_A	average pressure in paste plug
R_O	radius of die or orifice ($D_0/2$)
V	paste velocity in barrel or cylinder
V_O	paste velocity in die or orifice
α	constant in equation (1)
α^1	constant in equation (2)
β	constant in equation (1)
β^1	constant in equation (2)
ϕ	cone semi-angle in flowing paste
σ_O	die entry flow parameter (equations 1 and 2)
τ_O	die land flow parameter, or shear stress at wall of die or tube (equations 1 and 2)

REFERENCES

1. J. J. Benbow, and J. Bridgwater, T. A. Lawson, E. W. Oxley: *J. Ceramic Bulletin*, 1989, **68** (10), 1821.
2. D. R. Oliver: 'Theoretical and Applied Rheology', Moldenaers and Keunings eds, *Proc 11th Internat. Congress on Rheology*, Brussels, 1992, Elsevier Sci Publ. B.V. Vol. 2, 865.
3. M. A. Hepworth: *British Ceramic Proceedings 46, Advanced Engineering with Ceramics*, Inst. of Ceramics, 1990, 113.
4. J. J. Benbow and J. Bridgwater: *Tribology in Particulate Technology*, B. J. Briscoe ed., M. J. Adams Publ. Adam Hilger, Bristol & Philadelphia, 1987 86–88.
5. A. B. Metzner and A. P. Metzner: *Rheol. Acta*, 1970, **9** (2), 174.
6. D. R. Oliver and M. Whiskens: Inst. of Chem. Engrs. Research Event. Publ. Inst. Chem. Engrs. 1994, **2**, 767–769.

Biomaterials

Bioactive Glass-Ceramics

E. CLAXTON, R. D. RAWLINGS and P. S. ROGERS

Department of Materials, Imperial College of Science, Technology and Medicine, South Kensington, London SW7 2BP, UK

ABSTRACT

The development of a bioactive glass-ceramic incorporating ductile metal particles is discussed in this paper. The glass-ceramic known as Apoceram has been processed in the past by either casting and heat treatment or by powder and hot pressing routes. In order to improve the toughness of the material ductile metal inclusions, in the form of titanium, have been added but some difficulties were encountered due to a reaction at the particle-matrix interface. The aim of the current studies has been to reduce the fabrication temperature in order to limit the extent of the interfacial reaction. Various amounts of sodium silicate have been added to the parent glass and the effects of this on sintering, crystallisation and microstructure have been assessed. It was found that small additions of sodium silicate reduced the crystallisation temperature, without adversely affecting the sintering behaviour, thus alleviating the interfacial reaction problem. Alternative approaches for the production of bioactive glass-ceramics are also discussed briefly.

1. INTRODUCTION

Ductile metal particle reinforcement is an effective means of enhancing the mechanical properties of brittle ceramics, glasses and glass-ceramics for applications not requiring high hardness or high temperature capabilities. Consequently ductile metals (aluminium, Co–Cr alloy, silver, stainless steel and titanium) have been used to reinforce bioactive glass and glass-ceramics (GC) including Apoceram (see for example review[1]). Improvements in strength and toughness of particulate reinforced bioactive glass and glass-ceramic matrix composite have been observed in a number of systems, for example GC + Ti,[2–6] GC + Ag,[6] Bioglass + Ag,[7] Bioglass + 316 stainless steel.[8]

Clearly a major prerequisite of the reinforcement is that it should be biocompatible. Aluminium can be discarded for this reason and although Co–Cr alloy and stainless steel are biocompatible materials, they are found to disturb the process of osteogenesis, perhaps by a synergistic effect of the glass and the metal ions.[9] Titanium has low toxicity and does not interfere with the process of osteogenesis. It is well accepted by the body, is a well established biomaterial and is less expensive than silver. Thus it may be concluded that titanium is preferable as the reinforcement phase in bioactive glass-ceramic matrix composites.

Apoceram was originally developed as a cast material at Imperial College. It is a bioactive glass-ceramic containing apatite and wollastonite. The material is from the $Na_2O–CaO–Al_2O_3–SiO_2$ system and additions of P_2O_5 are made as a nucleating agent and for the formation of apatite.[10,11] The initial studies on Apoceram indicated the mechanical properties were promising[11,12] and the bioactivity was good.[13] In order to improve the strength and toughness of the monolithic material, powder routes and hot pressing have been utilised. These were carried out at a temperature at which sintering and crystallisation occurred consecutively. The toughness was, however, still inadequate for load-bearing implant applications.[14]

In an attempt to enhance the mechanical properties ductile titanium particles have been incorporated into an Apoceram matrix but there was only a slight improvement compared with the monolithic material.[6] A reaction layer, found to be a compound of titanium and silicon, of between 1–4 μm in thickness was formed between the particles and the matrix and was considered to prevent the attainment of the full potential of the titanium reinforcement to improve the mechanical performance of the composite. In order to gain further toughness in an Apoceram–titanium composite this reaction layer should be reduced.

There are a number of approaches that could be taken to try to optimise the benefits from ductile particle reinforcement of bioactive glass-ceramics, namely:

(i) coat the titanium particles to reduce the extent of the reaction,

(ii) use silver rather than titanium particles,

(iii) employ a bioactive glass-ceramic matrix which is silica-free, eg, one based on the $CaO–P_2O_5$ system,

(iv) adjust the composition of the silica-containing bioactive glass-ceramics in order to reduce the processing temperature and hence the extent of the interfacial reaction, and

(v) using any of the approaches (i) to (iv) to produce a functionally graded material with the amount of titanium increasing towards the interior of the component.

All these approaches are being actively pursued but not all are reported here. This paper will discuss briefly the use of silver particles (approach ii) but is mainly concerned with the results of an investigation into adding sodium silicate to Apoceram as a means of reducing the processing temperature by lowering the temperature of crystallisation without hindering the sintering process (approach iv).

2. EXPERIMENTAL PROCEDURE

The parent glass for Apoceram was prepared from the melt of the constituent oxides and fluoride at 1500°C. A frit, which was prepared by quenching the molten glass in iced water, was milled and the powder sieved to a particle size of less than 38 μm. Either 2.5 or 5 weight percentage of sodium silicate (molecular ratio $1Na_2O:2.58SiO_2$), was added as a 42 wt % solution in water (water glass) to the powdered glass; the resulting slurry was thoroughly mixed and allowed to dry. The dried powder was ground with an agate pestle and mortar to remove any aggregates due to drying. The composite was fabricated by mixing Apoceram glass with 2.5 wt % sodium silicate and 30 weight % sponge titanium. The latter was supplied by Active Metals Ltd. and had an approximate particle size of 50μm.

Samples were uni-axially hot pressed using the same parameters of pressure and time as had previously been used for Apoceram with 30 weight % titanium, these being, 4.5 MPa and 60 minutes respectively. In the case of the composite with an addition of sodium silicate the fabrication temperature was reduced from 1000°C to 875°C. It should be emphasised that these parameters have not been optimised for Apoceram with sodium silicate, rather they have been specifically chosen to establish if the reduction in temperature reduced the interface layer between the matrix and titanium. The conditions chosen, in particular the low pressure, are known to give the worst interface for Apoceram with titanium.[5,6]

X-ray diffraction analysis was carried out in order to establish the effect of sodium silicate on the crystalline phases formed and on the degree of crystallinity within the glass-ceramic. In order to carry out quantitative analysis an internal standard of MgO, in the form of periclase, was added to the powdered crystalline samples. Periclase was chosen as it gave very few diffraction peaks and these were at angles which did not interfere with those of the crystalline phases of the Apoceram. Calibration standards were produced for comparison purposes; the estimated error in the calculated proportions is approximately ±3 absolute %. The percentage of the crystalline phases was found by comparing the areas under the main apatite and wollastonite peaks with those from the periclase. Samples of Apoceram and Apoceram with 2.5 weight % sodium silicate were prepared by cold pressing and then heat treating in argon from room temperature to 1000°C at a heating rate of 200°C h^{-1}. Further quantitative analysis was carried out on monolithic samples of Apoceram and Apoceram with 2.5 weight % sodium silicate that had been hot pressed at 1000°C and 875°C respectively and held at 13MPa for 60 minutes.

Differential thermal analysis (DTA), was conducted using 30 mg samples of Apoceram parent glass with additions of sodium silicate. The samples were placed in platinum crucibles with an alumina reference and an inert argon atmosphere. Heating rates of 3, 5, 7, 10, 20, and 30°C min^{-1} were used and for each heating rate the exothermic peak temperatures of the two crystalline phases, apatite and wollastonite, were plotted. The resulting curves were extrapolated to a heating rate of 0°C min^{-1}. This allowed the isothermal crystallisation temperature to be determined. The activation energy for crystallisation, E, for apatite and wollastonite was calculated using the Augis and Bennett equation:[15]

$$\ln \frac{(T_p - T_o)}{\alpha} = \frac{E}{RT_p} - K_o \tag{1}$$

where T_p and T_o are peak and room temperature respectively, α is the heating rate, K_o is a frequency factor and R the universal gas constant.

Differential scanning calorimetry (DSC), at a heating rate of 10°C min^{-1}, was used to establish the enthalpy change associated with the crystallisation of the phases. The crucibles, reference and atmosphere were as for DTA. In all cases the work was compared with the Apoceram parent glass with no additions. The effect of the additions of sodium silicate on the sintering densification were evaluated using dilatometry at 10°C min^{-1} from room temperature to 1100°C.

Density measurements were carried out on bulk samples of Apoceram and Apoceram with 2.5 weight % sodium silicate. This was carried out on the same samples as used for quantitative X-ray analysis. The density of the bulk samples of the composites was also established; in each case the Archimedes' principle was employed.

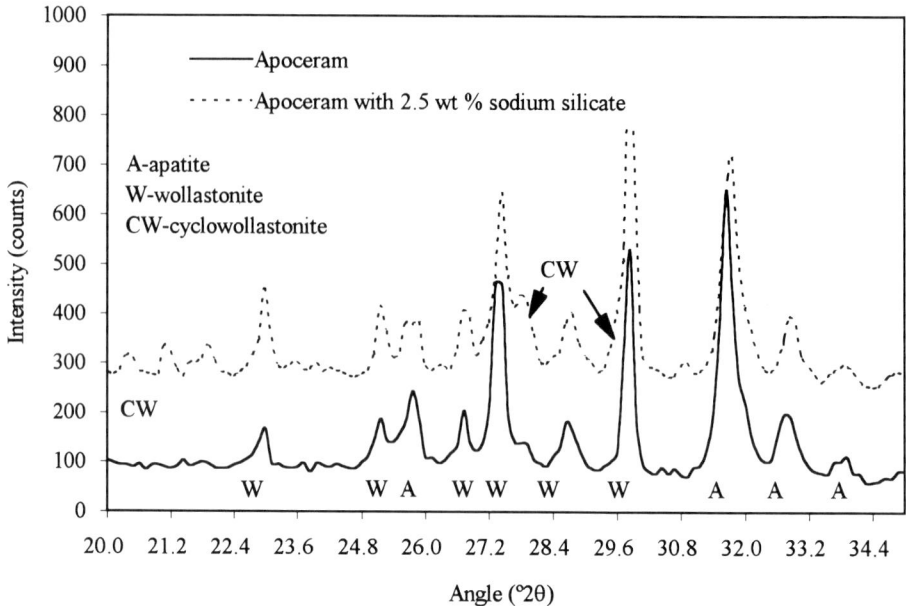

Figure 1.   X-ray diffraction traces showing the comparison between Apoceram and Apoceram with an addition of sodium silicate. (Cold pressed and heat treated from room temperature to 1000°C in argon)

3. RESULTS AND DISCUSSION

3.1 Monolithic Material: Effect on Crystallisation

The crystalline phases in Apoceram were found, by X-ray diffraction, to be a solid solution of hydroxy- and fluorapatite and parawollastonite. When sodium silicate was introduced these phases were still present and there was an extra crystalline phase, cyclowollastonite, as shown in Fig. 1. The diffraction patterns for the two apatite phases were very similar, as were the patterns for para- and cyclowollastonite, and hence only the proportions of apatite and wollastonite could be determined. Comparison of the areas under the apatite and wollastonite crystalline peaks and that of the periclase internal standard demonstrated that for cold pressed and heat treated Apoceram and Apoceram with 2.5 weight % sodium silicate the amount of apatite was reduced and that of wollastonite was unchanged as a result of the addition of sodium silicate (Table 1).

The DTA curve for Apoceram had two distinctive exothermic peaks; X-ray diffraction showed the lower temperature peak to be associated with the crystallisation of apatite and the other peak to wollastonite. The temperature at which crystallisation occurred was found to be reduced by making additions of sodium silicate to Apoceram. This was the case for both the apatite and the wollastonite at all heating rates employed, as shown by Fig. 2, and as a result the isothermal crystallisation temperature for both the crystalline phases has been lowered. The activation energy

Table 1. Proportions of the crystalline phases as determined by X-ray diffraction

Composition [fabrication conditions]	Weight % Apatite	Weight % Wollastonite
Apoceram [room temp–1000°C]	25	29
Apoceram with 2.5wt% sodium silicate [room temp–1000°C]	18	28
Apoceram hot pressed [1000°C–13MPa–60min]	21	28
Apoceram with 2.5wt% sodium silicate hot pressed [875°C–13MPa–60min]	27	28

for crystallisation of apatite was found to increase with increasing amounts of sodium silicate, as shown in Table 2. This is consistent with the quantitative X-ray analysis results in Table 1; the slower crystallisation kinetics due to the higher activation energy with a negligible change in K_o, meant that the equilibrium amount of apatite was not formed at the heating rates used in the thermal analysis work. However, hot pressing at 875°C for 60 minutes allowed the Apoceram with 2.5 weight % sample to form the full amount of apatite. In contrast to the silicate-containing samples, Apoceram that had been hot pressed at 1000°C had a similar amount of apatite as that heat treated from room temperature to 1000°C once the error of ± 3 % absolute is taken into account.

The values for the activation energy of wollastonite were found to be slightly reduced with additions of sodium silicate. It is considered that as there was only a slight change in activation energy, coupled with the high temperatures of the reaction, that the kinetics do not prevent the wollastonite reaction going to completion. This is in agreement with the fact that about 28% wollastonite was present in all samples irrespective of the thermomechanical history and sodium silicate content (Table 1).

Table 2. Activation energies for the crystallisation of apatite and wollastonite as determined from DTA data and equation 1

Composition	Activation Energy Apatite (kJ/mol)	Activation Energy Wollastonite (kJ/mol)
Apoceram	327	375
Apoceram with 2.5wt% sodium silicate	382	332
Apoceram with 5wt% sodium silicate	461	349

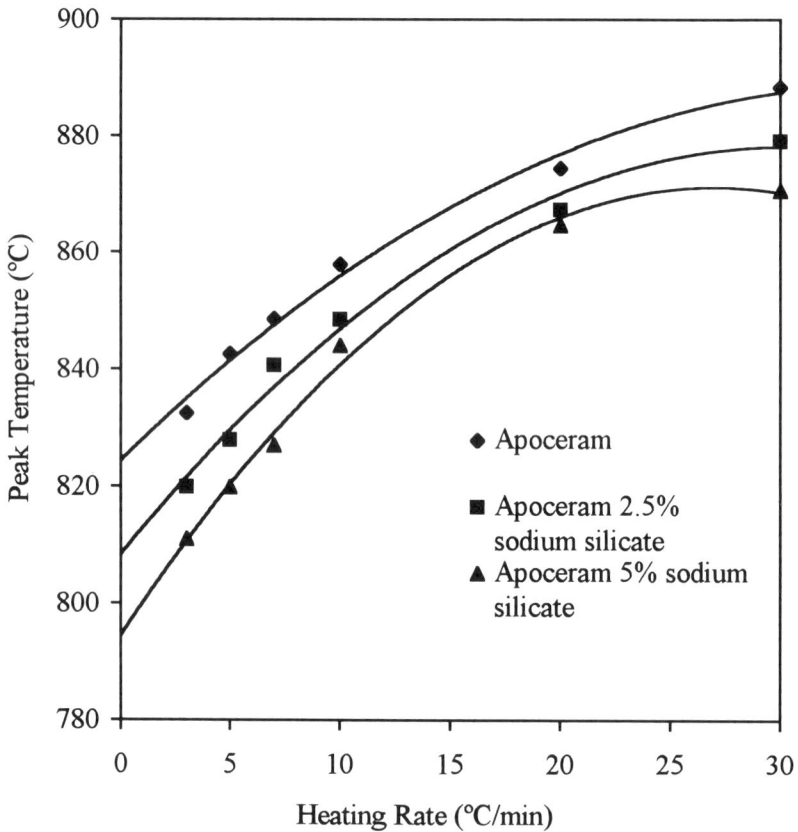

Figure 2a.   Effect of sodium silicate on the exothermic peak and isothermal crystallisation temperatures of apatite.

The reduction in the exothermic peak temperatures can also be seen in the DSC traces in Fig. 3. Although the peak associated with the formation of apatite has shifted to lower temperatures with additions of sodium silicate it can be seen that there is no other major change. Thus, other than a reduction in crystallisation temperature and an increase in activation energy, the apatite crystallisation process is not sensitive to the presence of sodium silicate. In the case of the crystallisation of wollastonite, the peak has not only shifted to a lower temperature with the addition of sodium silicate but the area under the peak has also been reduced. This may be attributed to the formation of both parawollastonite and cyclowollastonite as found by X-ray diffraction. The X-ray data also showed that the total amount of wollastonite present was unaffected by the presence of sodium silicate. The smaller wollastonite exothermic peak in the sodium silicate containing samples may be accounted for if cyclowollastonite is metastable and therefore has a lower enthalpy of formation than parawollastonite.

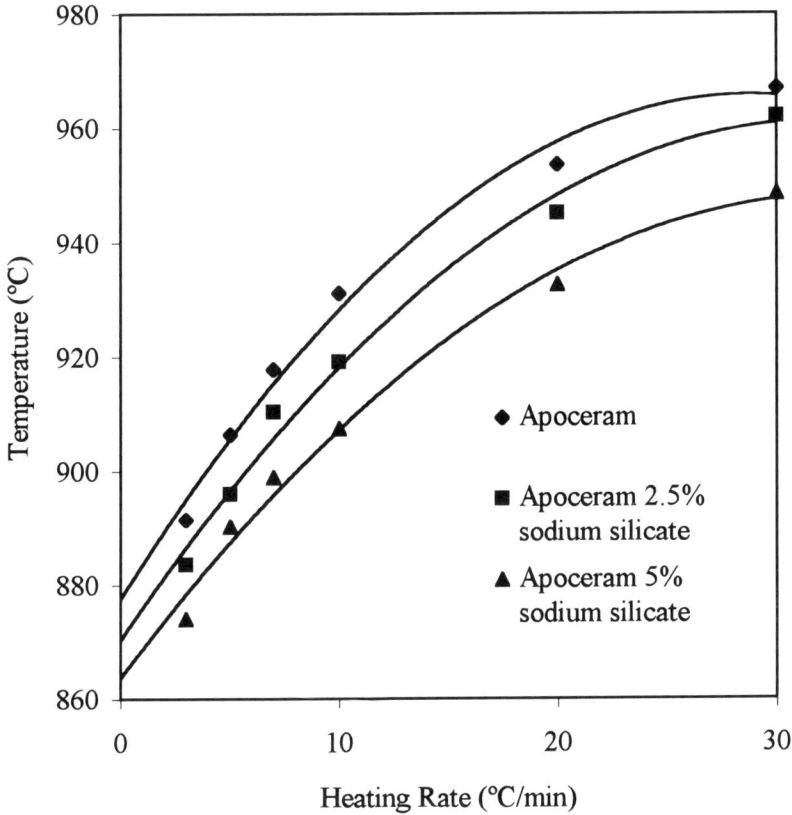

Figure 2b.   Effect of sodium silicate on the exothermic peak and isothermal crystallisation temperatures of wollastonite.

3.2 Monolithic Materials: Effect on Densification

The onset of sintering has not been affected by additions of 2.5 wt % sodium silicate, as shown in Table 3. Also the maximum rate of sintering was found to be at 800°C for Apoceram and Apoceram with a 2.5 wt % addition of sodium silicate and the end point for sintering was at a similar temperature. As a result it is concluded that 2.5wt % sodium silicate has not had an adverse effect on sintering. In contrast, when a 5 wt % addition of sodium silicate was made the temperature range over which sintering occurred was reduced; this reduction was mainly associated with sintering ending at a temperature some 15°C lower than for Apoceram. From DTA this can be attributed to the onset of crystallisation at lower temperatures. As a result of the adverse effect on the densification caused by the onset of crystallisation no further work was carried out on Apoceram with 5 wt % sodium silicate.

The density of Apoceram with 2.5 wt % sodium silicate cold pressed and heat treated was less than that of Apoceram (Table 4). This difference is attributed to the

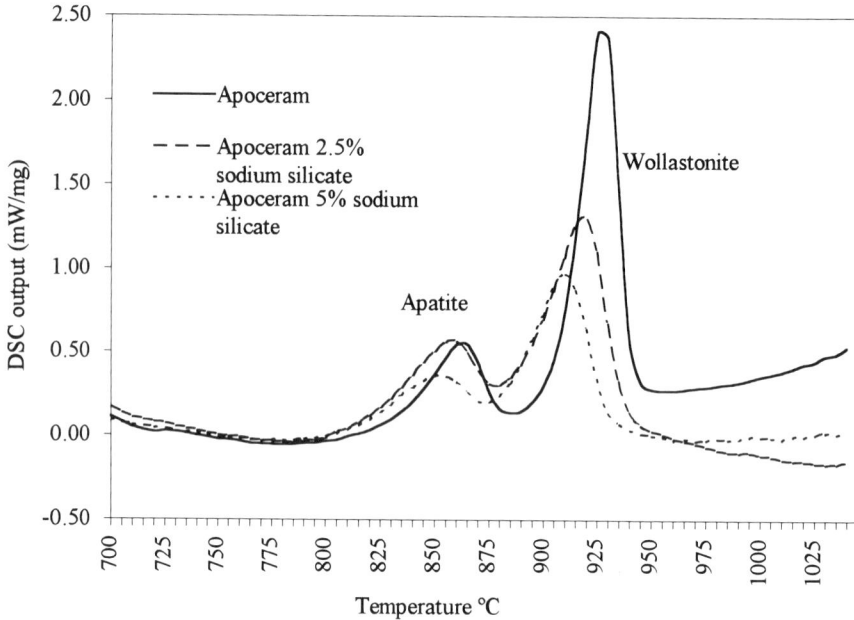

Figure 3.   DSC traces for Apoceram and Apoceram with additions of sodium silicate.

slightly smaller amount of apatite in the former, as discussed previously, as micro-scopy and porosimetry indicated that the porosity contents were low in both cases.

The fact that 2.5 wt % sodium silicate has not affected the sintering behaviour but has reduced the crystallisation temperatures has allowed Apoceram containing sodium silicate to be processed at 875°C instead of 1000°C used for Apoceram. In spite of the lower fabrication temperature of the sodium silicate containing mater-ial, Figs 4 and 5 show that there is little difference in the microstructure of the uni-axially hot pressed materials. The dark areas represent where glass has been etched away and in both microstructures it is difficult to distinguish the two crystalline phases although there is a tendency for the apatite fibrils to be larger than those of wollastonite. Using this criterion for distinguishing the crystalline phases it would appear that Apoceram with sodium silicate has slightly more apatite as was also

Table 3.   Effect of sodium silicate on the sintering characteristics as determined by dilatometry.

Composition	Sintering onset temp. (°C)	Sintering end temp. (°C)	Temp. of max. rate of sintering (°C)
Apoceram	670	850	800
Apoceram (with 2.wt% sodium silicate	670	845	800
Apoceram with 5wt% sodium silicate	675	835	820

Table 4. **Effect of sodium silicate on the densities of the monolithic and composite materials**

Composition [fabrication conditions]	Density (Mgm^{-3})
Apoceram [room temp–1000°C]	2.59
Apoceram with 2.5wt% sodium silicate [room temp–1000°C]	2.49
Apoceram hot pressed [1000°C–13MPa–60min]	2.69
Apoceram with 2.5wt% sodium silicate hot pressed [875°C–13MPa–60min]	2.64
Apoceram 30wt% Ti hot pressed [1000°C–4.5MPa–60min]	2.90
Apoceram 2.5wt% sodium silicate 30wt% Ti hot pressed [875°C–4.5MPa–60min]	3.06

indicated by quantitative X-ray analysis for the hot pressed samples. The porosity contents of both were very small and hence the densities of the hot pressed samples were higher than and did not differ so much as the cold pressed materials.

3.3 Composites

Preliminary experiments on hot pressing Apoceram–titanium composites containing sodium silicate at 875°C have shown a greatly reduced reaction between the constituents, as shown by the scanning electron micrographs of Figs 6 and 7. It should be

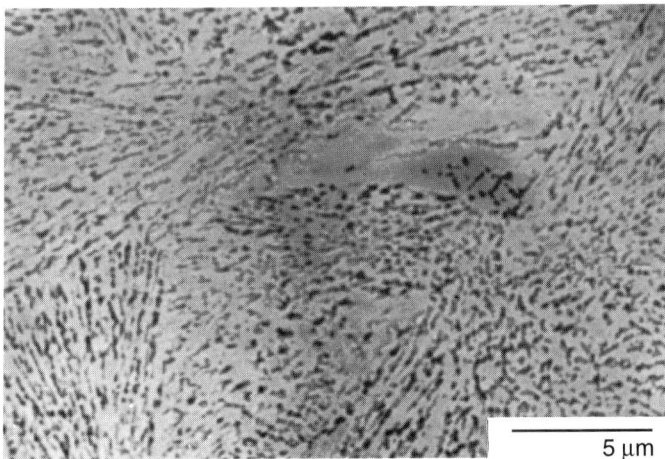

5 µm

Figure 4. Secondary electron image of hot pressed Apoceram (1000°C–13MPa–60min)

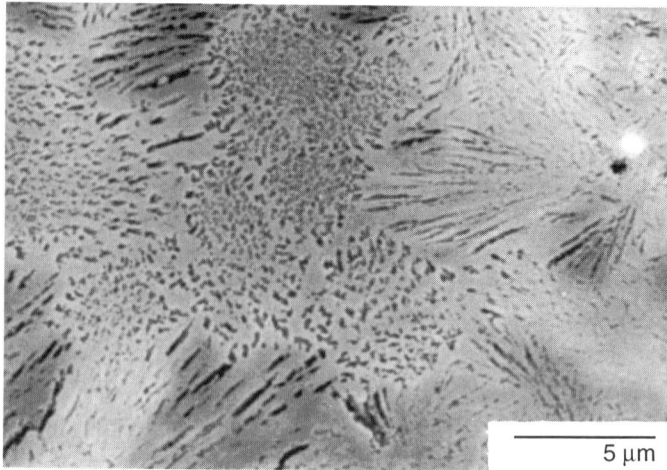

Figure 5.   Secondary electron image of hot pressed Apoceram with 2.5 wt% sodium silicate (875°C–13MPa–60min).

remembered that the hot pressing conditions of both these samples were chosen to emphasise the adverse microstructural features associated with the interfacial reaction. Optimisation of the conditions improves the interface but sodium silicate additions always result in a better interface. Furthermore, the matrix of the sodium silicate containing material is finer and more homogeneous, in fact more similar to that of the monolithic material; a consequence of these improvements in the microstructure is an increase in the density as shown in Table 4.

Silver particulate reinforcement of Apoceram may be achieved by mixing the parent glass with silver particles and hot-pressing as previously described for the titanium–Apoceram composites. Alternatively Ag_2O, which decomposes at around

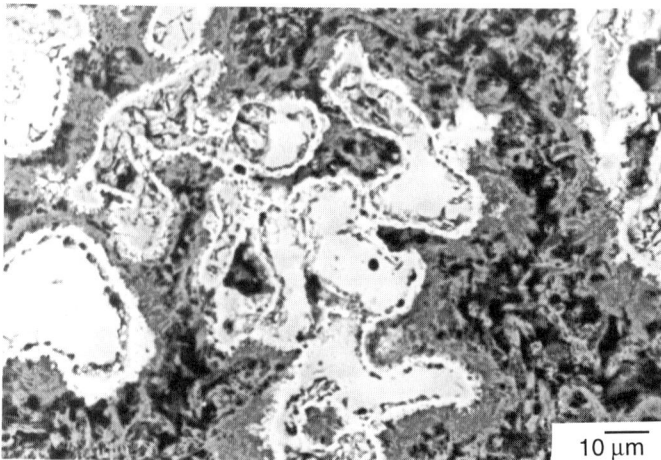

Figure 6.   Backscattered electron image of hot pressed Apoceram 30 wt% titanium (1000°C–4.5MPa–60min) showing the reaction layer between matrix and particle.

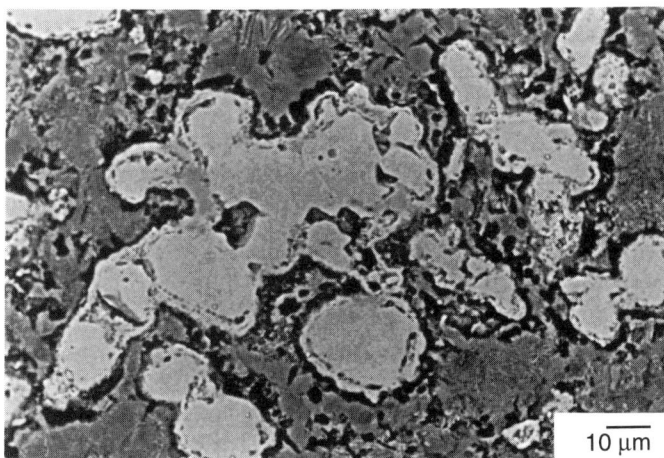

Figure 7. Backscattered electron image of hot pressed Apoceram with 2.5wt% sodium silicate 30wt% titanium (875°C–4.5MPa–60min) with improved microstructure.

420°C, may be mixed with the parent glass. In either case problems arise if the processing temperature is greater than the melting point of silver (961°C) as is the case for Apoceram. Clearly the reduction of the processing temperature to 875°C by the addition of 2.5wt% sodium silicate should facilitate the production of silver containing composites; experiments, by the authors, have shown this to be the case.

Although silver is more expensive than titanium, if the silver content is minimised, commensurate with obtaining acceptable mechanical properties, silver reinforced Apoceram may be a viable proposition for small critical components. Indeed, silver foil is readily available and therefore the possibility exists of producing laminate structures.

4. CONCLUSIONS

The crystallisation temperatures for apatite and wollastonite have been successfully reduced by making small additions of sodium silicate. This has not caused any adverse effects on the sintering behaviour. As a result Apoceram-titanium composites with an addition of sodium silicate have been hot pressed at a temperature 125°C lower than previously employed for Apoceram-titanium. This has given rise to a finer and more homogeneous matrix with less porosity. The interface between the titanium particles and the matrix has also been improved and can still be optimised.

REFERENCES

1. R. D. Rawlings: *Clinical Materials,* 1993, **14**, 155–179.
2. M. J. Lodeiro: MSc Thesis, Imperial College, London University, 1993.

3. T. B. Troczynski and P. S. Nicholson: *J. Amer. Ceram. Soc.*, 1991, **74**, 1803–1806.

4. E. Devaux: MSc Thesis, Imperial College, London University, 1994.

5. B. A. Taylor, R. D. Rawlings, and P. S. Rogers: *Bioceramics*, 1994, **7**, 255–260.

6. B. A. Taylor: PhD Thesis, Imperial College, London University, 1994.

7. T. B. Troczynski and P. S. Nicholson: *J. Amer. Ceram. Soc.*, 1990, **73**, 164–166.

8. P. Ducheyne and L. L. Hench: *J. Mater. Sci.*, 1982, **17**, 595–606.

9. E. Schepers, P. Ducheyne and M. De Clercq: *Biomed. Mater. Res.*, 1989, **23**, 735–752.

10. P. R. Carpenter, M. Campbell, R. D. Rawlings and P. S. Rogers: *J. Mater. Sci. Letts.*, 1986, **5**, 1309–1312.

11. C. B. Ponton, R. D. Rawlings and P. S. Rogers: SERC Report, Department of Materials, Imperial College, London, 1990.

12. R. D. Rawlings, P. S. Rogers and P. M. Stokes: Materials Science Monographs, 1987, **39**, *Ceramics in Clinical Applications*, P. Vincenzini, ed., Elsevier Science Publishers B.V., Amsterdam, 73–82.

13. L. A. Wolfe: *J. Dent. Res.*, 1989, **68**, 872.

14. H. Alanyali: PhD thesis, Imperial College, London, 1992.

15. J. A. Augis and J. E. Bennett: *J. Thermal Anal.*, 1978, **13**, 283–292.

Prospects for Zirconia in Total Joint Replacements

J. W. STUART, I. C. ALEXANDER and A. W. PRYCE

Morgan Materials Technology Ltd., Bewdley Road, Stourport-on-Severn, Worcestershire DY13 8QR, UK

ABSTRACT

Over the last 20 years, the introduction of ceramics as biomaterials for implant devices has been dramatic. Yttria-stabilised zirconia materials are increasingly being used for applications such as femoral heads for total hip replacements. The move away from metal femoral heads to ceramic materials is mainly because of their superior wear characteristics which makes them well suited for younger and more active patients.

The physical and mechanical properties of a commercially available 3 mol% yttria stabilised zirconia are presented. The effect of exposure to a simulated physiological environment has been investigated as part of an on-going study of these materials. After immersion in simulated body fluid at 37°C for 300 days, there was a slight change in the crystal structure on the surface but there was no discernible loss of strength. Ultimate compression tests have been carried out on different femoral head designs. The results showed that zirconia heads had about twice the strength of comparable alumina heads.

As we move towards a new century, with the continuing development of zirconia as a bio-material, we predict a continued increase in its role for implant components. This could improve the quality of life for patients and reduce the need for revision operations.

1. INTRODUCTION

Every year, in the UK over 40,000 people have a hip replacement operation. This routine procedure is improving the quality of life for people with damaged or worn out hips and is considered one of the great successes of modern orthopaedic surgery. In this paper we consider the role of, and the future of, zirconia in total hip replacements.

Most hip replacement systems are based on that designed by the orthopaedic surgeon John Charnley around 30 years ago. It was a single solid piece made from forged or cast stainless steel with a tapering stem to fix into the femoral bone and a spherical head which rotated in a polyethylene cup fixed in the pelvic bone. There has now been a move away from using a single piece prosthesis towards a modular system which gives the surgeon the flexibility to choose, in the operating theatre, the most appropriate stem and femoral head.

A range of materials are available for hip replacement components. In the UK, metal (for example CrCoMo alloy) femoral heads are widely used. However, metal has certain disadvantages. One of the main concerns is the wear characteristics of the femoral head against the acetabular cup, which is typically ultra high molecular weight polyethylene (UHMWPE). Any wear debris can be detrimental to the success of the implant and can lead to loosening of the prosthesis. The raised edges associated with damage, such as a scratch, on a metal surface will increase the wear of the UHMWPE cup. The non-ductile surface of a ceramic, however, is not prone to this type of roughening, resulting in reduced wear and, therefore, fewer associated problems.

2. CERAMIC FEMORAL HEADS

Alumina has been very successful as a femoral head material and is widely used particularly in Europe. Conforming to the International Standard ISO 6474 rev. 2, alumina implant material has excellent biocompatibility and superior wear properties compared to metal. However, despite the experimental and clinical evidence supporting the use of alumina for femoral heads, its selection has been influenced by the small (<0.02%) number of failures. Also, there is the desire to produce a wider range of ceramic head designs to match the diversity of metal heads. This has led to a quest for an improved ceramic and the move towards zirconia which has twice the mechanical strength of alumina and a higher fracture toughness whilst maintaining comparable wear properties. These superior properties enable the manufacture of smaller diameter heads (<26 mm) and heads with long neck lengths.

In this paper, the properties of a biograde 3 mol% yttria stabilised tetragonal zirconia polycrystal (3 mol% Y-TZP) are presented. There have been several years of clinical use of this zirconia with no failures to date. However, there have been some concerns about the use of this material, namely the inherent radioactivity and the long term in-vivo stability of the crystal structure.

By using high purity raw materials this first concern is minimised as the effective body dose from a 3 mol% Y-TZP femoral head is nearly 200 times less than the limit as recommended by the International Commission on Radiological Protection.[1]

The stability, over extended time periods, of the tetragonal structure of Y-TZP in-vivo can be assessed by measuring the effect of a simulated physiological environment on the structure and properties of the material. This is discussed in this paper.

3. WEAR PERFORMANCE

The main consideration in the wear process is the wear of the UHMWPE acetabular cup by the femoral head. It is the polyethylene wear debris which mainly causes concern and can lead to loosening of the prosthesis. Wear is influenced by the physical properties of the femoral head material. A hard, tough material is less prone to scratching and pull out caused by localised fracture. A small grain size will produce a stronger, tougher ceramic. Surface smoothness will have an obvious effect on wear. The surface finish obtainable on a polished ceramic is as good as that obtained on a metal. In a comparable reciprocating pin-on-flat wear test against UHMWPE, the wear with a CrCoMo alloy was approximately 30 times greater than for either a commercial grade zirconia or alumina.[2]

Table 1.   Physical and Mechanical Properties of 3 mol% Y-TZP (Morgan Matroc Zyranox zirconia)

Parameter	Zyranox (typical)	ASTM draft #	ISO/CD 13356	ISO 6474 rev 2 grade alumina
Bulk density (g cm⁻³)	6.07	>6.00	>6.00	3.95
Ave grain size MLI (µm)	0.35	<0.5	<0.6	1.53
Rockwell hardness	87	–	–	91
Vickers hardness (Hv)	1,300	1,200	–	–
Mean strength (MPa)	810	>800	>900	396
Elastic modulus (GPa)	220	>200	–	394
Monoclinic phase content	<1%	<5%	–	N/A

4. MECHANICAL PROPERTIES

The physical and mechanical properties of a 3 mol% Y-TZP (Morgan Matroc Zyranox zirconia) are shown in Table 1. The properties are compared with draft ASTM and ISO standards and typical values of high purity alumina in accordance with ISO 6474 rev 2.

Two points to note from this table are the mean flexural strength (as measured by 4 point bending) which is about twice that of alumina, and the elastic modulus of 220

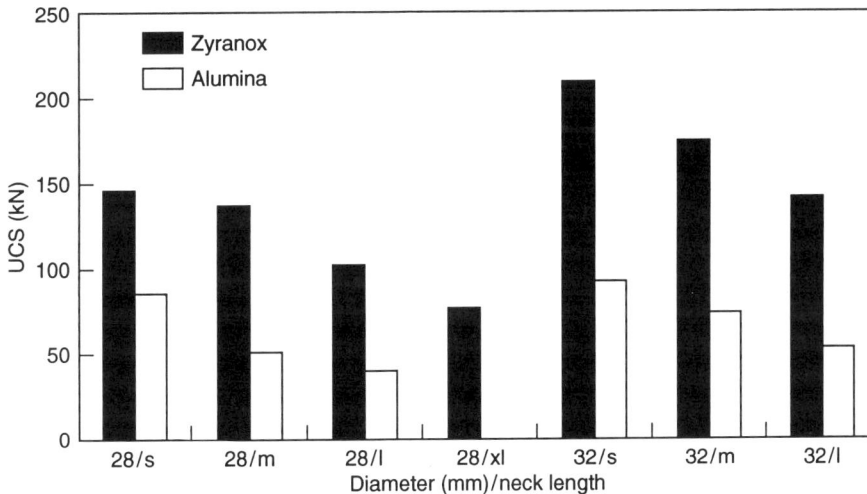

Figure 1.   Ultimate compressive strength of zirconia and alumina femoral heads

GPa which is close to the modulus of the metal stem (typically 200 GPa). This modulus match reduces the mechanical mismatch between the two materials which could cause loosening and wear debris.

Figure 1 shows the ultimate compressive strength of femoral heads made from Zyranox zirconia and alumina, mounted on titanium spigots. The heads were 28mm or 32mm in diameter with short (s), medium (m) or long (l) neck length design. The extra long (xl) neck design is only possible with zirconia because of its superior mechanical properties. These results show that zirconia heads have about twice the compressive strength of alumina heads.

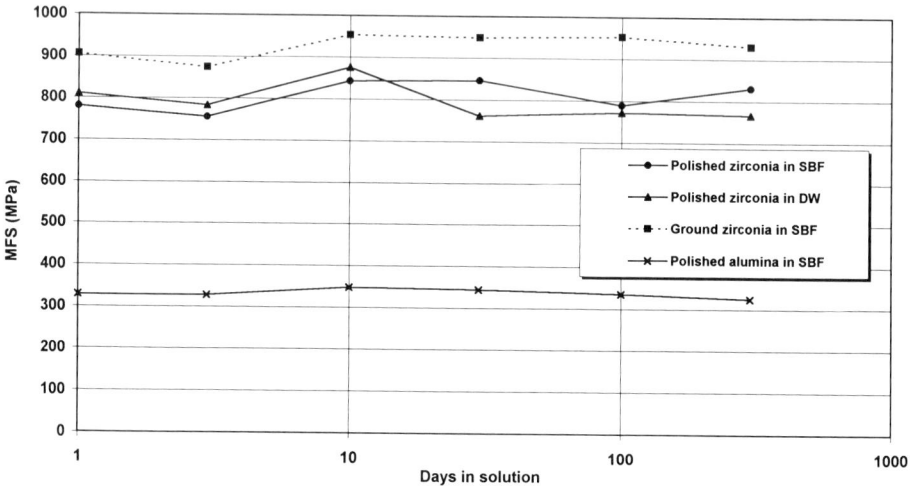

Figure 2.   Mean flexural strength of zirconia and alumina after immersion in solution at 37°C

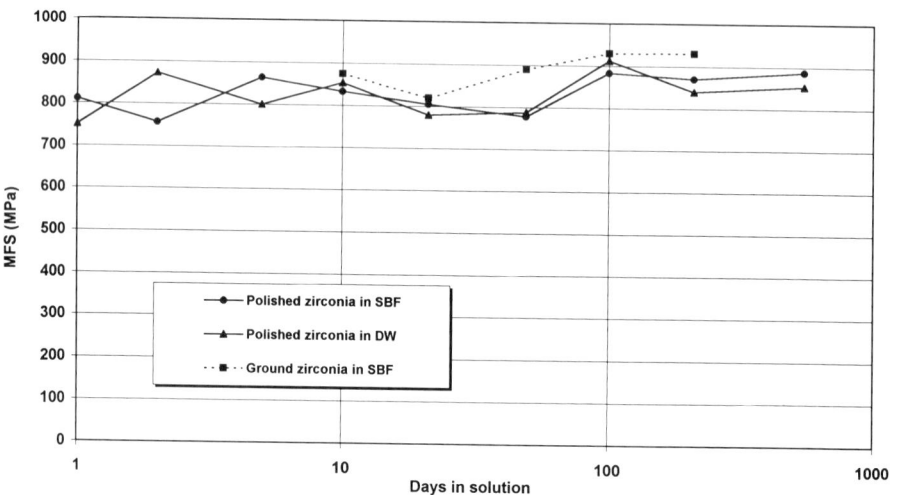

Figure 3.   Mean flexural strength of zirconia after immersion in solution at 60°C

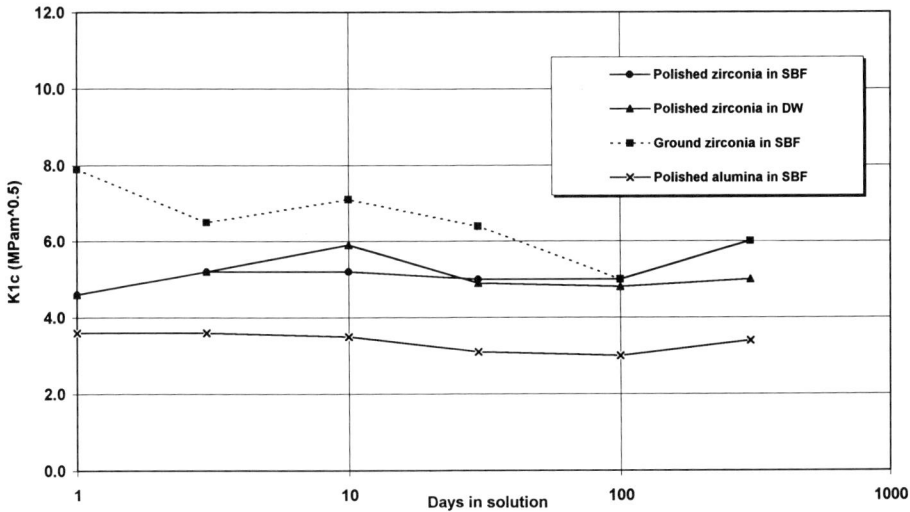

Figure 4.   Fracture toughness of zirconia and alumina after immersion in solution at 37°C

5. IN-VITRO STABILITY

The toughening mechanism in Y-TZP is a result of the transformation of the meta-stable tetragonal structure to a stable monoclinic structure in a region of high stress, for example at a crack tip. The concern that this transformation would take place over time in the body, and therefore reduce the toughness of the zirconia, has been addressed by a study of the long term effects of a simulated physiological environment on Zyranox zirconia.

Figure 2 shows the mean flexural strength (by 4 point bending) of zirconia and alumina after immersion in simulated body fluid (SBF) and distilled water (DW) at 37°C. Again, it should be noted that the strength of the zirconia is more than double that of the alumina. The values for the surface ground samples are higher than the polished samples. This is probable a result of the small amount of transformation on the surface which occurred during grinding. The increase in volume associated with regions of monoclinic microstructure could result in a compressive stress in the surface which would affect the strength. This transformed layer is removed during polishing. The microstructure of the sample surfaces have been examined by XRD and are discussed later.

The important points to note are that there was no reduction in the strength of these materials after 300 days in solution and that there was very little difference between the effect of the distilled water and the simulated body fluid.

Similar results were obtained at 60°C for zirconia (Fig. 3).

The 3 point bend indent fracture toughness values (K_{1C}) of zirconia and alumina samples immersed in solution at 37°C are shown in Fig. 4. The surface of the sample bars were indented with a Vickers hardness diamond and then the bars were broken

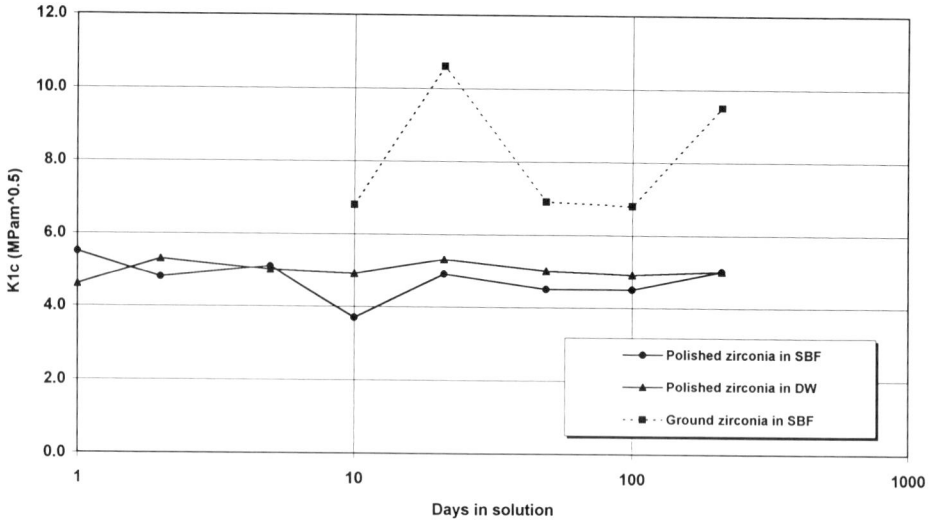

Figure 5.   Fracture toughness of zirconia after immersion in solution at 60°C

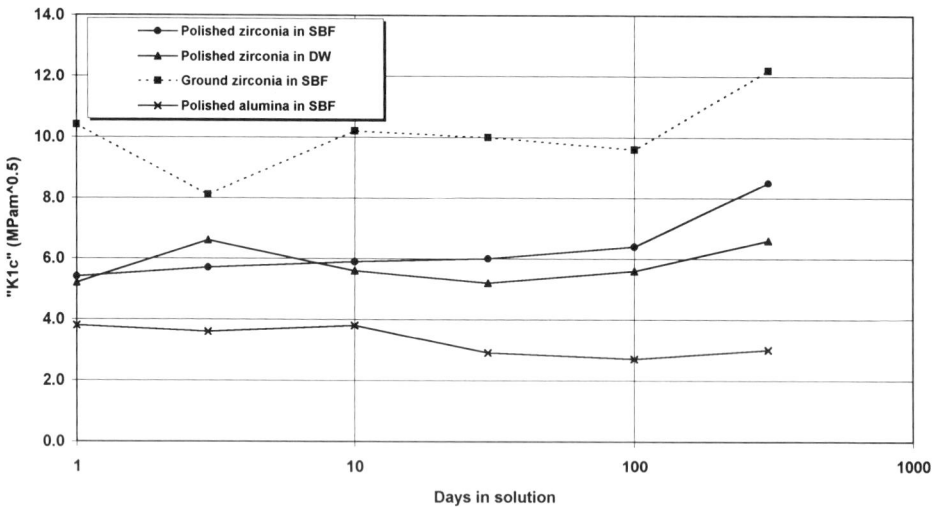

Figure 6.   Effect of immersion in solution at 37°C on a surface defect in zirconia and
alumina

in 3 point bending. These results show the higher fracture toughness of the zirconia
and the similarity of the effect of simulated body fluid and distilled water. As with the
strength results, there was no reduction in fracture toughness for polished zirconia
after 300 days. The ground samples have a higher fracture toughness initially but this
reduces with time to the same value as for the polished samples by 100 days (note
that for 100 and 300 days the points for ground and polished samples coincide).

The results obtained for zirconia at 60°C are shown in Fig. 5. The data for polished
samples were similar to those at 37°C. The data for ground samples were more

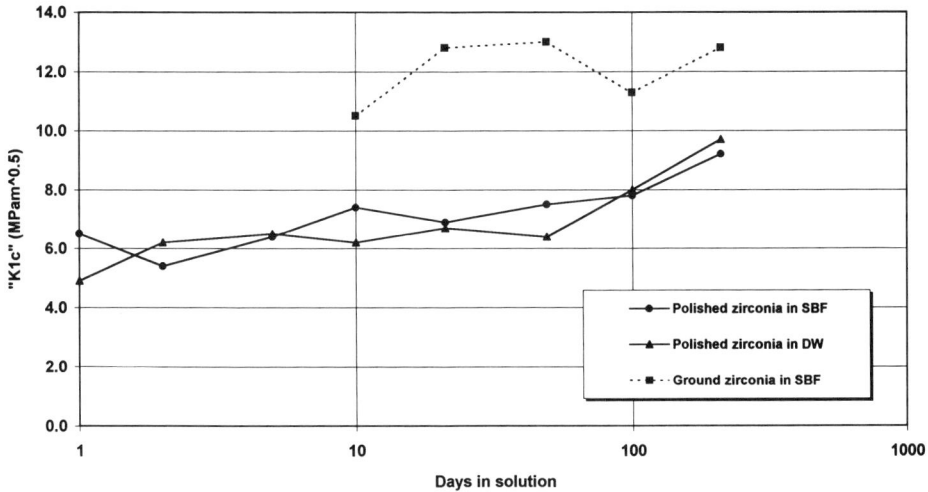

Figure 7.   Effect of immersion in solution at 60°C on a surface defect in zirconia

irregular, probably because of the effect of the surface ground layer on what is a surface sensitive measurement.

The effect of immersion on a surface defect was investigated by using a test similar to the fracture toughness measurement except that samples were indented before being placed in the solution. After each soak interval the sample bars were broken in 3 point bend and a toughness value was calculated using the K_{1C} formula.

The results shown in Fig. 6 show that there was very little change until after 100 days in solution. The measured values for zirconia then increased indicating a reduction in the stress associated with the indent, possibly due to crack blunting. This was even more marked at 60°C as shown in Fig. 7.

This stress relieving at the indent cracks may be expected as part of the toughening characteristics of a stabilised zirconia. However, the importance of this in terms of implant applications is that there is no crack growth evident as a result of the simulated physiological environment.

6. SURFACE ANALYSIS

The same sample, polished or ground, was periodically removed from the solution and examined by XRD for the relative amounts of monoclinic and tetragonal phase in the surface microstructure. XRD traces from the polished and surface ground samples, having been immersed in simulated body fluid at 37°C, are shown in Figs 8 and 9. After 300 days in solution there was a small amount (<1%) of monoclinic in the polished surface. However, this is not associated with any reduction in the mechanical properties.

As stated previously, there was initially some monoclinic in the surface of the ground sample due to the grinding action and it can be seen that this transformed

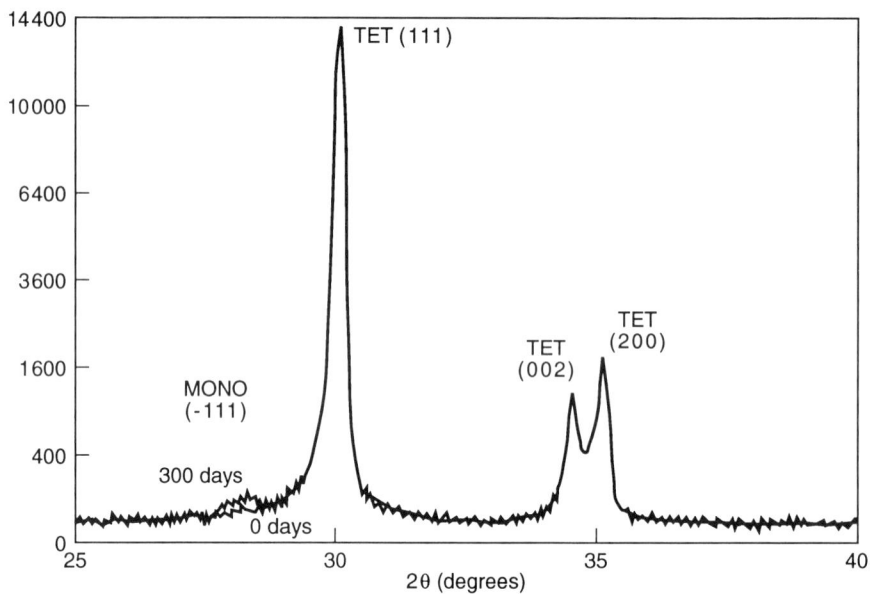

Figure 8.   XRD of polished zirconia after immersion in SBF at 37°C

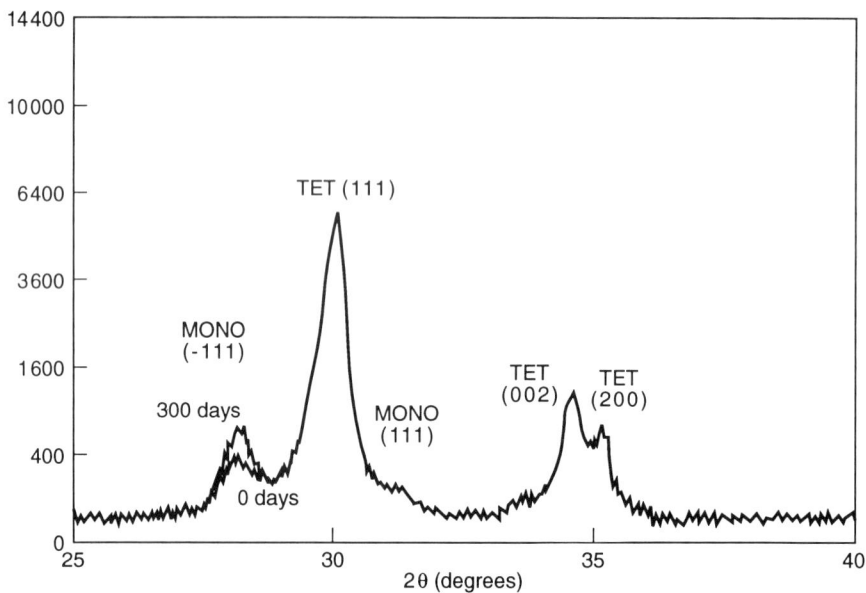

Figure 9.   XRD of surface ground zirconia after immersion in SBF at 37°C

layer is removed by polishing (compare Figs 8 and 9). After 300 days in simulated body fluid there was a noticeable increase in the amount of monoclinic phase in the ground sample. However, this cannot be correlated with any drop in strength and even with the slight drop in fracture toughness, the surface ground zirconia, which now had a value similar to the polished zirconia, was still tougher than alumina.

7. MOVING INTO THE 21ST CENTURY

Zirconia has been used for femoral heads in hip replacement operations for several years with no head failures. The in vitro tests show no detrimental changes in physical properties to date. However, as zirconia is being used for long term implant applications, the small changes in the surface structure, as measured in vitro, cannot be ignored. Work is continuing to monitor the effects of a simulated physiological environment with the view to understanding the processes involved. The current immersion trials will continue up to 3000 days (8 years) and there are plans to develop second generation zirconias with better mechanical properties which are even more stable to transformation.

As more clinical data becomes available and the ASTM and ISO International Standards for 3 mol% Y-TZP are finalised, zirconia will become more established as an implant biomaterial. With its superior mechanical properties it has the design flexibility for a wide range of femoral head designs and sizes. Other applications which can benefit from the use of 3 mol% Y-TZP, or second generation biograde zirconias, include knee joints, shoulders, phalangeal joints and spinal implants.

REFERENCES

1. ICRP Publication 60: *1990 Recommendations of the International Commission on Radiological Protection*, Pergamon Press, 1990.
2. V. Saikko: *J. Wear*, 1993, **166**, 169–178.

Preparation and Sintering of Carbonate Hydroxyapatite

J. BARRALET, S. BEST and W. BONFIELD

IRC in Biomedical Materials, Queen Mary and Westfield College, Mile End Road,
London E1 4NS, UK

ABSTRACT

This paper describes the effects of preparation conditions on the morphology of carbonate hydroxyapatite precipitates and the influence of furnace atmosphere during subsequent heat treatment on the density and hardness of the material after sintering. Carbonate hydroxyapatite was prepared using a reaction between calcium nitrate, tri-ammonium orthophosphate and sodium bicarbonate over a range of temperatures and reactant concentrations. A carbonate hydroxyapatite prepared using one particular set of reaction conditions was formed into pellets using a slip casting technique and sintered at temperatures between 1000°C and 1300°C for four hours in four different atmospheres: air, dry carbon dioxide, wet carbon dioxide and wet air. The composition and crystallography of the sintered apatites were characterised by X-ray diffraction (XRD) and Fourier transform infrared spectroscopy (FTIR) and the density and micro hardness of the materials were determined. A dry carbon atmosphere was found to result in the poorest densification characteristics while wet and dry air atmospheres resulted in samples of greater than 95% density. The use of wet carbon dioxide led to the production of 100% dense carbonate hydroxyapatite characterised by the formation of translucent discs during sintering.

1. INTRODUCTION

Improvements are constantly being sought in the biological and mechanical performance of materials for bone replacement. A great deal of interest has been directed towards the application of hydroxyapatite (HA),[1,2,3] an osseoinductive calcium phosphate material which resembles the mineral component of bone. However, human bone mineral differs in composition from stoichiometric HA ($Ca_{10}(PO_4)_6(OH)_2$) in that it contains additional ions, of which carbonate is the most abundant species (approx 5%).[4] There is now growing appreciation of the potential benefits of the use of carbonate hydroxyapatite (CHA) as a bone replacement material since it approaches more closely than HA the composition of the natural tissue. As the production and characterisation of sintered CHA has not been extensively reported in the literature, an investigation was initiated into the preparation and sintering behaviour of this material. To achieve the optimum mechanical properties of a sintered ceramic, it is necessary to eliminate porosity and minimise grain size. Removal of porosity in sintered HA has been previously achieved by means of hot isostatic pressing[5] (HIP) and microwave sintering.[6] This paper documents the production of fully dense CHA at atmospheric pressure by careful control of the furnace atmosphere.

2. MATERIALS AND METHODS

2.1 Precipitation

A precipitation reaction based on that used by Nelson and Featherstone (1982)[7] was used for the preparation of hydroxyapatite and carbonate hydroxyapatites with varying degrees of substitution. Carbonate apatites may be prepared aqueously by one of two methods; the Direct Method where the calcium ion solution is added to the phosphate solution or the Inverse Method where the order of addition is reversed. In this study the Inverse Method was used. A 130mM solution of AR (analytical grade) tri-ammonium orthophosphate at pH ≥ 9 was dripped into a continuously stirred 210mM solution of AR calcium nitrate 4-hydrate over a period of approximately two hours. Double distilled water (DDW) and analytical grade reagents were used throughout the preparation. Five reaction temperatures were selected, 3°C, 25°C, 45°C, 70°C, and 90°C. The 3°C temperature range was achieved by immersing the reaction vessel in an ice bath, while other temperatures were maintained by the use of a thermostatically controlled hot plate. Up to seven concentrations of AR sodium bicarbonate between 10mM and 640mM were added to the phosphate solution at each temperature. All of the resulting precipitates were aged for 24 hours.

2.2 Characterisation

The crystallite size and morphology resulting from the precipitation reaction was determined with transmission electron microscopy. Copper grids with carbon support films were dipped into suspensions of the precipitate and were observed using a JEOL 1200 EX2. The precipitates were examined in bright field mode and using electron diffraction at magnifications of up to 150,000x and an accelerating voltage of 100kV.

The carbon content of the carbonate apatite powder samples was measured over the range of bicarbonate ion additions in the reaction. Carbonate content was determined using a Control Equipment Corporation Model 240 XA CHN elemental analyser. Calcium, phosphorus and sodium measurements were obtained with a Perkin Elmer Plasma 40 Emission Spectrometer. The reproducibility of the method was found to be better than 2% of the measured value.

X-ray diffraction patterns of carbonate hydroxyapatite powders were collected using a Siemens D5000 diffractometer using Cu K_α radiation, $\lambda = 154.18$pm with a graphite monochromator and a step size of 0.03° with a count time of 2 seconds per step. The data was analysed on Diffrac AT program. The phases present in the samples were determined by comparing the patterns with stick patterns from standard JCPDS files.

2.3 Sintering

A CHA precipitated at 3°C containing 3.2weight% carbonate, was slip cast into pellets of approximately 20mm in diameter. After drying, the green density was calculated using Archimedes principle. A Carbolite STF tube furnace was used to sinter pellets at 1000°C, 1150°C, 1250°C and 1300°C, in four different atmospheres: air, commercial purity grade CO_2, wet (10–20mg water per litre of gas) and wet CO_2. Moisture was added to the gases by passing them through DDW at a constant flow rate of 1.5lmin^{-1}. The sintering regime was fixed at a heating/cooling rate of 2.5°Cmin^{-1}, with a four hour dwell.

After sintering sample densities were measured using Archimedes' principle. Samples were then sectioned, polished and etched using 10% orthophosphoric acid. The microstructure of the sintered discs was studied using scanning electron microscopy, performed on gold coated samples using a JEOL 6300 microscope with an accelerating voltage of 20kV. The hardness of each specimen was measured using a Shimadzu microhardness tester with a 1kg load. Ten measurements were averaged to give a mean hardness measurement.

FTIR spectra of the green body and apatites sintered in the four atmospheres at 1250°C were obtained after drying for twenty four hours under vacuum with desiccant using the photoacoustic method in a Nicolet 800 spectrometer. XRD was performed for the samples after sintering in the various atmospheres at 1150°C.

3. RESULTS

3.1 Characterisation of Precipitates

HA crystallites, precipitated in the absence of bicarbonate ions at 90°C and 70°C, were found to be acicular in morphology. The crystallites resulting from reactions at 45°C and 25°C appeared to have an ellipsoidal morphology while those prepared at 3°C were virtually spheroidal in shape. The size of the crystals decreased with temperature from needles of the order of 800nm in length down to 10nm diameter crystallites for the 3°C reaction, as shown in Fig. 1.

Figure 2 shows the variation in precipitate morphology with bicarabonate ion concentration. It is evident that, as the bicarbonate ion concentration increased, the morphology of the apatites precipitated changed from acicular to spheroidal, with a marked decrease in size. At lower reaction temperatures the effect of bicarbonate ion concentration on crystallite size was less marked.

Table 1..   Effect of sintering temperature and atmosphere on carbonate hydroxyapatite density. (Mgm^{-3}), (n.d. not determined)

	AIR	DRY CO_2	WET CO_2	WET AIR
1150°C	3.12	3.06	3.13	n.d.
1250°C	3.12	3.07	3.13	3.14
1300°C	3.10	3.06	3.13	n.d.

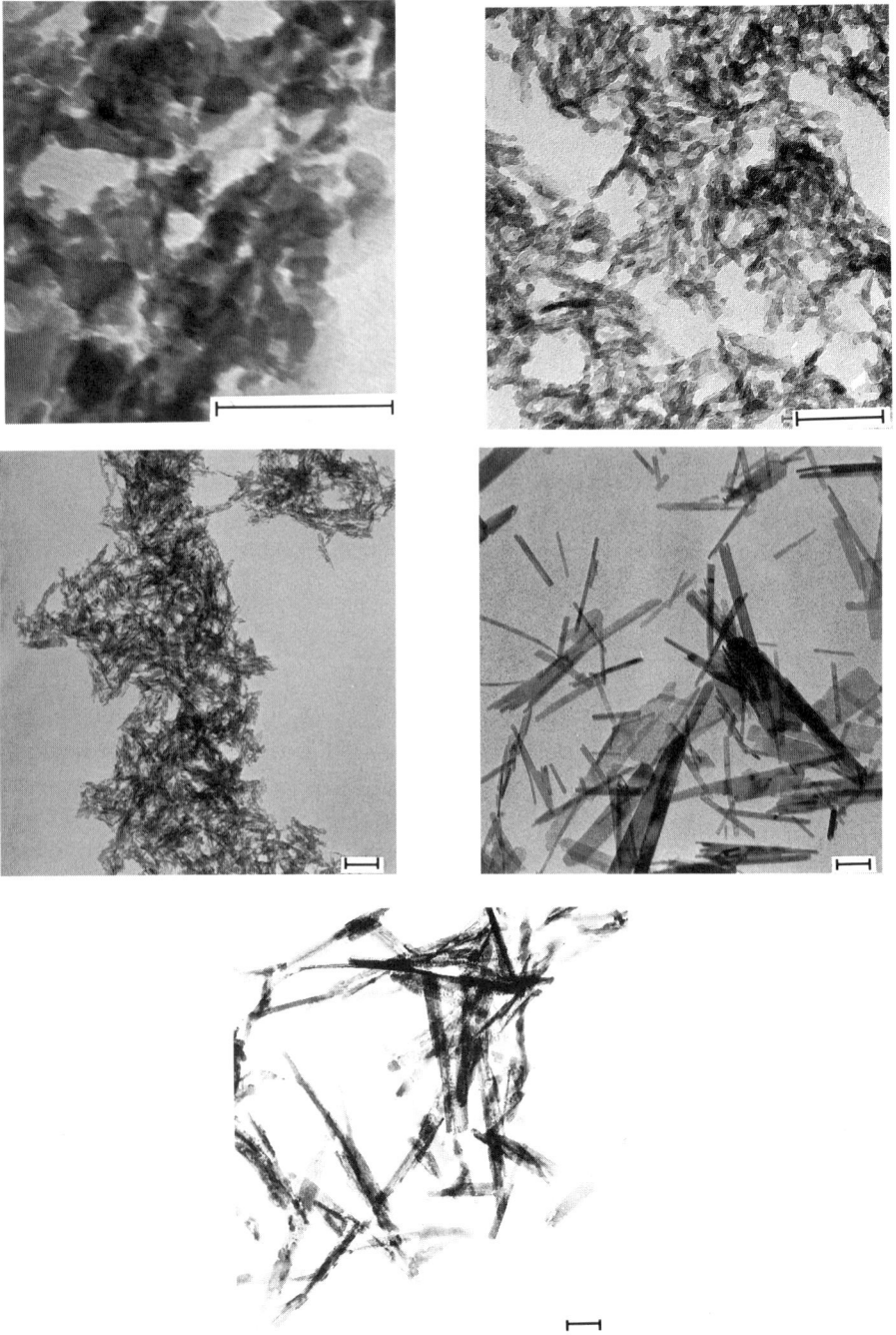

Figure 1.   Effect of precipitation temperature on the size and morphology of hydroxyapatite precipitates at a) 3°C, b) 25°C, c) 45°C, d) 70°C and e) 90°C. All marker bars are 100nm.

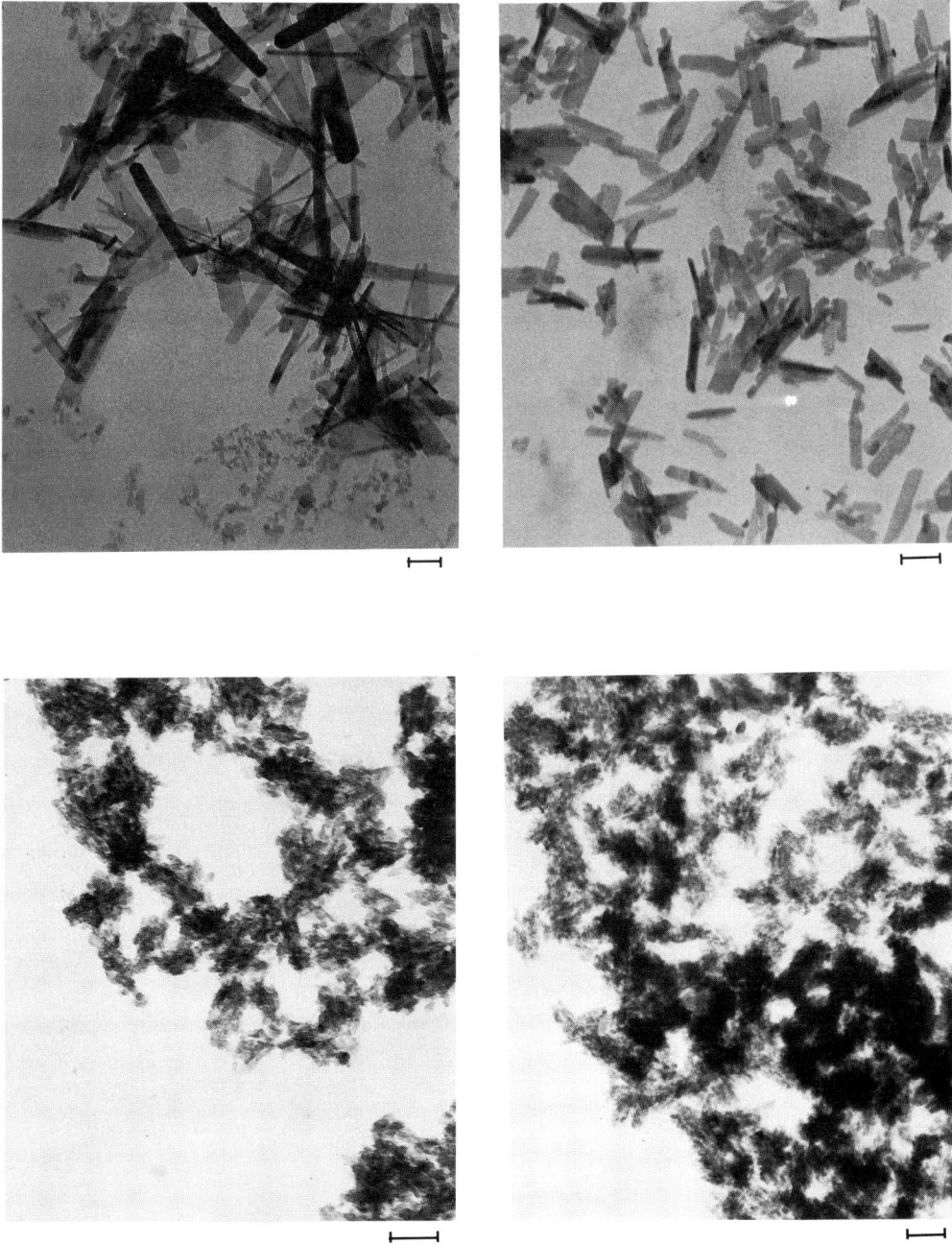

Figure 2. Effect of bicarbonate ion concentration on the size and morphology of carbonate hydroxyapatite precipitates at 70°C. a) 0mM, b) 40mM, c) 160mM and d) 320mM. All marker bars 100 nm.

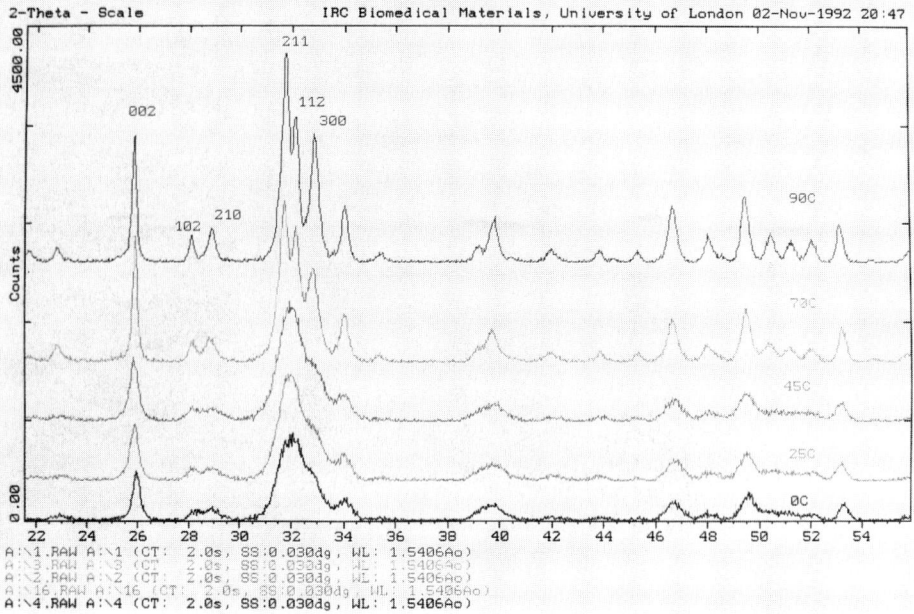

Figure 3. XRD patterns showing broadening due to decrease in particle size with decrease in reaction temperature.

3.2 Sintering

The slip casting method adopted for this set of experiments resulted in green bodies with a density of approximately 40% of the theoretical. Table 1 shows the effect of sintering temperature and atmosphere on the density achieved for CHA.

The microstructures of CHA specimens sintered at 1250°C in the four different atmospheres are shown in Fig. 4. All the samples except those sintered in dry carbon dioxide appeared have densified well after sintering at 1150°C. For the samples sintered in a dry carbon dioxide atmosphere, approximately 8% porosity was measured after sintering at 1150°C. It was observed that no additional densification occurred where higher sintering temperatures were used and in fact these conditions simply resulted in pore agglomeration. Small amounts of intergranular porosity were evident in CHA samples sintered in wet and dry air, but the microstructures of those sintered in wet carbon showed no porosity. In addition, the CHA samples sintered in the wet carbon dioxide atmosphere were translucent whereas samples sintered in all other atmospheres were opaque (Fig. 6).

3.3 Characterisation of the Sintered Materials

The microhardness values for the CHA specimens sintered under the various conditions are shown in Fig. 5. It can be seen that CHA samples sintered in dry carbon

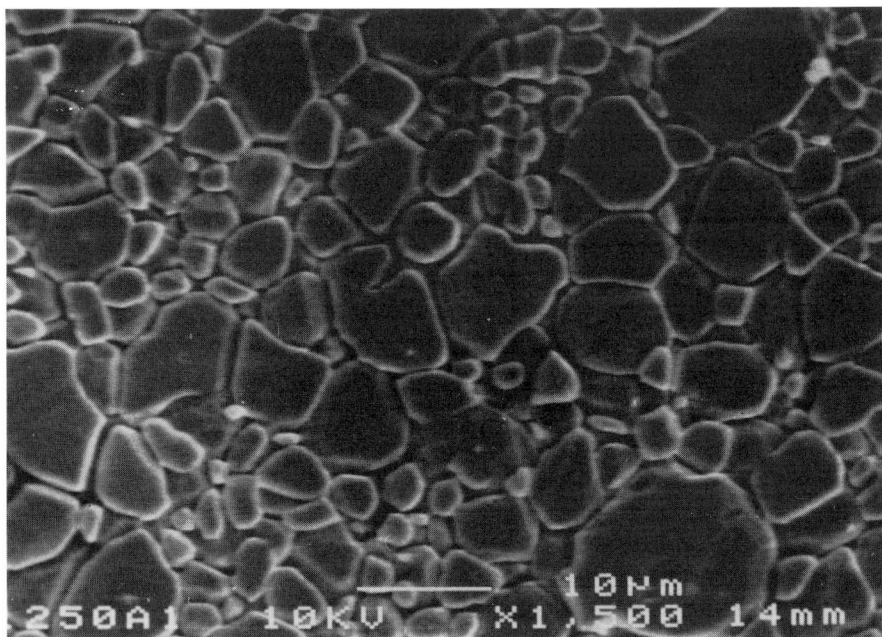

Figure 4a. Microstructures of CHA sintered at 1250°C in air.

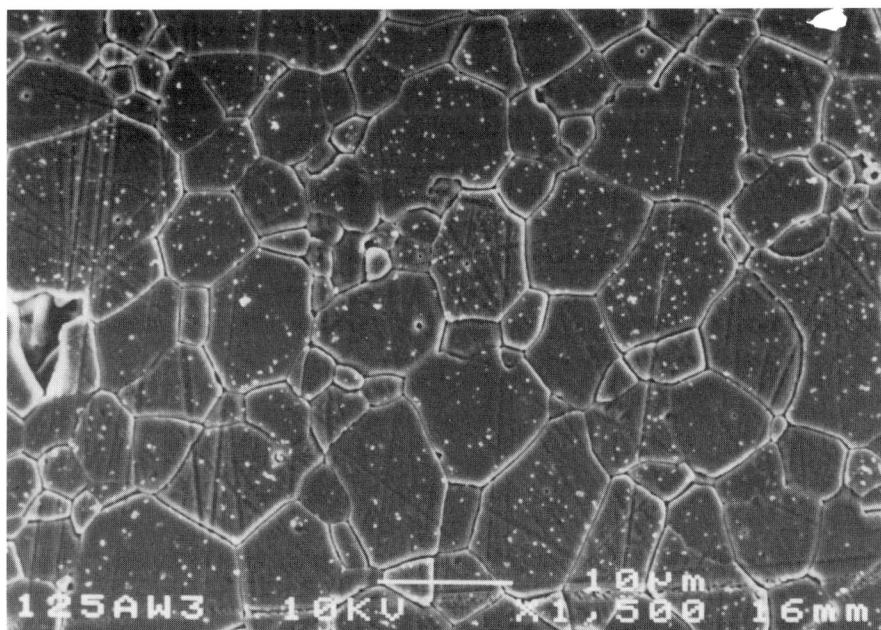

Figure 4b. Microstructures of CHA sintered at 1250°C in wet air.

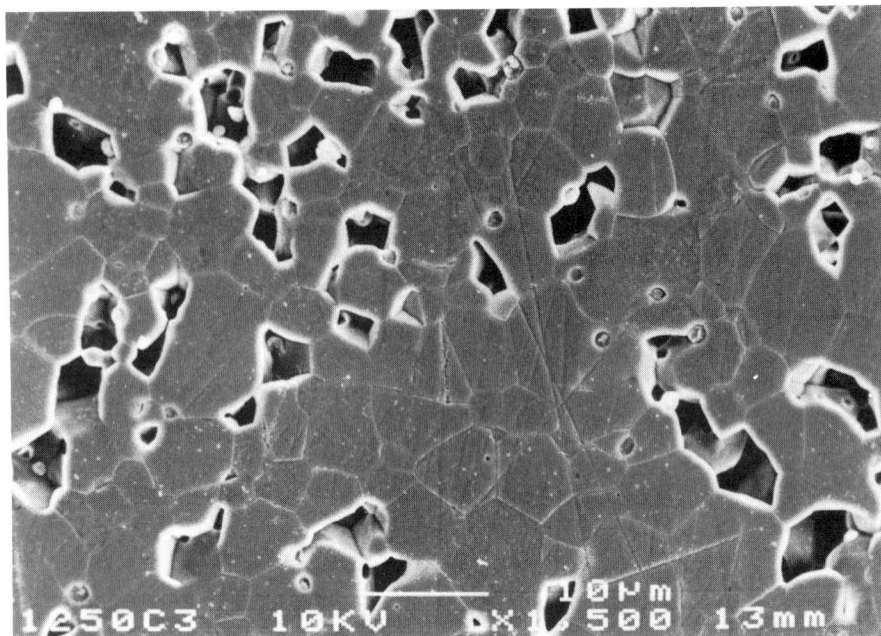

Figure 4c. Microstructures of CHA sintered at 1250°C in dry carbon dioxide.

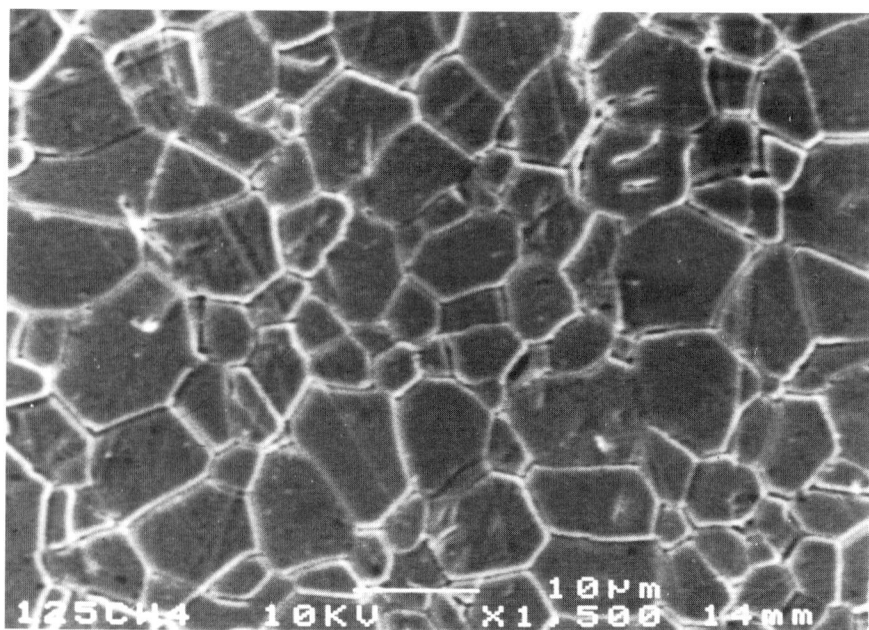

Figure 4d. Microstructures of CHA sintered at 1250°C in wet carbon dioxide.

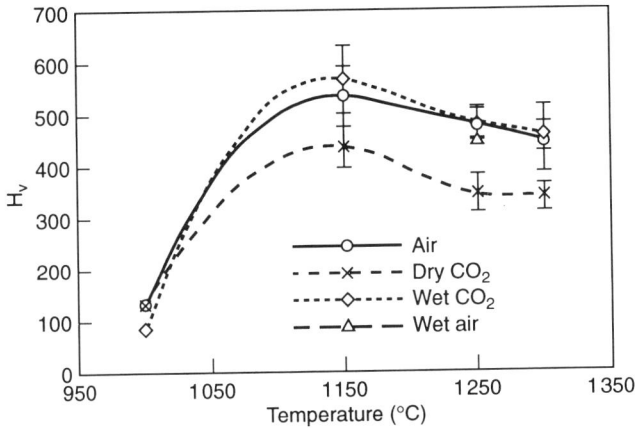

Figure 5.   Effect of furnace temperature and atmosphere on hardness.

dioxide displayed a signficantly lower hardness than the other samples over the whole range of sintering temperatures. FTIR spectra indicated that CHA specimens sintered in air atmospheres showed no carbonate peaks after sintering at 1250°C, whereas samples sintered in wet atmospheres gave enhanced OH absorption compared with their 'dry' counterparts. The XRD pattern of CHA specimens sintered in wet carbon dioxide at 1150°C (1.5wt% carbonate) confirms the material to be a single phase of apatite, as illustrated in Fig. 7. The microstructure of this material is shown in Fig. 8.

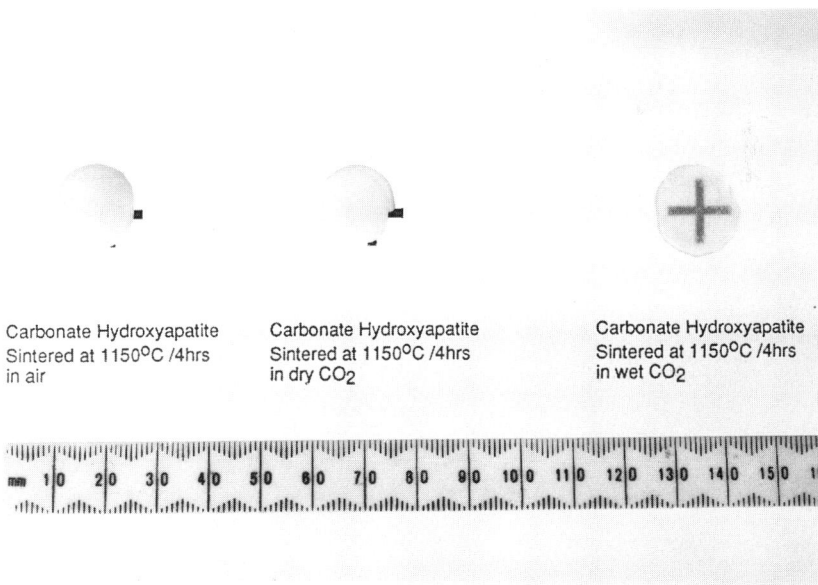

Carbonate Hydroxyapatite
Sintered at 1150°C /4hrs
in air

Carbonate Hydroxyapatite
Sintered at 1150°C /4hrs
in dry CO2

Carbonate Hydroxyapatite
Sintered at 1150°C /4hrs
in wet CO2

Figure 6.   Illustration of translucent nature of CHA sintered in wet carbon dioxide at 1150°C (1.5wt% CO_3^{2-})

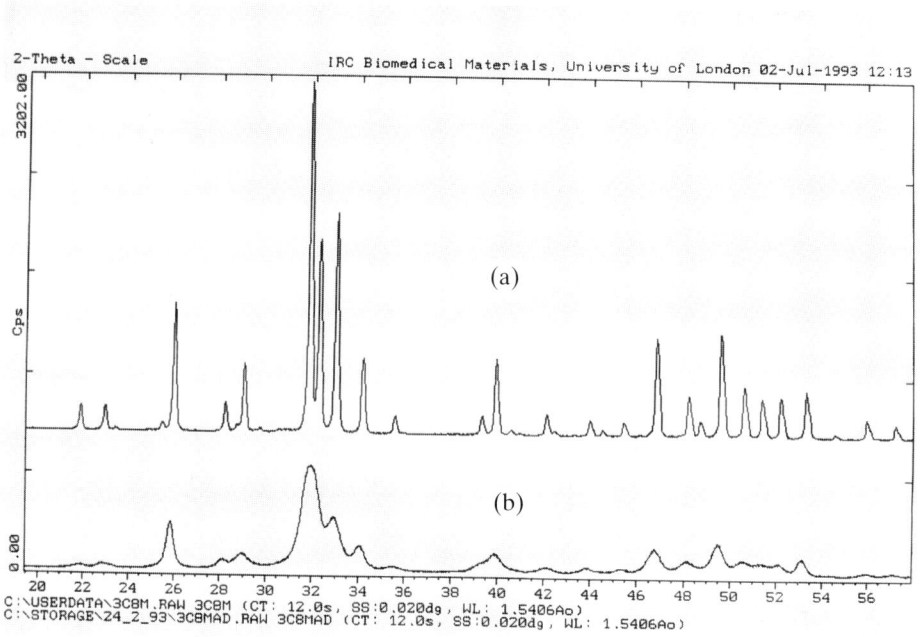

Figure 7. XRD patterns of: A CHA sintered in wet CO_2 at 1150°C and B green CHA.

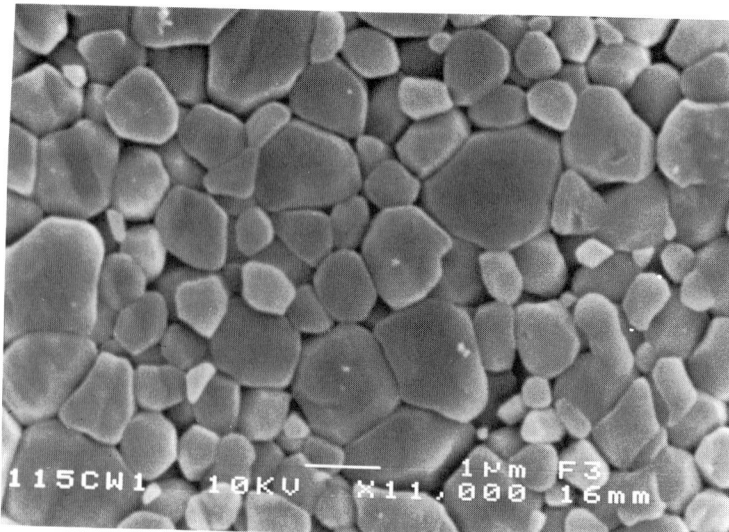

Figure 8. Microstructure of CHA sintered in wet CO_2 at 1150° (× 3 300)

4. DISCUSSION

It was found that by adjusting the reaction conditions, the size, composition and morphology of the CHA crystals could be altered. Although the effects of carbonate on apatite morphology have been reported by other workers using similar systems,[8,9] the influence if tenperature does not appear to have been noted previously.

The sintering results indicated that the partial pressure of water in the sintering atmosphere strongly influenced the densification process of CHA and that a carbon dioxide sintering atmosphere appeared to promote grain growth. The porosity of CHA sintered in dry carbon dioxide was reflected in the reduction in hardness values for this sintering environment. Full densification was achieved by sintering in a moist carbon dioxide atmosphere, as indicated by the translucent nature of apatite sintered in these conditions and the opacity of all other samples. However, it is not the purpose of this paper to discuss the mechanisms of sintering[10,11] of CHA in moist CO_2 as these will be discussed elsewhere.

5. CONCLUSIONS

Both reaction temperature and bicarbonate ion concentration strongly influence the morphology and sintering characteristics of carbonate hydroxyapatite. Due to the volatility of the carbonate in the material, special atmospheres need to be used during sintering and in this work the most suitable atmosphere was moist carbon dioxide. From the investigations, a novel method has been devised for the production of fully dense carbonate hydroxyapatite using pressureless sintering. The technique has potential, and it is cheaper than hot isostatic pressing and the densification temperature can be lower than for both of these alternative methods.

REFERENCES

1. M. Jarcho, C. H. Bolen, M. B. Thomas, J. Bobick, J. F. Kay and R. H. Doremus, *J. Mater Sci.*, 1976, **11**, 2027–2035.
2. J. G. J. Peelen, B. V. Rejda and K. DeGroot, *Ceramurgica*, 1976, **2**, 71.
3. G. DeWith, H. J. A. van Dijk and N. Hattu, *Proc. Brit. Ceram. Soc.*, 1981, **31**, 181.
4. F. C. M. Driessens, *Bioceramics of Calcium Phosphate*, CRC Press, Bucca Raton, Florida, USA, 1983.
5. H. Aoki, *Science and Medical Applications of Hydroxyapatite*, Takayama Press System Centre Co. Inc. JAAS, 1991, 101–106.
6. D. K. Argrawal, Y. Fang, D. M. Roy and R. Roy, *Microwave Processing of Materials III*, MRS, Pittsburgh, Pensylvania, USA, 1992, 231–236.
7. D. G. A. Nelson and J. D. B. Featherstone, *Calcif. Tiss. Int.*, 1982, **34**, S69–S81.
8. R. Z. LeGeros, O. R. Trautz, J. P. LeGeros, E. Klein and W. P. Shirra, *Science*, 1967, *155*, 1409–1411.
9. S. Shimoda, T. Aoba, E. C. Moreno and Y. Miake, *J. Dent. Rest.*, 1990, **69**, 1731–1740.
10. L. G. Ellies, D. G. A. Nelson and J. D. B. Featherstone, *J. Biomed Mats. Res.*, 1988, **22**, 541–553.

11. C. Rey, M. Frèche, J. C. Heughebaert, J. L. Lacout, A. Lebugle, J. Szilagyi and M. Vignoles, *Bioceramics 4* Butterworths, Guildford, UK, 1991, 57–64.

ACKNOWLEDGEMENTS

The authors of this paper acknowledge the financial support of the Science and Engineering Research Council in the provision of a Research Studentship for J. Barralet.

Glass-Ionomers: Prospects for the 21st Century

J. W. NICHOLSON

Biomaterials Department, Dental Institute, King's College School of Medicine and Dentistry, Denmark Hill, London SE5 9RW, UK

ABSTRACT

Glass-ionomer cements are prepared by reaction of acid-decomposable glass powders with polymeric acids in aqueous solution. They have been used in clinical dentistry for almost 20 years, but continue to undergo development, for example with metal-reinforcement or as modified materials capable of curing partially by photopolymerization. Outside dentistry, they are being developed for use in craniofacial reconstruction surgery. They have also been used recently in the fabrication of artificial ear ossicles. In general, they exhibit excellent biocompatibility, particularly when in direct contact with bone. Unfortunately, inappropriate use can cause problems; in recent cases in France, glass-ionomer cement led to massive build-up of aluminium in the blood and urine of patients resulting in epileptic seizures. Prospects will be discussed for the development of improved cements for these wider medical uses that will be less dangerous in the hands of inexperienced clinicians.

1. INTRODUCTION

Glass-ionomers became available in the last quarter of the twentieth century as materials for use in clinical dentistry. However, following the discovery of their excellent biocompatibility, and with the growing need for materials showing such compatibility in a wide range of clinical situations, their use has recently been extended. It currently includes artificial ear ossicles and bone-substitute plates for craniofacial reconstruction, and experimental studies have included use in orthopaedics, ENT surgery (for fixation of cochlea implants and sealing defects in the skull), and augmentation of the alveolar ridge of edentulous patients. More recently, the range of possibilities in clinical dentistry has been expanded through the development of resin modified (light-curable) versions of these cements. All of this clinical interest suggests that glass-ionomers will continue as clinically important materials well into the next century. The background to this confident prediction will be explored in the current paper.

2. THE CHEMISTRY OF GLASS-IONOMER CEMENTS

Glass-ionomers fall into the class of material known as acid-base cements.[1] Their setting involves neutralisation of acid groups on a water-soluble polymer with a powdered, solid base. This latter is a special calcium aluminosilicate glass which also contains fluoride, an important feature, since it causes the cement to release clinically useful amounts of this ion and thereby to prevent the development of secondary caries around restorations.[2] The glasses are bases in the sense that they are proton-acceptors, even though they are not soluble in water.

As the cements set, water becomes incorporated into the material, and there is no phase separation. In fact, water has been identified as having a number of roles: (i) it is the solvent for the setting reaction, since without it, the polymeric acid would be unable to exhibit its full properties as an acid, (ii) it is one of the reaction products, (iii) it acts as both co-ordinating species to the metal ions released from the glass and as hydrating species at well-defined sites around the polyanion, and finally (iv) it may act as plasticiser and reduce the rigidity of the bulk polymeric structure.[3]

A number of factors are known to influence the speed of the setting reaction and the final strength of the cement. These include molar mass of the polyacid concentration of the acid solution, powder: liquid ratio, and the presence of chelating agents, such as (+) – tartaric acid,[4] reduces the setting time and increases the compressive strength of the cement once set.

Glass-ionomers undergo gradual maturation processes which are poorly understood. For example, in cements prepared from poly(acrylic acid), compressive strength gradually rises over the first 3 months or so of the cement's life to a maximum value some magnitude greater than the value at 24 hours. The ratio of bound to unbound water increases, this being defined as the ratio of water that may be removed by chemical desiccation (eg by storage for 24 hours over silica gel at elevated temperature) to that which is retained in the cement during this treatment. Finally, translucency also changes, gradually becoming greater and more like natural tooth material as the cement ages.[4] The setting reactions in glass-ionomers are as follows:

(i)   Decomposition of the glass under the influence of the aqueous polyacid, leading to the release of Ca^{2+} and Al^{3+} ions. These latter are probably released in the form of complex oxyanions containing several atoms,[5] a structure which reflects the form they have occupied within the glass prior to acid attack.[6]

(ii) Rapid reaction of the Ca^{2+} ions with the polyacid chains, followed by slower reaction of Al^{3+} species gradually released from the anionic complex. This reaction displaces water from some of the hydration sites,[7] and leads to some ionic crosslinking of the polyacid chains, both of which effects lead to insolubilisation of the polymer and stiffening of the material.

(iii) Gradual hydration of the inorganic fragments also released in step (i) to yield a matrix of increasing strength, greater resistance to desiccation, and improved translucency.[8]

These steps are illustrated in Figure 1.

Improvements in strength and durability of glass-ionomer cements have been sought by such means as the inclusion of finely divided silver alloy or of a silver cermet formed from the glass plus silver in a fusion process.[9] Fibres have also been used to reinforce experimental cements.[10]

A disadvantage of glass-ionomers is that they are sensitive to moisture in the early stages following placement.[4] This may result in either the washing out of reacting ions from the immature cement by saliva or, in patients who tend to breathe through

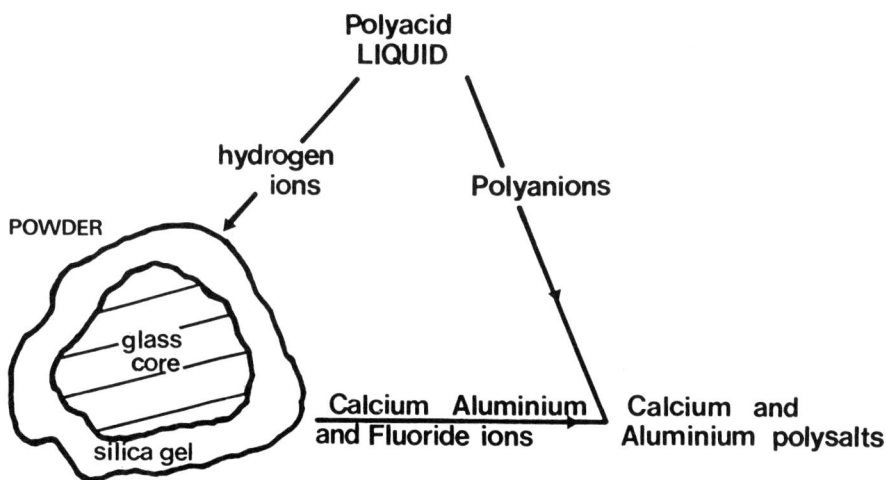

Figure 1.   Schematic representation of the setting of glass-ionomer cements.

the mouth, in desiccation and arrest of the setting reaction. Both are undesirable, and to overcome the problem, dentists are advised to cover a freshly placed cement with an imperious layer of varnish or petroleum jelly.[4]

Glass-ionomers are able to form true adhesive bonds to dentine and enamel,[4] and for this reason have found a wider range of applications than other dental cements. These uses may be summarised as follows:[11–13]

(i)   Sealing of occlusal pits and fissures.
(ii)   Filling of occlusal fissures with early signs of caries.
(iii)   Restoration of deciduous teeth.
(iv)   Restoration of cavities on the adjacent surfaces of neighbouring teeth (class III cavities) which are in low-stress situations, typically on the lingual face.
(v)   Restoration of erosion cavities along the gum line (class V cavities).
(vi)   Repair of defective margins in restorations.
(vii) Minimal cavity repair.[4,14] Here, the concept is that, since caries is mainly a disease of the dentine, only a minimal amount of enamel needs to be removed, just sufficient to allow access to the carious dentine. The carious dentine is then excavated, and the region filled with glass-ionomer cement. Since it is adhesive, it holds the enamel shell together, and thus preserves the sound enamel.

3. RESIN-MODIFIED GLASS-IONOMERS

These materials, the majority of which are visible-light cured, are hybrids that involve the incorporation of polymerizable components into an acid-base glass-ionomer cement. They were first described in the late 1980s.[15] The use of visible light to cure these materials, at least as far as the initial development of structure is concerned, limits the depth of individual layers of cement that can be used, because of limitations in the extent to which light can penetrate these materials. Typically, this depth is of the order of 2–3mm, a feature which restricts the use of these materials to certain limited areas, such as cavity lining or incisal edges.

Resin-modified glass-ionomers consist of a complex mixture of components,[16] including poly(acrylic acid) or a graft-copolymer of poly(acrylic) in which a photo-curable side chain has been added, photocurable monomers, such as hydroxyethyl methacrylate (HEMA), as well as the calcium aluminosilicate glass and water. These materials set by a number of competing reactions and have complex structures. There is evidence of slight swelling of the cured cement in aqueous media,[17] but to date, only one account of this leading to a catastrophic clinical failure, when a tooth filled with an inappropriately formulated material which underwent phase separation prior to use was split by the osmotic pressure in a HEMA rich cement.[18] In general, however, clinical indication for their use has been promising, with good adhesion,[19] acceptable fluoride release[20] and excellent aesthetics,[21] being reported. Clinical use has included deciduous[21] as well as fully developed adult teeth.[22] It is, however, too soon to have data on long term durability or aesthetics.

4. BIOCOMPATIBILITY

The environment inside the body is both hostile and extremely sensitive. To select materials for use in this environment requires that attention be paid to both the physical function and the biocompatibility of the candidate substance. Biocompatibility is defined as the ability of a material to perform with an appropriate host response in a specific application.[23] It is thus distinct from inertness, which would imply no response from the host. Moreover, it is not a single phenomenon, but rather is a collection of processes involving different but interdependent mechanisms of interaction between a material and the tissue. It is also specific to a particular application and location in the body.

Glass-ionomer cements are generally bland towards oral tissues, and as restorative materials show only mild pulpal irritation at a similar level to zinc polycarboxylate and zinc phosphate cements.[24] This reaction is so mild that glass-ionomers can generally be used as intracoronal restoratives without a lining or a base,[25–27] although in some human studies, moderate inflammatory responses in the pulp were reported.[28,29] However, the use of certain glass-ionomers extracoronally, as luting cements, has been shown to be associated with pulpal hypersensitivity.[30,31] The particular cements that cause this are the so-called anhydrous cements, which are formulated by mixing glass and polymer powders, and activating the setting process by the addition of the appropriate amount of water. The reason that these cements cause pulpal sensitivity was initially thought to be due to the slow dissolution of the poly-acid, which was assumed to maintain the local pH at low levels for longer than in conventionally formulated glass-ionomers.[32] However, a study of pH change in setting cements showed that the anhydrous cements underwent a slightly more rapid neutralisation than conventional ones, and that their setting profile was almost the same as that of anhydrous zinc polycarboxylate.[33] No pulpal sensitivity has ever been reported for this latter material. It thus seems likely that pH is not the cause of the reported sensitivity.

Biological studies have shown that different glass-ionomers differ in their ability to develop and sustain a marginal seal that exclude bacteria from the region close to the pulp.[34–36] Moreover, beneath certain glass-ionomer restorations, including experimental crowns luted with this material, there have been found bacteria in an active state of metabolism. These have always had associated with them moderate pulpal inflammation.[34–36] It thus seems that glass-ionomers are not directly responsible for this adverse pulpal reaction, but in certain brands, through not forming an adequate seal, are responsible indirectly, since bacteria can be admitted to cause the adverse biological effect.

Outside dentistry, the general effect of glass-ionomers on cells has been tested, and in general shown to be minimal. For example, cytotoxicity against Hela cells has been studied,[37] with the finding that, while there was some evidence of toxicity in freshly prepared cements, in cements that had matured for at least 24 hours, there was little or no toxic response. This correlates well with what is known about the setting chemistry of these cements, that they are sensitive to early exposure to moisture, and will release a variety of water-soluble species in the first few hours after mixing.

More recently, osteoblast biocompatibility of a particular glass-ionomer has been studied.[38] This cement is prepared from maleic/acrylic acid copolymers, and is being considered for a variety of surgical procedures involving direct contact with bone. It gave no evidence of cytotoxicity towards osteoblasts,[38] despite leaching aluminium that was found to have entered the individual osteoblast cells. This initially surprising finding was attributed to detoxification through the formation of complexes with silica, also released from the cement, in a manner previously described in the literature.[39] In a separate study, it was also shown that bovine osteoblast cells attached themselves rapidly to glass-ionomer surfaces, particularly by comparison with other biomaterials, a feature aided by the good wettability and ionic character of the cement surface.[40]

Biocompatibility, in both cell cultures and in animal models, was studied using a series of glass-ionomers, both commercial and experimental, and including one resin modified (light curable) material.[41] The study involved both determination of acute cytotoxicity using contact techniques with cultured fibroblasts, and toxicity following extraction. The effect of implanting smooth rods of set cement into the femurs of adult hooded rats was also determined. Results varied widely between materials and evaluation techniques. However, some general conclusions can be drawn. Firstly, glass-ionomers that cure entirely by an acid-base reaction showed very good biocompatibility in each test regime, with good bone growth in the animal models, and good cell viability in the culture techniques.[41] Secondly, the resin modified cement studied showed consistent and significantly inferior biocompatibility, regardless of which test method was used. It was slightly less bad following extraction, suggesting that the poor biocompatibility arose due to a water-soluble substance that was readily leached out. No conclusions were drawn from this concerning the identity of the toxic species,[41] but a knowledge of the constituents of the cement plus their biological effects would strongly imply that HEMA was responsible.

For autocuring glass-ionomers, three features have been identified as contributing to the good biocompatibility.[42] They are:

(a) the minimal setting exotherm, typically no more than 3–4°C even in quite large samples of the cement;[43]
(b) the speed of neutralisation, which removes the potentially irritating acid;
(c) the substances leached from the cement are either benign or beneficial towards the tissue in which the cement is placed. On this last point, the species found to be leached are calcium, sodium, aluminium, silica, phosphate and fluoride. No organic species arising from the polymer have ever been found to be leached from the cement. Of these, silica is probably benign; certainly in the form and location of the release, no harmful side effects have ever been found.[43] Calcium, sodium and phosphate all find a multitude of physiological uses in the body, and seem acceptable at the levels released. Fluoride in low doses is helpful for teeth and bones, and again seems acceptable at the levels released.

The only species released for which there is concern is aluminium. Cell culture studies, as we have seen, suggest that aluminium is detoxified in some way, presumably via formation of complexes with silica, and no cytoxic effects have been shown, even when aluminium has been shown to have entered the cells.[38] However, there have been problems in patients, due to aluminium toxicity.[44] In France, two patients who had had translabyrinthic otoneurosurgery and bone reconstruction with a glass-ionomer cement became ill (at one month and two months respectively following surgery). Their symptoms include subacute coma, epileptic seizures, and in one case, stupor, mutism and grasping.[44] Aluminium levels were extremely high in the lumbar cerebro-spinal fluid, serum and urine, and in all fluids, fell following removal of the cement and treatment with sequestering agent (desferrioxamine 0.5–2.0g day^{-1} for 2 months). Symptoms, however, persisted, and following the reporting of these cases, the use of the glass-ionomer bone cement was stopped in France by order of the Direction Générale de la Santé on 31 March 1994.

Quite how these symptoms appeared in these patients, given the results from cell culture and other studies, is not clear. However, ions are known to be washed out of a cement if it is prematurely exposed to aqueous liquids, and it may be that the particular surgical procedures resulted in premature exposure of the setting cement to the body fluids. Certainly, as subsequent sections of this paper will show, many patients have been treated in a variety of ways by glass-ionomer cement in direct contact with bone, and no comparable problems have ever arisen before.

5. MEDICAL APPLICATIONS OF GLASS-IONOMER CEMENTS

Glass-ionomer was originally studied as a potential replacement for poly (methyl methacrylate) bone cement, a material which is not entirely satisfactory.[45–47] Its use has been associated with a variety of problems, including tissue sensitivity and necrosis as a result of its large polymerisation exotherm. Moreover, it is not suitable for use in reconstructive surgery following cancer or radiation treatment of diseased bone.

Initial biological observations were made using baboons as the animal model.[45] In these animals, glass-ionomer was shown to be stable in bone contact, and to show no signs of surface dissolution. New bone growth was promoted and calcification observed on the surface of the cement, features that were ascribed to fluoride release from the cement. Osteogenesis is known to be stimulated by fluoride at appropriate levels, a feature that is used in fluoride therapy for osteoporosis.[48] However, there is an optimum level of fluoride release,[49] since large doses are known to be toxic, as has been demonstrated in an enclosed in vitro situation.[50–52]

In another study, Brook et al[52,53] studied glass-ionomer cement as a possible material for augmenting the alveolar ridge in edentulous patients. They confirmed the earlier findings of good bone biocompatibility, as cements formed intimate bioactive bonds with bone cells and became fully integrated into the bone. In this behaviour, glass-ionomer cement compared favourably with other materials, and was, for example, better than hydroxyapatite, and bone bonded directly to the glass-ionomer without the formation of an intervening layer.

These favourable biological properties have been widely exploited in ear, nose and throat surgery. For example, Ramsden et al[54] have reported on a 1-year study in which a newly available commercial glass-ionomer, IONOS cement, was used in a variety of procedures, including fixation of a prosthetic chochlea implant, sealing the Eustachian tube and sealing other defects in the skull which would otherwise have allowed cerebro-spinal fluid to leak through the middle ear and Eustachian tube to the nose or nasopharynx.

The cement was presented to the clinician in a capsule containing a mixed powder of glass and acrylic/maleic acid copolymer. This was mixed with water by shaking, after which it took 5 minutes to set to a reasonable hardness. This preliminary study indicated that the material is safe, non-toxic and does not affect either the organ of corti or the facial nerve. These findings contrast with those of Renard et al,[44] who found severe aluminium toxicity following use of this same commercial cement in acoustic neurone surgery. As previously stated, the reasons for this latter result, contrasting as it does with results from numerous other studies, are not clear. Certainly, there are other accounts of the use of this material in ENT surgery.[55,56] Cochlea implant fixation has been widely studied; for example, Lehnhardt[56] has carried out a clinical trial involving more than 237 devices, mainly in children, using the IONOS cement for the fixation of the electrode and the receiver of the implant, and has encountered not a single adverse reaction.

Repair of the ossicular chain using preformed glass-ionomer ossicles has been widely reported.[55,57,58] Such implants have now been placed in more than 2000 operations world wide, with excellent results. Glass-ionomer shows good compatibility and an extremely low rejection rate by patients.

Finally, glass-ionomers have been used in maxillo- and cranio-facial reconstruction surgery[56,57] In this technique, solid implants are made for the major part of the implant, each implant being custom made to precisely fit the contours of the patient's skull or jaw line. These large pieces, cured outside the body to develop full mechanical properties prior to placement, are then fixed in place by a glass-ionomer paste, which cures in situ. As in other accounts, this material is reported to have good biocompatibility, and there remains healthy bone growth right up to the edge of the implant. To date, relatively few of these custom-made implants have been placed, but already these are indications that this will prove a successful surgical procedure for the small group of patients for whom they are required.

6. FUTURE PROSPECTS

Glass-ionomers are now well established in clinical dentistry, and recent developments are likely to extend and enhance their use in this field. Their adhesion under clinically demanding conditions, fluoride release and aesthetics make them highly favoured materials, and with the growing emphasis on aesthetics in restorative dentistry, coupled with the increasing proportion of the population who remain at least partially dentate until well into old age, they seem likely to retain their position.

Outside dentistry, despite the recent set-back in France, they seem likely to find increasing use. Their bone-contact biocompatibility is excellent, and provided they can be presented to the clinician in a fail safe form, their use in areas such as craniofacial, maxillofacial, acoustic neurone and orthopaedic surgery seems likely to grow. Unlike all other possible materials glass-ionomer cement is biocompatible and will bond well to bone. It has thus opened up a new dimension in reconstructive bone surgery, and in the future is likely to allow a whole range of new surgical techniques. As a biomaterial it is certainly one to watch in the 21st Century.

REFERENCES

1. A. D. Wilson and J. W. Nicholson: *Acid-Base Cements: Their biomedical and industrial applications,* The University Press, Cambridge, 1993.
2. M. L. Swartz, R. W. Phillips and H. E. Clark: *J Dent Res,* 1984, **63**, 158.
3. J. W. Nicholson: *Chem Soc Rev,* 1994, **23**, 53.
4. A. D. Wilson and J. W. McLean: *Glass-ionomer cement,* Quintessence Publishers, Chicago, 1988.
5. E. A. Wasson and J. W. Nicholson: *Br Polym J,* 1990, **23**, 179.
6. R. G. Hill and A. D. Wilson: *Glass Technol,* 1988, **29**, 150.
7. A. Ikegami and N. Imai: *J Polym Sci,* 1962, **56**, 133.
8. E. A. Wasson and J. W. Nicholson: *J Dent Res,* 1993, **72**, 481.
9. E. A. Wasson: *Clin Mater,* 1993, **12**, 181.
10. C. W. B. Oldfield and B. Ellis: *Clin Mater,* 1991, **7**, 313.
11. J. W. McLean and A. D. Wilson: *Aust Dent J,* 1977, **22**, 31.
12. J. W. McLean and A. D. Wilson: *Aust Dent J,* 1977, **22**, 120.
13. J. W. McLean and A. D. Wilson: *Aust Dent J,* 1977, **22**, 190.
14. J. W. McLean: *J Californian Dent Assoc,* April 1986, **20**.
15. S. B. Mitra: Eur Patent Appl, 0323120A2, 1989; J. M Antonucci, J. E. McKinney and J. W. Stansbury: US Patent Appl, 160, 856, 1988.
16. H. M. Anstice: *Chem & Ind,* 1994, 899.
17. H. M. Anstice and J. W. Nicholson: *J Mater Sci; Mater in Med,* 1992, **3**, 447.
18. J. W. Nicholson and H. M. Anstice: *Trends in Polym Sci,* 1994, **2**, 272.

19. S. B. Mitra: *J Dent Res,* 1991, **70**, 72.
20. H. Forss: *J Dent Res,* 1993, **72**, 1257.
21. T. P. Croll and C. M. Killian: *Quintessence Int*, 1993, **24**, 561.
22. T. P. Croll: *Quintessence Int,* 1993, **24**, 109.
23. D. E. Williams: *Definitions in Biomaterials,* Elsevier, Amsterdam, 1987.
24. C. G. Plant, P. J. Knibbs, R. S. Tobias, A. S. Britton and J. W. Rippin: *Br Dent J,* 1988, **165**, 54.
25. J. W. McLean: *Br Dent J*, 1984, **157**, 432.
26. I. Mjor, I. Nordahl and L. Tronstad: *Endod Dent Traumatol,* 1991, **7**, 59.
27. D. A. Felton, C. F. Cox, M. Odom and B. E. Kanoy: *J. Pros Dent,* 1991, **65**, 704.
28. I. R. Cooper: *Int Endod J,* 1980, **13**, 76.
29. G. C. Plant, R. M Browne, P. J. Knibbs, A. S. Britton and T. Sorahen: *Int Endod J,* 1984, **17**, 51.
30. G. Christensen: *J Am Dent Assoc,* 1990, **120**, 69.
31. G. H. Johnson, L. V. Powell and T. A. Deroven: *J Am Dent Assoc,* 1993, **124**, 39.
32. D. C. Smith and N. D. Ruse: *J Am Dent Assoc,* 1986, **112**, 654.
33. E. A. Wasson and J. W. Nicholson: *J Dent,* 1993, **21**, 122.
34. R. S. Tobias, C. G. Plant, J. W. Rippin and R. M. Browne: *Endod Dent Traumatol,* 1989, **5**, 242.
35. C. H. Pameijer and H. R. Stanley: *Am J Dent,* 1988, **1**, 71.
36. C. G. Plant, R. S. Tobias, J. W. Rippin, J. W. Brooks and R. M. Browne: *Dent Mater,* 1991, **7**, 217.
37. B. L. Dahl, and L. Tronstad: *J Oral Rehab,* 1976, **3**, 19.
38. U. Meyer, D. H. Szulczewski, R. H. Barkhaus, M. Atkinson and D. B. Jones: *Biomaterials,* 1993, **14**, 917.
39. J. D. Birchall, C. Exley, J. S. Chappell and M. J. Phillips: *Nature,* 1989, **338**, 146.
40. U. Meyer, D. H. Szulczewski, K. Moller, H. Heide and D. B. Jones: *Cells and Materials,* 1993, **3**, 129.
41. P. Sasanaluckit, K. R. Albustany, P. J. Doherty and D. E. Williams: *Biomaterials*, 1993, **14**, 906.
42. J. W. Nicholson, J. H. Braybrook and E. A. Wasson: *J Biomater Sci Polym Ed,* 1991, **4**, 277.
43. R. K. Iler: *The Chemistry of silica*, John Wiley & Sons, New York, 1979.
44. J. L. Renard, D. Felton and D. Bequet: *Lancet,* 2 July 1994, **344**(63), 8914.
45. L. M. Jonck, C. J. Grobbelaar and H. Strating: *Clin Mater,* 1989, **4**, 85.
46. L. M. Jonck, C. J. Grobbelaar and H. Strating: *Clin Mater,* 1989, **4**, 201.
47. L. M. Jonck and C. J. Grobbelaar; *Clin Mater,* 1990, **6**, 323.
48. H. M. Frost: *Orthop Clin N America,* 1981, **12**, 649.
49. R. T. Turner, R. Francis, D. Brown, J. Garand, K. S. Hannan and N. H. Bell: *J Bone Mineralogy Res,* 1989, **4**, 477.
50. H. Kawahara, Y. Imanishi and H. Oshime: *J Dent Res,* 1979, **58**, 1080.
51. S. Hetem, A. K. Jowett and M. W. Ferguson: *J Dent,* 1989, **17**, 155.
52. I. M. Brook, G. T. Craig and D. J. Lamb: *Clin Mater,* 1991, **7**, 295.
53. I. M. Brook, G. T. Craig and D. J. Lamb: *Biomaterials,* 1991, **12**, 179.
54. R. T. Ramsden, R. C. D. Herdman and R. H. Lye: *J Laryngol and Otol,* 1992, **106**, 949.
55. G. Babighian: *J Laryngol and Otol,* 1992, **106**, 954.
56. W. Zollner and C. Rudel: 'Glass Ionomers: The Next Generation', P. Hunt (ed.), *International Symposium in Dentistry PC*, Philadelphia, 1994, pp57–60.
57. G. Geyer and J. Helms: *Transplants and Implants in Otology II*, 1991, **165**.
58. J. Helms and G. Geyer: *Eur Arch Otorhinolaryngol,* 1991, **250**, 253.

Magic Angle Spinning NMR of Bioglass Materials

D. HOLLAND, M. W. G. LOCKYER and R. DUPREE

Department of Physics, Warwick University, Coventry, CV4 7AL, UK

ABSTRACT

Glasses from the Na_2O–CaO–SiO_2–P_2O_5 system have been prepared, covering the inert, soluble and bioactive regions of the phase diagram. ^{23}Na, ^{31}P and ^{29}Si NMR has been performed on these glasses and the spectra obtained have been used to develop a model of the glass structure which is compatible with the observed behaviour in body fluid. The Na^+ and Ca^{2+} ions were found to associate preferentially with the phosphate tetrahedra in the glasses but there was no observable differentiation between Na^+ and Ca^{2+}. The remaining metal ions then distribute themselves within the silicate network. In the bioactive region this network was found to consist of Q^3 and Q^2 type silicons. The change in the chemical shifts of the NMR resonances from these two sites, as a function of composition, was interpreted on the basis of a preferred association of Ca^{2+} with Q^2 and Na^+ with Q^3. The structural units produced by this association control the chemical properties of the glasses and thus the degree of activity.

1. INTRODUCTION

'Bioglass'[1] has been the subject of many studies aimed at understanding the mechanism of formation of the glass-tissue interface[2–10] but there has been no investigation of glass structure which might give some insight into the prerequisites for bioactivity. Possible implant materials have been classified in terms of their response to body fluid. If the implant is toxic then the surrounding tissue dies, if the implant is non-toxic but soluble then gradually the neighbouring tissue replaces it, or finally, if the implant material is non-toxic and insoluble, the host tissue encapsulates the implant, generally with fibrous tissue that has little strength at the interface.[11] Bioglass elicits a different response, the formation of a strong chemical bond across the tissue/implant interface.

The mechanism of formation of this interface is thought to consist of 5 stages.[12]

1) ion exchange between the implant surface and the surrounding physiological environment with Na^+ being replaced by H^+ or H_3O^+.
2) break up of the silicate network to form silanols
3) condensation and partial repolymerisation of these silanols to form a SiO_2 rich gel layer at the interface.
4) migration of Ca^{2+} and PO_4^{3-}, through the open gel structure, to the interface region to form a CaO–P_2O_5 rich film on the gel layer which develops further by absorption of calcium and phosphate ions from the physiological solutions.
5) crystallisation of the amorphous calcium phosphate into a mixed hydroxyl, carbonate and fluoro-apatite layer – the OH^-, CO_3^{2-} and F^- anions being derived from the host tissue.

The composition dependence of the bioactivity of glasses within the soda-lime silicate (+ P_2O_5) system is illustrated in Fig. 1, in which the phase diagram can be considered in four regions.

1) Compositions within region A form strong bonds with living tissue. This bioactivity decreases as the boundaries of the region are approached.
2) Compositions from region B are inert and they become encapsulated by fibrous material from the host.
3) Compositions within region C are resorbable, disappearing completely after 10 to 30 days of implantation.
4) Region D is non glass-forming.

We have employed ^{29}Si, ^{23}Na and ^{31}P MAS NMR in the examination of the structure of a series of Na_2O–CaO–SiO_2–(6 wt% P_2O_5) glasses and in this paper we try to relate the behaviour of glasses in these regions to the observed structure. Simple binary silicate glasses, such as the sodium silicate system, have been studied extensively using MAS NMR.[13] However, application of the technique to the soda lime silicate system has been limited. Dupree, Holland and Williams[14] published results of ^{29}Si and ^{23}Na MAS NMR of a Na_2O $4SiO_2$ glass where CaO was substituted for Na_2O. The ^{29}Si shifts for silicon in Q^3 and Q^4 became more negative, as the amount of CaO was increased, from -90 to -93 ppm and -105 to -108 ppm respectively. (Q^n refers to a silicon with n bridging oxygens to other silicons and 4-n non-bridging oxygens). The corresponding ^{23}Na peak position was also shifted from -3 ppm in the base glass to -15 ppm when the ratio of Na_2O to CaO was 2:3. In another series of glasses with nominal composition $^x/_2$ Na_2O $^x/_2$ CaO $(100-x)SiO_2$ the ^{23}Na peak position became less shielded, -13 ppm to -4 ppm, as the total modifier content was raised from 20 to 40 mol%. XPS data from similar glasses (Veal et al[15]) suggested that only 80% of the Ca^{2+} behaves as network modifiers, contrary to the ^{29}Si MAS NMR which binary distribution predicted by composition.

2. EXPERIMENTAL

2.1 Glass Preparation

The compositions prepared are presented in Table 1 and are also illustrated in the ternary (Na_2O–CaO–SiO_2) phase diagram of Fig. 1. To mimic bioglass, a constant 6 weight% of P_2O_5 (around 2.6 to 2.7 mol%) was included. 0.05 mol% $MnCO_3$ was added to each batch to decrease the ^{29}Si relaxation time. The two tie lines represent CaO substitution for Na_2O (AA) and for SiO_2 (BB). The stoichiometric amounts of analytical grade sodium carbonate, calcium carbonate, sodium dihydrogen orthophosphate and Limoges quartz were dry milled for 16 hours and then melted, between 1250 and 1300°C with a hold for 1 hour, in a Pt/Rh crucible. The glass was poured into a graphite coated steel mould at room temperature and annealed at 450°C for 6 hours prior to slow cooling (Hench et al[1]). After annealing, the samples were examined by XRD and stored in a vacuum desiccator until required.

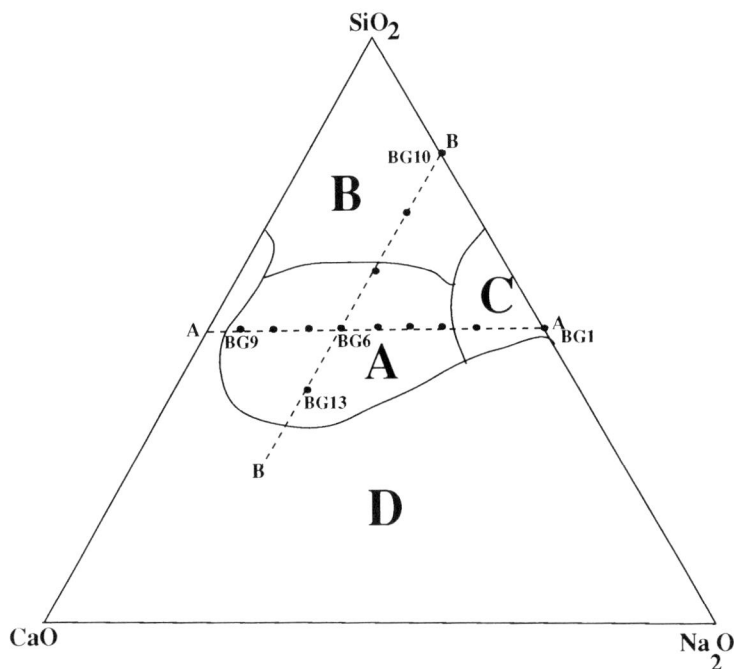

Figure 1.   Ternary phase diagram (in weight %) illustrating the compositions prepared in this study, compared with their biological response. All compositions also contain a constant 6 wt% of P_2O_5.

The concentrations of SiO_2, Na_2O and CaO, in one of the glasses BG4, were determined by gravimetric analysis (BS 2649[16]) and the concentration of P_2O_5 was obtained by difference. The composition was found to be within experimental error of the nominal composition, hence all glasses in this study are discussed in terms of their nominal composition.

2.2 Nuclear Magnetic Resonance

Spectra were acquired on a Bruker MSL 360 (8.45T) spectrometer, at 71.5, 145.8 and 95.3 MHz for ^{29}Si, ^{31}P and ^{23}Na respectively. For the ^{29}Si spectra, the powdered sample was spun at 3–3½ kHz. TMS was used as the reference and a 2μsec pulse length (π/6) and 60 sec delay were employed. The ^{23}Na spectra were recorded in a high speed Doty probe, using a 1μsec pulse length and 1 sec delay. 1 molar NaCl solution was the reference. ^{31}P spectra required spinning speeds in excess of 10 kHz to allow separation of the spinning sidebands. A pulse length of 4μsec and a delay of 1 sec were employed and 85% H_3PO_4 solution was used as the reference.

Table 1.   Nominal compositions, in mol% (m/o) and weight % (w/o), of the glasses prepared in this system.

Sample	Na$_2$O		CaO		SiO$_2$		P$_2$O$_5$	
	m/o	w/o	m/o	w/o	m/o	w/o	m/o	w/o
BG1	47.9	47.0	–	–	49.5	47.0	2.7	6.0
BG2	37.9	37.6	10.5	9.4	48.9	47.0	2.6	6.0
BG3	33.0	32.9	15.6	14.1	48.7	47.0	2.6	6.0
BG4	28.2	28.2	20.8	18.8	48.4	47.0	2.6	6.0
BG5	23.2	23.5	25.8	23.5	48.2	47.0	2.6	6.0
BG6	18.6	18.8	30.8	28.2	48.0	47.0	2.6	6.0
BG7	13.9	14.1	35.8	32.9	47.7	47.0	2.6	6.0
BG8	9.2	9.4	40.7	37.6	47.5	47.0	2.6	6.0
BG9	4.6	4.7	45.6	42.3	47.3	47.0	2.6	6.0
BG10	19.0	18.8	–	–	78.4	75.2	2.6	6.0
BG11	18.8	18.8	10.4	9.4	68.1	65.8	2.6	6.0
BG12	18.7	18.8	20.7	18.8	58.0	56.4	2.6	6.0
BG13	18.5	18.8	40.8	37.6	38.1	37.6	2.6	6.0

3. RESULTS

X-ray diffraction showed that all samples were amorphous, with the exception of the final compositions of each tie line. BG9 and BG13 contained α–CaSiO$_3$ and a combination of α–Ca$_2$SiO$_4$ and β–CaSiO$_3$ respectively in addition to a glass phase. Tables 2(a) and 2(b) contain the chemical shifts of ^{29}Si and ^{31}P, peak positions of ^{23}Na and linewidths of all resonances for the two series of compositions. Figures 2 and 3 illustrate representative ^{29}Si spectra obtained from the compositions along tie

Table 2a.   The spectral parameters of compositions lying along AA. ^{29}Si shifts labelled ▼ are composed of overlapping Q^2 and Q^3 resonances.

Sample	^{23}Na		^{31}P		^{29}Si	
	peak posn (± 2 ppm)	FWHM (± 1 ppm)	shift (± 0.2 ppm)	FWHM (± 0.2 ppm)	shift (± 0.2 ppm)	FWHM (± 0.2 ppm)
BG1	–8.7	43.0	15.6	4.2	–76.7	7.8
					–86.3	9.8
BG2	–9.1	42.2	13.3	7.5	–78.9	8.0
					–86.1	9.8
BG3	–8.0	41.5	11.2	7.5	–80.3 ▼	15.6
BG4	–10.6	42.7	9.5	8.1	–81.2 ▼	16.5
BG5	–10.6	43.4	7.7	7.5	–81.6 ▼	15.4
BG6	–9.5	40.6	6.4	7.5	–81.1 ▼	15.2
BG7	–9.1	40.0	4.2	7.5	–82.0 ▼	14.8
BG8	–9.5	38.4	4.7	7.5	–82.7 ▼	16.2
BG9	–3.9	29.9	2.8	6.5	–83.2 ▼	15.4
			–8.4	n.m.		

Table 2b.   The spectral parameters of compositions lying along BB. ^{29}Si shifts labelled ▼ are composed of overlapping Q^2 and Q^3 resonances.

Sample	^{23}Na		^{31}P		^{29}Si	
	peak posn ($\pm$ 2 ppm)	FWHM ($\pm$ 1 ppm)	shift ($\pm$ 0.2 ppm)	FWHM ($\pm$ 0.2 ppm)	shift ($\pm$ 0.2 ppm)	FWHM ($\pm$ 0.2 ppm)
BG10	−15.4	41.8	13.8	4.7	−92.2	10.3
			2.4	5.3	−106.9	12.5
BG11	−13.6	38.4	2.7	10.5	−92.1	14.2
			−2.6	5.7	−104.9	12.5
BG12	−12.6	39.3	6.2	7.5	−89.0 ▼	14.2
BG6	−9.5	40.6	4.7	7.5	−81.1 ▼	15.2
BG13	−1.7	28.6	1.8	8.7	−69.3	9.0
			−12.1	n.m	−78.5	6.9
					−89.0	6.3

line AA and BB and examples of the ^{31}P and ^{23}Na spectra recorded are given in Figs 4 and 5 respectively.

4. DISCUSSION

4.1 NMR

4.1.1 CaO substitution for Na$_2$O

In tie line AA, CaO was substituted for Na$_2$O on a weight % basis so the ratio of modifier to silicon increased slightly from 0.97 to 1.06 on going from BG1 to BG9. The NMR data are summarised in Table 2(a). BG1 contains no CaO and the ^{31}P spectrum (Fig. 4) contains a single peak at 15.6 ppm, close to that of crystalline Na$_3$PO$_4$.[17] Hence it appears that the phosphorus is closely associated with sodium ions which have effectively been removed from their network modifying role in the silicate network. 2.7 mol% P$_2$O$_5$ would remove 8.1 mol% of Na$_2$O to form a charge balanced 'Na$_3$PO$_4$-like' species. Assuming that the remaining sodium ions are evenly distributed within the silicate framework, the Binary model would predict a Q^2 to Q^3 ratio of approximately 3:2, consistent with the observed intensity ratio for the ^{29}Si resonance in sample BG1 (Fig. 2) where the spectrum contains two overlapping resonances, centred on −86.3 (Q^3) and −76.7 (Q^2) ppm, each with a pair of associated spinning sidebands. The ^{23}Na spectra are very similar for all of the AA series and a typical spectrum (BG3) is shown in Fig. 5.

In sample BG2, 10 weight% of Na$_2$O has been substituted by CaO and this affects both the ^{31}P and ^{29}Si spectra. The ^{31}P resonance (not shown) has both a less positive shift and a greater half-width than in BG1, although it is still characteristic of an orthophosphate unit. Increased half-width implies a greater range of environments and, with the less positive shift, suggests that calcium ions are also associated with the phosphorus. The reported shifts for ^{31}P in Ca$_3$(PO$_4$)$_2$ are 0.0 ppm[17] and 3.0 ppm.[18]

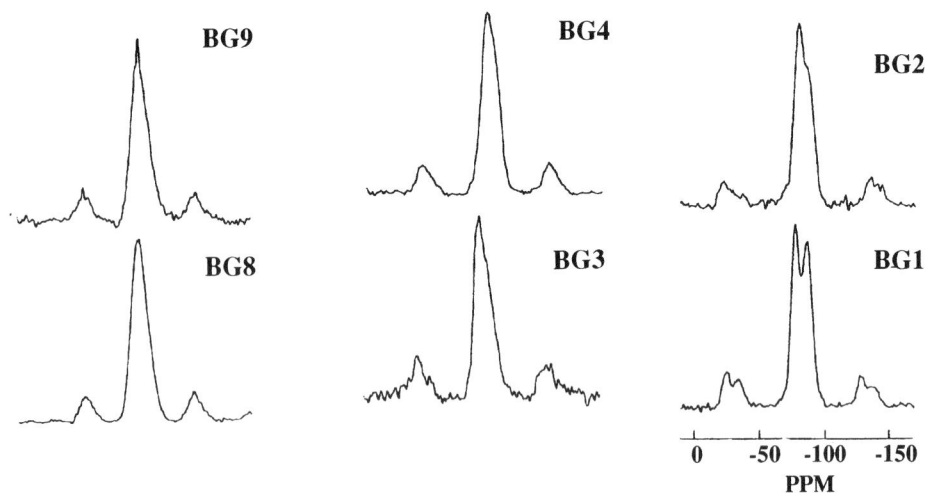

Figure 2.   The ^{29}Si NMR spectra recorded from compositions lying along AA, corresponding to a progressive substitution of Na$_2$O by CaO.

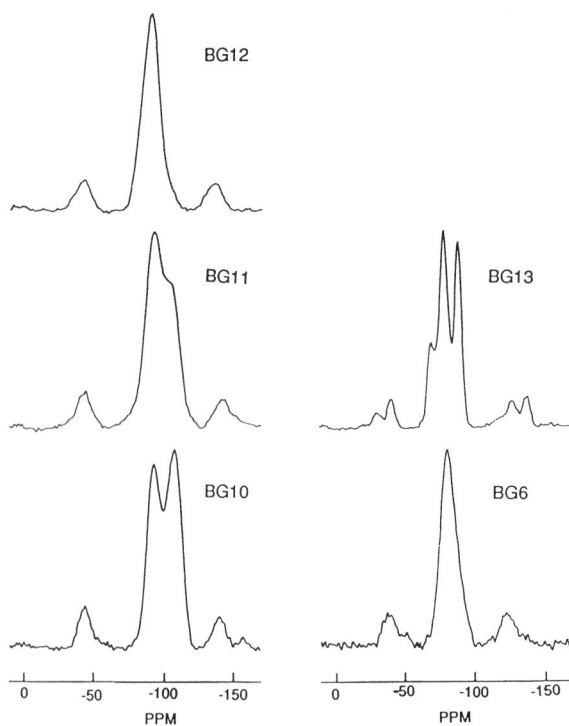

Figure 3.   The ^{29}Si NMR spectra recorded from compositions lying along BB, corresponding to a progressive substitution of SiO$_2$ by CaO.

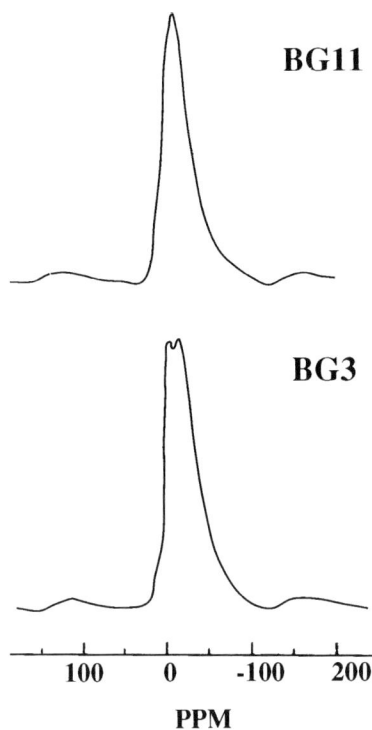

Figure 4.   An example of ^{23}Na MAS NMR spectra from each of the two tie lines.

This pattern is continued throughout the CaO for Na_2O substitution series and Fig. 6 shows the change in the ^{31}P shift as the molar fraction of modifier that is CaO is increased. The near straight line relationship indicates that there is no preferential association with either sodium or calcium ions and these will therefore be present in the ratio given by the overall glass composition. The [29]Si spectrum contains both Q^2 and Q^3 resonances, in nearly the same ratio as for BG1 (the total modifier does change because substitution is in weight %)) but the degree of overlap is greater (Fig. 2).

Fitting of the spectrum to two Gaussians gives shifts of −86.0 (Q^3) and −78.9 ppm (Q^2) with similar linewidths to the BG1 [29]Si resonances. Engelhardt and Michel[19] quote Q^2 shifts for $Na_2O.SiO_2$[20] and $CaO.SiO_2$[21] glasses of −76 ppm and −81.5 ppm respectively. Dupree et al[22] report a Q^3 shift of −86.0 ppm and Q^2 shift of −75.7 ppm for a 45.0 Na_2O–55.0 SiO_2 glass composition. This indicates that the Q^3 species in these glasses has non-bridging oxygens neutralised by sodium ions and the Q^2 species has a combination of calcium and the remaining sodium ions, producing a Q^2 shift between that reported for $Si(OSi)_2(O. . .Na)_2$ and $Si(OSi)_2(O. . .\frac{1}{2}Ca)_2$.

This merging of the Q^3 and Q^2 resonances continues as more CaO is substituted for Na_2O and the [29]Si spectrum obtained from BG3 has a single resonance at −80.3 ppm. However, this is nearly twice as wide as the individual peaks in BG1 and BG2 and the Binary model requires the presence of both Q^2 and Q^3. The calcium ions are again associated with the Q^2 species, the greater electronegativity of the Ca^{2+} ion

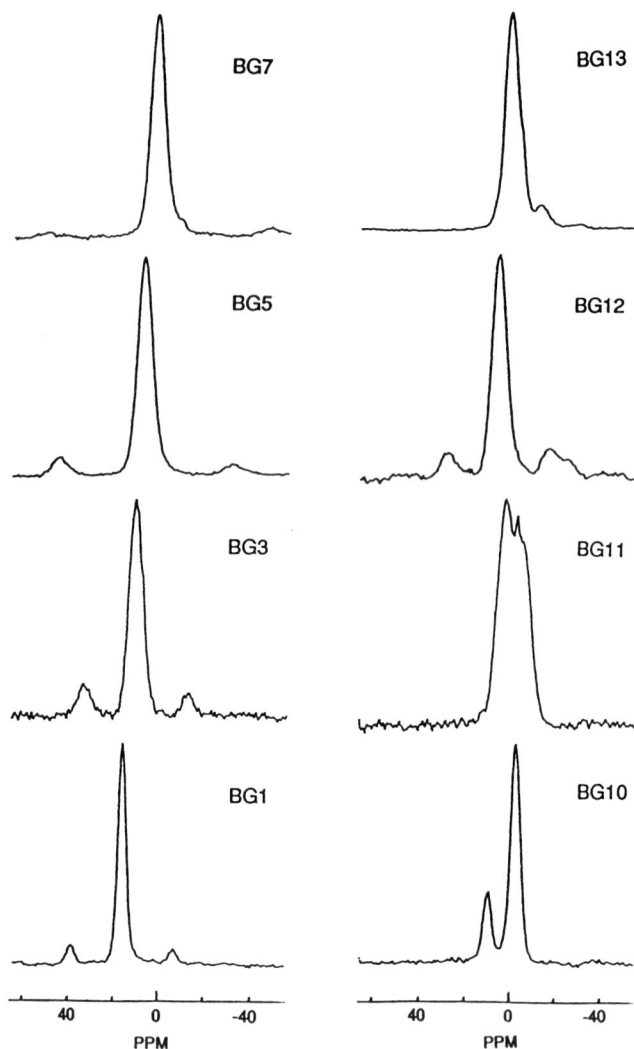

Figure 5.   Examples of ^{31}P MAS NMR spectra from compositions lying along both AA and BB.

shifting the Q^2 position to the high field in comparison to a sodium associated Q^2 shift. At this composition the sodium ions must be split between both silicon species, but are probably the sole modifiers of the Q^3 units.

In glasses BG4 to BG8 this single [29]Si line moves to a slightly more negative shift with increasing CaO substitution. Thus it appears that, throughout this series of glasses, the Ca^{2+} ions are associated with Q^2 units whereas the Na^+ ions are more and more only associated with the more polymerised Q^3 species, giving calcium metasilicate-like environments and sodium disilicate-like environments. No visible evidence was found for phase separation but it may occur on a microscopic level. The predicted Q^2:Q^3 ratio changes from 1.55 for BG1 to 3.8 for BG9 as a result of the use

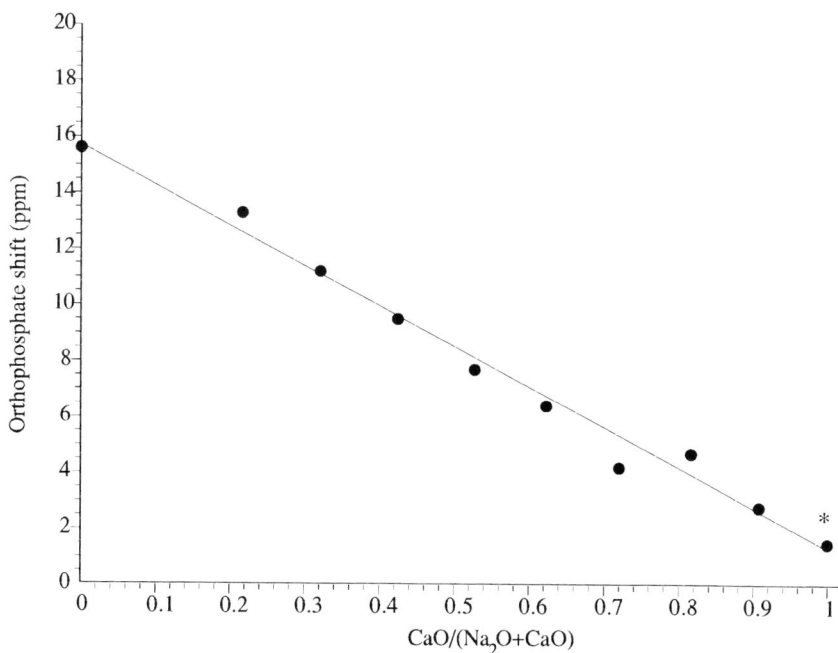

Figure 6.   A plot of the measured ^{31}P chemical shift against CaO content as a fraction of total modifier oxide for compositions along AA. Errors are insignificant on this scale. The asterisked point is an average of the reported values of the ^{31}P chemical shift for $Ca_3(PO_4)_2$ and corresponds to complete replacement of Na_2O by CaO.

of weight % substitution. Thus the peak position becomes less influenced by the Q^3 contribution.

The final composition studied in series AA, BG9, contained crystalline α–$CaSiO_3$ as well as glass. This is evident from XRD and also in the ^{29}Si spectrum which contains a narrow peak at –83.2 ppm (α–$Ca_3Si_3O_9$ has a shift of –83.5 ppm[19]). The presence of this phase is indirect evidence for the presence of a calcium metasilicate-like environment within this series of glasses.

The changes in the Q^2 and Q^3 shifts are plotted in Fig. 7 which attempts to illustrate the changes occurring to the silicon environment as Na_2O is replaced by CaO. It assumes that:

a) Na^+ and Ca^{2+} are equally likely to associate with the phosphate species in the glass and can be discounted from interaction with the silicate network in amounts proportional to their compositional presence.

b) Ca^{2+} preferentially associates with Q^2 silicons and therefore the Q^2 resonance should move progressively towards the shift observed for $CaO.SiO_2$ glass whilst the Q^3 shift should remain similar to that observed in the glass with no CaO substitution.

All Q^2 will be neutralised by Ca^{2+} when the ratio of CaO to total modifier reaches approximately 0.63 (allowing for removal by the phosphate species). After this, Ca^{2+}

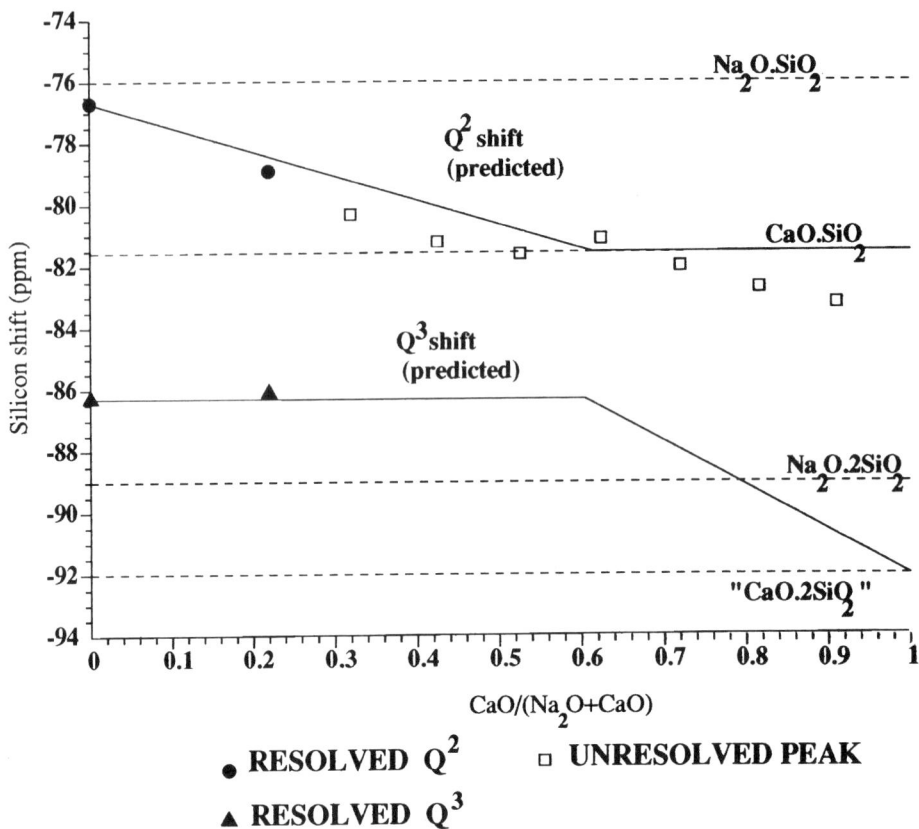

Figure 7.   Peak position of the [29]Si resonance versus CaO content as a fraction
of total modifier oxide for compositions along AA. (Errors lie within the symbols).
The predicted shifts are deduced on the basis of preferred association of
Ca^{2+} with Q^2.

will associate with Q^3 and the shift of this species will tend towards that typical of
$CaO.2SiO_2$ glasses. Shift data for calcium disilicate glass is not available but we have
taken it to be approximately –92ppm by comparison with other alkaline earth silicates. The predicted change in shifts for the two species is indicated by the solid lines
in Fig. 7. The effect of superimposing the two resonances is, as observed in this work,
to produce, initially, two resolvable peaks which converge on the position of the
dominant $CaO.SiO_2$ – like Q^2 at a substitution factor of 0.63 and then which move to
more negative values as the Q^3 environment becomes influenced by Ca^{2+}. This move
to more negative shifts is not very pronounced, since the peak position is increasingly
dominated by the Q^2 resonance. Figure 7 would also predict that the width of the
composite peak should go through a minimum when the CaO to total modifier ratio
goes through 0.63. Table 2a shows that the peak width does in fact decrease slightly
on going through BG7.

Table 3. Increasing depolymerisation of the silicate network as a result of the substitution of CaO for SiO$_2$ along tieline BB.

Sample	CaO/SiO$_2$	Relative Amounts of Q^n Species				
		Q^4	Q^3	Q^2	Q^1	Q^0
BG10	0	0.49	0.51			
BG11	0.15	0.32	0.68			
BG12	0.36		1.0			
BG6	0.64		0.30‡	0.70‡		
BG13†	1.07			0.50		0.13†
				0.37†		

† α Ca$_2$SiO$_4$ and β CaSiO$_3$ present
‡ calculated from Binary model.

4.1.2 CaO substitution for SiO$_2$
^{29}Si spectra

Tie line BB goes from a high silica content glass, through the 'inert' region and then into the bioactive composition range (Fig. 1) Since CaO is replacing the network former SiO$_2$, there is a change in the Q distribution as shown in Table 3.

The ^{29}Si MAS spectrum of BG10 (Fig. 3) contains two separate resonances at –92.2 ppm and –106.9 ppm which can be assigned to Q^3 and Q^4 respectively. In BG11, the replacement of 10 weight% of SiO$_2$ by CaO increases the depolymerisation of the silicate network and reduces the relative intensity of the Q^4 resonance giving a Q^3:Q^4 ratio of approximately 5:2. In BG12 the Binary model predicts largely Q^3 with only 6% Q^2 and the ^{29}Si MAS NMR spectrum fits this prediction well, the width and a shift of –89.0 ppm being typical of Q^3.

BG6 was discussed in the previous section. Here the Q^2 and Q^3 species are in the ratio 2.7:1 and overlap to give a single peak. The Q^3 structural unit is associated with Na$^+$ and the Q^2 with Ca^{2+}.

The final composition, BG13, must be close to the glass forming boundary and thus the sample is partially devitrified. The crystal phases are, from XRD, α–Ca$_2$SiO$_4$ and β–CaSiO$_3$. The ^{29}Si NMR spectrum contains three resonances, –69.3, –78.5 and –89.0 ppm (the latter two with associated spinning sidebands). They have reported shifts of –70.3 and –89.0 ppm[21] consistent with the observed –69.3 and –89.0 ppm in the spectrum whilst the main resonance, at –78.5 ppm originates from a Q^2 residual glass phase containing largely Na$^+$.

^{31}P spectra

The Na$_2$O-only (BG10) has resonances from an orthophosphate-like species at 13.8 ppm and a pyrophosphate-like species at 2.4 ppm. BG11 also contains two resonances, at 2.7 and –2.6 ppm. At lower spinning speeds, small sidebands can be seen, associated with the narrow resonance at –2.6 ppm. This asymmetry and shift suggests a pyrophosphate unit associated with sodium. Ca$_3$(PO$_4$)$_2$ has ^{31}P shifts in the range 3.0 to 0.0 ppm[17,18] and thus the broader resonance at 2.7 ppm is most likely from phosphorus in a calcium orthophosphate environment, though the width of this resonance indicates that this particular species has a greater range of environments than the pyro species.

BG12 contains only a single resonance centred on 6.2 ppm which is between the shifts of Na_3PO_4 and $Ca_3(PO_4)_2$ and, as in the AA tie line, it appears as though both types of modifier ion are associated with the orthophosphate unit. The ^{31}P spectrum of BG6 is also assigned to orthophosphate with a combination of modifier ions similar to that of BG12.

BG13 contains two separate resonances, from phosphorus in ortho- and pyro-arrangements with predominantly calcium cations responsible for charge neutralisation. The width of these lines suggests that the phosphorus is in the glass phase.

^{23}Na spectra

The ^{23}Na spectra of these glasses all consist of a single broad resonance (Fig. 4) shows BG11). The peak positions do however move to less negative positions as the total amount of modifier is increased – consistent with the observations on soda-lime – silica glasses.[14]

5. CONCLUSIONS

5.1 NMR

MAS NMR has shown that for the compositions with no CaO, BG1 and BG10, the distribution of non-bridging oxygens in the silicate network is close to that calculated for a binary distribution. The P_2O_5 present in these compositions increases the relative amount of the higher Q^n species by removal of sodium ions from the network to form 'Na_3PO_4 – like' structural units. Replacing some Na_2O with CaO produces a more negative ^{31}P shift, a trend that is linear as the amount of substituting CaO increases, indicating that the phosphorus shows no preference for the type of cation it removes from the silicate network. The ^{29}Si resonances from Q^2 and Q^3 species merge as the Na_2O is substituted by CaO, indicating preferential association of calcium ions with Q^2 species and sodium ions with Q^3 species. Consequently glasses in series AA consist of a combination of environments similar to calcium meta-silicate, sodium disilicate and mixed sodium/calcium orthophosphate. The presence of $CaSiO_3$ as a crystal phase in BG9 indirectly supports this hypothesis.

In tie line BB the silicate network becomes more depolymerised as the substitution of CaO for SiO_2 proceeds. The initial mixture of Q^4 and Q^3 is replaced by Q^3 alone and then Q^3 and Q^2 as detailed in Table 3. The ^{31}P spectra of BG12 and BG6 show an orthophosphate environment with a mix of charge neutralizing cations as seen in the AA series.

5.2 Glass Structure and Bioactivity

Tie line AA traverses from the 'Dissolution' region into the 'Bioactive' region and the MAS NMR results can explain the difference in behaviour in the following manner.

a) Dissolution

  BG1 glass contains only sodium modifier and the structural species are similar to sodium meta and disilicate, which are highly soluble. Substitution of 10 mol%

Na_2O by CaO (BG2) produced a convergence of the two separate ^{29}Si MAS NMR resonances due to Q^2 and Q^3, however the fairly soluble sodium disilicate structure remained predominant.
b) Bioactive

Further additions of CaO produced glasses with structures composed of a combination of sodium disilicate and calcium metasilicate species. The presence of the latter species controls the dissolution of the former. It is suggested that silica gel layer formation is via a process of restricted dissolution of the sodium disilicate species, followed by condensation of silanols to form the silica gel. Then the calcium and phosphate ions can diffuse through the gel layer to form, initially, an amorphous calcium phosphate layer.

Tie line BB traverses from the inert region, through the bioactive region and towards the glass forming boundary.
a) Inert

Glass BG10 has a structure comparable to that of sodium tetrasilicate, which is of low solubility and, consequently an inert biomaterial. BG11 is also still too highly polymerised to undergo the controlled dissolution required for bioactivity.
b) Bioactive

BG12 lies near the boundary of the 'bioactive' region and consequently has a lower degree of bioactivity than composition BG6 nearer the centre of region A.[1,9] The sodium disilicate species in BG12 can undergo a dissolution/condensation process to form the gel layer but the greater difficulty of calcium ion diffusion from the calcium disilicate species slows the rate of formation of the calcium phosphate. BG6 has the necessary combination of Q^2 and Q^3.
c) Non-glass forming

BG13 is compositionally in the bioactive region but, under the conditions of glass preparation used here, its partial devitrification would place it on the glass forming boundary.

The determination of the local structural units in these glasses has given some insight into their response as implant materials.

ACKNOWLEDGEMENTS

We wish to thank the SERC for funding for equipment and for provision of a studentship (MWGL).

We also wish to thank the editors of the *Journal of Non-Crystalline Solids* for permission to reproduce data and figures from M. W. G. Lockyer, D. Holland and R. Dupree: 'NMR investigation of the structure of some bioactive and related glasses', *J. Non-Cryst. Solids*, 1995, **188**, 207–219.

REFERENCES

1. L. L. Hench, T. K. Greenlee, R. J. Splinter and W. C. Allen: *J. Biomed. Res. Symp.*, 1971, **2**, 117.
2. O. H. W. C. Andersson, K. H. Karlsson and K. Kangasniemi: *J. Non-Cryst. Solids*, 1990, **119**, 290.

3. J. Christoffersen, M. R. Christoffersen, W. Kiblczyc and F. A. Andersen: *J. Crystal Growth,* 1989, **94**, 767.

4. A. E. Clark Jr., C. G. Pantano and L. L. Hench: *J. Am. Ceram. Soc.,* 1976, **59**, 37.

5. Y. Ebisawa, T. Kokubo, K. Ohura and T. Yamamuro: *J. Mat. Sci.: Materials in Medicine,* 1990, **1**, 239.

6. U. M. Gross, J. Brandes, V. Strunz, I. Bab and J. Sela: *Biomed. Mat. Res.,* 1981, **15**, 291.

7. U. Gross: *C.R.C. Critical Reviews in Biocompatability*, vol. 4, CRC Press, 1988, 155.

8. L. L. Hench and A. E. Clark Jr., *Biocompatibility of Orthopaedic Implants*, vol. 2, D. F. Williams, ed., CRC Press, 1982.

9. P. Li and F. Zhang: *J. Non-Cryst. Solids*, 1990, **119**, 112.

10. M. Ogino and L. L. Hench: *J. Non-Cryst. Solids,* 1980, **38/9**, 673.

11. L. L. Hench and J. Wilson: *Science*, 1984, **226**, 630.

12. L. L. Hench: *J. Am. Ceram. Soc.,* 1991, **74**, 1487.

13. R. Dupree, D. Holland, P. W. McMillan and R. F. Pettifer: *J. Non-Cryst. Solids*, **68**, 399.

14. R. Dupree, D. Holland and D. S. Williams: *J. Phys. (Paris)*, 1985, **46**, C8–119.

15. B. W. Veal, D. J. Lam, A. P. Paulikas and W. Y. Ching: *J. Non-Cryst. Solids,* 1982, **49**, 309.

16. British Standard BS 2649:PART 1:1988. Analysis of glass, Part 1. Glasses of the soda-lime-magnesia-silica type. B.S.I. 1988.

17. I. L. Mudrakovskii, V. P. Shmakkova and N. S. Kotsarenko, *J. Phys. Chem. Solids,* 1986, **47**, 335.

18. G. L. Turner, K. A. Smith, R. J. Kirkpatrick and E. Oldfield: *J. Mag. Res.*, 1986, **70**, 408.

19. G. Engelhardt and D. Michel: *High Resolution Solid State NMR of Silicates and Zeolites*, Wiley, 1987.

20. R. Dupree, D. Holland and M. G. Mortuza: *Phys. Chem. Glasses*, 1988, **29**, 18.

21. A-R. Grimmer, A-R. Mägi, A-R. Hahnert, H. Stade, A. Samoson, H. Wieker, W. and E. Lippamaa: *Phys. Chem. Glasses*, 1984, **25**, 105.

22. J. B. Murdoch, J. F. Stebbins and I. S. E. Carmichael, *Am. Mineral*, 1985, **70**, 332.

Electrical Ceramics

Advances in Ceramic Soft Magnets

F. R. SALE and J. FAN

Manchester Materials Science Centre, University of Manchester and UMIST,
Grosvenor Street, Manchester, M1 7HS, UK

ABSTRACT

Despite the maturity of the soft ferrite industry much vital, basic research and product development is still occurring worldwide. Selected aspects of this research, development and production are presented here with a view to the prediction of the requirements that must be met for advances in soft ferrites for the 21st Century. The areas of particular application that are considered briefly are power switching supplies, multilayer chip applications and yoke rings for TV deflection units. Developments in these three areas are shown to need a further understanding of both the effects of additives and impurities, and particularly their interaction, as well as improved raw materials and improved processing methods.

1. INTRODUCTION

The soft ferrite industry is mature, being more than 50 years old. Following on from the early work of Hilpert,[1] who published the first systematic study of the chemical and magnetic properties of a number of binary iron oxides, Forestier,[2] Hilpert and Wille,[3] Kato and Takei,[4] Kawai[5] and Snoek[6] gave the scientific and technological basis of practical ferrites in 1945. Following all this early work the ferrite industry became rapidly established in Europe and Japan. From these beginnings the production of soft ferrite in Japan, United States, Europe and Southeast Asia reached a value of $1.25 billion in 1993,[7] a figure which is predicted to increase to $1.875 billion by the year 2000 (which will include production in P.R. China). This increase in production will see a move in the manufacturing base. Today, Japan dominates world production, however, as progress to the 21st Century occurs there will be increased production in P.R. China, Southeast Asia and the United States.[7,8] As examples, TDK have set up a new facility in Oklahoma and China Steel have set up the new plant of Himag Magnetics in Taiwan. Both Philips and TDK have installed facilities in P.R. China, which will increase the role of P.R. China in the supply of soft ferrites.[7]

The major markets that are predicted to cause increases in soft ferrite usage may be listed as automotive, telecommunications, lighting, noise suppression in electronics, and switched mode power supplies. The automotive requirements will arise from two types of applications. General usage of ferrite magnets in automotives will continue to rise as more devices needing electric motor power and sensors are added to greater numbers of cars. However, of potentially greater importance is the drive to establish electric vehicles, which are likely to involve substantial amounts of soft ferrite. Within the telecommunications field the growth of equipment is estimated at

10% per year. Soft ferrite yoke rings will be an important part of this growth as a result of large screen televisions, high definition televisions and the ever-growing demands of new markets in developing countries as well as replacements in the industrialised countries. In this field soft ferrites will also be used in miscellaneous transformers and power supplies.

The use of soft ferrites in lighting is driven by the ever-present desire to reduce energy usage. Thus increased use of compact fluorescent lamp bulbs and electronic ballast will bring about increased need for soft ferrites. In 1993, 134 million compact fluorescent bulbs were sold,[7] which was an increase of 23% over the sales of 1992 and the worldwide electronic ballast market is projected to be 100 million units by 1999.[7]

Soft ferrite cores for noise suppression will increase into the 21st Century as a result of tighter regulations in the electronics industry and the growth of that industry. In the field of switched mode power supplies the decrease in size and increase of operating frequency will place important technical requirements, with respect to power loss, upon the soft ferrites which are used. A major drive here is the change from one central power source for a system to the use of distributed power supplies. It is of significant importance that the frequency of operation of switched mode power supplies has increased from around 20 kHz in the late 1970s to 50–120 kHz by the middle 1980s and on to 500 kHz–2 MHz in the 1990s. This increase in operating frequency has increased dramatically the contribution of relaxation losses to the overall power losses of the ferrites, which up to frequencies of the order of 500 kHz are dominated by eddy current and hysteresis losses.

2. POWER LOSSES IN SOFT FERRITES

For power switching supplies with reduced power losses at ever-increasing frequencies, new materials based on Mn–Zn ferrite have been developed over recent years by controlling microstructure and nano-structure. These developments have been driven by the desire to miniaturise devices[9] and further drives will occur as further miniaturisation, and hence higher operating frequency, becomes necessary. The largest component in a switched mode power supply is the main transformer. Consequently, the miniaturisation of the switched mode power supply depends upon the transformer and manufacturers are trying to develop flat-shaped transformers for surface mounting such that thinner and lighter power supplies can be produced which are more efficient, more noise-proof and more reliable.[10] To achieve such reduction in size the core loss of the ferrite needs to be reduced. The important factors associated with power losses in soft ferrites, that have been identified over a number of years, are purity of raw materials, ferrite composition and particularly the roles of additives as well as quality control of the overall manufacturing process. It is, however, the microstructural control that is perhaps the most important. These same considerations apply with the use of soft ferrites as yoke rings in deflection units in televisions particularly with the advent of high definition TV (HDTV) which requires to operate at higher frequencies than those used previously to give the improved resolution.

In general terms the overall (core) loss of a ferrite may be expressed as an increasing function of hysteresis loss, eddy current loss and relaxation loss e.g.

$$P_c = f(P_h, P_e, P_r) \tag{1}$$

In the frequency range below that at which the operating frequency approaches the frequency of domain wall displacements, the latter loss P_r is negligible. However, at frequencies > 500 kHz the relaxation loss, P_r, may be prohibitive at high induction levels.

These three losses may be considered further by reference to the following:

$$P_h = W_h f \tag{2}$$

where P_h is the hysteresis loss,
   W_h is the energy equivalent to the area of d.c. B-H loop under the same maximum flux density as that used in power loss measurement
and f is the frequency.

$$P_e = cd^2 f^2 B_m{}^2 /_\rho \tag{3}$$

where c is a shape constant related to the eddy current circuit
   d is the dimension of the eddy current circuit (often taken as grain size)
   B_m is the maximum flux density
   f is the frequency
and ρ is the resistivity

As indicated above relaxation losses occur as a result of domain wall displacements[11]. According to the Globus model[12,13] the relationship

$$f_r(\mu_{si} - 1) = 3/4 \ \frac{(4\pi M_s)^2}{\pi^2 \beta D} \tag{4}$$

holds, where
   μ_{si} is the initial static permeability
   f_r is the relaxation frequency
   $4\pi M_s$ is the saturation magnetization
   D is the average grain size
and β is the damping coefficient

Hysteresis loss may be reduced by increase in chemical homogeneity, the development of uniform large grains, high density in the sintered products and the reduction of grain boundary stress. Eddy current loss requires as small a grain size as possible and as high a resistivity as possible. These two requirements are often met with the production of high resistivity grain boundaries by the addition of appropriate oxides to the basic ferrite mix. Relaxation losses may be minimised by small grain size and saturation magnetization.

Table 1.   Effects of grain size and additions on power losses in Mn-Zn ferrites[12]

	Hysteresis Loss	Eddy Current Loss	Relaxation Loss
Small grain size	Increase	Decrease	Decrease
Additions (CaO, SiO$_2$)	Increase	Decrease	Decrease

It is therefore apparent that there are conflicting microstructural requirements for the minimisation of each type of loss. Small grain size, which increases resistivity by maximising the grain boundary surface area, increases the relaxation frequency but decreases permeability and increases coercive force. Additions of oxides, and the presence of impurities such as CaO and SiO$_2$, also increase the grain boundary resistivity, increase the relaxation frequency, decrease the permeability and increase the coercive force. As a result both small grain size and grain boundary layers may help to minimise eddy current and relaxation losses but will have the opposite effect on hysteresis loss.

The problem for the future in power switching applications and other high frequency applications will therefore be the minimisation of relaxation losses which, as indicated above, become important as the operating frequency approaches the frequency of domain wall displacement (relaxation frequency of wall displacement). At these higher frequencies the conventional high resistivity grain boundaries used to minimise eddy current loss are not as effective and it is important to rely upon the resistivity of the ferrite grains themselves.[11] If increases in intrinsic resistivity of the ferrite are not possible, then thicker grain boundary layers may be needed.[11]

The conflicting microstructural requirements for the minimisation of power loss have been summarised by Lebourgeois et al,[12] as given in Table 1.

3. MICROSTRUCTURAL AND PROCESSING REQUIREMENTS

3.1 Power Switching Supplies

Over the past five years many materials, often based on Mn–Zn ferrites, have been developed to give reduced power losses in switched mode power supply applications. These developments have all aimed at the control of both microstructure and nanostructure of the ferrite, which in turn have led to the miniaturisation of the devices as indicated in the previous section.[9,15] It is well established that the electromagnetic properties of the ferrites are strongly dependent upon compositional effects, chemical homogeneity, microstructure and nanostructure.[14,15] Despite this extended knowledge the full interrelationships of these factors are not understood and further study will be required into the 21st Century with reference to advances made so far in the materials for switched mode power supplies and to requirements that are foreseen for the future.

The control of sintering atmosphere, a topic which has been studied since the 1940s and has been shown to be vital in order to achieve optimum magnetic properties, requires a more complete assessment in relation to both initial permeability and power loss characteristics of the ferrites.[9,10] These properties are related to the degree of oxygen non-stoichiometry in the ferrite and to stresses present within the crystal lattices, topics which again have received much study. Nevertheless, new techniques for characterisation have become available for structural studies and closer study leads to a better understanding of the fundamental inter-relationships of atmosphere, oxygen non-stoichiometry, lattice distortion and magnetic properties. For instance, recent work on Mn–Zn ferrites has shown that the eddy current loss is particularly susceptible to variation in oxygen content of the sintering atmosphere whereas hysteresis loss is almost independent of the oxygen content.[9]

The importance of grain boundary chemistry is a vital subject of study, particularly in terms of its control of grain boundary resistivity, and hence eddy current losses, as indicated in the previous section. Much work has been reported over the last 20 years on the effects of additives such as SiO_2 and CaO (both intentional and as impurities) and TiO_2, SnO_2, V_2O_5 etc. It is interesting, however, that the roles of double additions such as TiO_2 and SnO_2, which may promote low temperature sintering and also give decreases in hysteresis loss by reducing the magnetic flux density and coercive force, are not yet fully explained. Similarly, the presence of amorphous phases, rich in SiO_2 and CaO, at grain boundaries has been shown to be important because of lattice distortion caused in the grain boundary region[16] as well as because of the inherent high resistivities of these phases. However, it is the interaction of SiO_2 and CaO with deliberate additives, or simply the ratio of these impurities, which may be the most important aspects of study for future advances. Recent work by Taylor et al[17] has shown that it is often not the total amounts of CaO and SiO_2 that are present which determine the overall effect on magnetic properties, but it is the ratio of these impurities. Work by Mahloojchi and Sale[18] has shown that in MgZn ferrites excess SiO_2, which is not associated with CaO, removes MgO from the ferrite and produces MgO-SiO_2 layers (often of enstatite composition) at grain boundaries. Of crucial importance is the recent work of Otsuka et al[19] who have reported the effects of 0.2 wt% additions of Al_2O_3, HfO_2, Nb_2O_5, SnO_2, Ta_2O_5, TiO_2, V_2O_5 and ZrO_2 on the resistivity and eddy current loss in a MnZn ferrite. These electrical properties were reported for samples containing different base levels of the impurities CaO and SiO_2, as shown in Tables 2 and 3 below. The dramatic effects caused by the change of SiO_2 and CaO levels respectively from 0.005 wt% and 0.005 wt% (labelled no SiO_2–CaO)

Table 2. Effect of additives on Pe at 1MHz, 50mT, R.T. and dc-ρ at R.T. (containing no SiO_2–CaO) after Otsuka et al[19]

Additives	–	Al_2O_3	HfO_2	Nb_2O_5	SnO_2
Pe (KW/m³)	4800	4000	4600	2500	4200
ρ (Ωcm)	0.8	1.0	0.8	2.7	1.0

Additives	Ta_2O_5	TiO_2	V_2O_5	ZrO_2
Pe (KW/m³)	1300	3600	3100	1300
ρ (Ωcm)	13.1	1.3	3.1	5.0

Table 3.   Effect of additives on Pe at 1MHz, 50mT, R.T. and dc-ρ at R.T. (containing SiO$_2$–CaO) after Otsuka et al[19]

Additives	–		Al$_2$O$_3$	HfO$_2$	Nb$_2$O$_5$	SnO$_2$
Pe (kW/m^3)	720		710	390	670	720
ρ (Ωcm)	390		540	3150	1750	470

Additives	Ta$_2$O$_5$	TiO$_2$	V$_2$O$_5$	ZrO$_2$
Pe (kW/m^3)	460	580	490	550
ρ (Ωcm)	2140	2140	1100	1600

to 0.035 wt% and 0.08 wt% (labelled containing SiO$_2$–CaO) are clearly self-evident and are explained by complex interaction at grain boundaries. It is interesting to note the different behaviours reported for additions of ZrO$_2$, TiO$_2$ and HfO$_2$. The dramatic increase in resistivity and decrease in eddy current loss which occurs with HfO$_2$ was explained by reaction with CaO at grain boundaries, however, with TiO$_2$ additions neither Ti nor Ca were found to segregate to the grain boundaries and so no high resistivity grain boundary phases resulted. It is evident that much work is required before such complex potential interactions at grain boundaries may be understood. In these studies it will be of vital importance to ensure that the processing method does not result in inconsistent observations because of a failure to produce chemically homogeneous samples. As a result it may be necessary to use non-conventional, solution-based preparative techniques for such studies. Citrate-gel processing is one such method which has been shown to be particularly applicable to the production of soft ferrites.[20,21]

3.2 Multilayer Chip Components

Many multilayer ferrite chip components produced over the last five years or so have silver internal windings.[22] The multilayer components are produced by the successive screen printing of silver and ferrite paste which is followed by a co-firing operation to obtain the required densities in both the silver and the ferrite layers by sintering. Ni–Cu–Zn ferrite is used as the ferrite because of its high resistance and good performance at high frequencies and because of its relatively low sintering temperature. Silver is adopted for the winding material because it has a generally low reactivity with the ferrite and has the lowest electrical resistivity amongst commercially available materials. A major drawback of this fabrication route is that silver may migrate from the winding into the ferrite during co-firing. The initial permeability and its temperature dependence and Q-factor, which are important properties for such chip components, have been shown to be strongly dependent upon both microstructure and stress. Excessive Ag migration in Ni–Cu–Zn ferrite has been shown to cause compressive stresses at grain boundaries, which in turn cause a deterioration of magnetic properties.[23] The control of Ag migration is, therefore, a problem of continuing importance. In particular, the presence of anion additives such as chlorine and sulphur, which may promote silver migration, needs further study and explanation. The use of Ag–Pd as winding material has been studied as a means of allowing sintering temperatures in excess of 960°C to be used.[22] However, increases in Pd content cause increases in inductance and decreases in Q factor because of an increase in DC resistance. It is evident that further advances are required in both the winding material and also in the ferrite, particularly in the reduction of sintering temperature. Increased CuO contents in the Ni–Cu–Zn ferrite bring about decreases in sintering temperature but also worsen magnetic properties. Sintering aids such as Li_2O and V_2O_5 can be used judiciously to suppress Ag migration during sintering. A small amount of Ag migration may also cause increased densification but it gives deterioration of magnetic properties such as the temperature dependence of μ_i.[24] It is evident that future advances may arise from the use of advanced chemical processing for the manufacture of highly sinterable ferrite powders,[20] particularly as the quantities used in such multilayer components lend themselves to specialised production processes, whereas the bulk use of tonnage material for yoke ring applications cannot easily take advantage of such superior but more costly powders. In an attempt to understand the important parameters concerning the wetting of ferrite surfaces by electrode materials sessile drop measurements have been undertaken.[25] The uses of such an experimental technique will need to be extended to cover all potential electrode materials.

The problems of interface reactions between the dielectric and the ferrite in multilayer components are also of critical importance. Some studies on Ni–Cu–Zn ferrites and Ti-based dielectrics have been made in relation to the enhanced grain growth that is observed at the interface regions and the effects that such growth has on properties.[23] However, much further work is necessary to understand the mechanisms of reaction and their effects for a large number of potential ferrite/dielectric composites.

3.3 Yoke Rings for Deflection Units

One of the major bulk usages of soft ferrites is undoubtedly as yoke rings in deflection units for televisions. Usually the ferrites are based on the MgZn or MnZn systems, although NiCuZn has been used recently for high frequenty applications needed in HDTV and high resolution display monitors[26] because of the need for high resistance to reduce ringing currents. NiZnCu ferrites containing additions of Mg and Ti have been shown to have power losses some 50% of those associated with non-doped materials. These recent applications at high frequency are re-focusing attention on to MgZn ferrites, which in their iron-deficient state, also possess extremely high resistivity ($\sim 10^6$ Ω m^{-1}) albeit with a lower initial permeability of 300–500 at $\leq$ 10 kHz, 0.1 mT at 25°C.

This bulk tonnage use of ferrites means that it is unlikely that advanced ceramic processing methods such as gel processing, freeze drying, hydrothermal synthesis etc. will ever be used on the bulk industrial scale for economic reasons. Goldman[27] has recently reviewed the supply of modern raw materials for ferrite processing and

Table 4.   Mn–Zn ferrite, comparison of magnetic properties for commercial and gel-derived material[30]

Property	Commercial (specification)	Gel-derived
μ_i	900 ± 25% (≤10 kHz, 0.1mT) 25°C	2400
T_c	≥150°C	>170°C
H_c	≤40 A/m	20.4 A/m
Power loss	≤100 mW/cm³ (16 kHz, 100mT) 25°C	19 mW/cm³

indicates that there are applications such as recording heads and microwave ferrites where the iron oxide raw material needs to be of high quality (low impurities, high reactivities) and may cost several dollars per pound, whereas for yoke ring applications (and radio antenna and low grade hard ferrites) a lower quality of iron oxide costing several cents per pound is available and suitable. Nevertheless, the search for higher grade materials at low, economic costs is a necessity for future yoke ring applications.

Of the non-conventional processing routes, co-precipitation and co-spray roasting are the methods which are of most commercial importance.[27] Co-precipitated materials are available and are being used in special recording heads. Co-spray roasting has been installed in Japan and the USA for internal use by ferrite component manufacturers. Of most relevance to the yoke ring industry is the work of Ruthner[28,29] who has recently reported the production of presintered ferrite powders (MnZn, NiZn and also strontium hexaferrite) using a vertical furnace in which the powders react whilst undergoing a short thermal treatment in a 'quasi-free fall' manner.

The advantages to be gained by microstructural control of standard yoke ring ferrites are very clear. Table 4 shows a comparison of data determined for citrate gel processed and conventionally processed MnZn ferrite for yoke ring applications.[30] The interesting observation is that whilst initial permeability is increased from 900 to 2400 the power loss at 16 kHz, 100 mT at 25°C is reduced from ≤ 100 mW cm⁻³ to < 20 mW cm⁻³. Both these increases arise from the control and uniformity of grain size and purity.[30] Similarly, for MgMnZn ferrites the control of processing available for typical yoke ring ferrite compositions with the citrate gel route has given 30% reductions in power loss at 64 kHz, 100 mT at 100–125°C.[31] These increases in property indicate the levels in improvement that are available with today's compositional specifications and show achievable targets for any alternative, economic, processing method.

4. CONCLUSIONS

It is evident that despite the maturity of the soft ferrite industry there are many challenges that must be met as the 21st Century is approached. These challenges exist in a variety of applications that range from the relatively small usage of ferrites in electronic applications to the bulk usage of ferrites in the television industry as yoke rings.

In summary, these challenges must be met by increases in purity of raw materials, a full understanding of the effects of interactions of impurities with deliberate additions to the ferrites, and by the development of new processing methods that give sinterable, controlled powders at a range of costs which must suit the individual different areas of application.

ACKNOWLEDGEMENTS

The authors would like to acknowledge the financial support of EPSRC which has made possible some of the work reported here.

REFERENCES

1. S. Hilpert: *Bes. Detsch. Chem. Ges.*, 1909, **42**, 2248.
2. H. Forestier: *Ann. Chim.*, *Xe Série, IX*, 1982, 316.
3. S. Hilpert and A. Wille: *Z. Phys. Chem.*, 1932, **B18**, 291.
4. V. Kato and T. Takei: *J. Inst. Elec. Engrs.* Japan, 1933, **53**, 408.
5. N. Kawai: *J. Soc. Chem. Ind.* Japan, 1934, **37**, 392.
6. J. L. Snoek: *Physica*, Amsterdam, 1936, **3**, 463.
7. K. S. Talbot: *Am. Ceram. Soc. Trans.*, 1995, **47**, 133.
8. T. Abraham: *Am. Ceram. Soc. Bull.*, 1994, **73**, 62.
9. E. Otsuki, S. Yamada, T. Otsuka, K. Shoji and T. Sato: *J. Appl. Phys.*, 1991, **69**, 5942.
10. J. A. T. Taylor, S. T. Reczek and A. Rosen: *Am. Ceram. Soc. Bull.*, 1995, **74**, 91.
11. T. Mochizuki: *Proc. 6th Int. Conf. Ferrites, ICF-6*, 1992, Japan Soc. Powder and Powder Met., Tokyo, 53.
12. R. Lebourgeois, C. Deljurie, J. P. Ganne, P. Perriat, B. Lloret and J. L. Rolland: *ibid*, 1169.
13. M. Guyot and V. Cagan: *JMM Mat*, 1982, **27**, 202.
14. Y-S. Kim and S-J. Kwon: *Proc 6th Int. Conf. Ferrites, ICF-6*, 1992, Japan Soc. Powder and Powder Met., Tokyo, 37.
15. E. Otsuki and S. Yamada: *Am. Ceram. Soc. Trans.*, 1995, **47**, 147.
16. S. Okamoto, T. Nomura and T. Ochiai: Erekutoroniku-Seramikusu, Winter, 1986, 41.
17. S. M. O. Taylor, W. M. Dawson and F. R. Sale: *Third Euro-Ceramics – Vol. 2, Properties of Ceramics*, P. Duran and J. F. Fernandez eds, Faenza Editrice Iberica, Castellón de la Plana, Spain, 1993, 653.
18. F. Mahloojchi and F. R. Sale: *Proc. 6th CIMTEC, High Technology Ceramics*, P. Vincenzini ed., Materials Science Monographs, Elsevier, Netherlands, 1987, 1038.
19. T. Otsuka, E. Otsuki, T. Sato and K. Shoji: *Proc. 6th Int. Conf. Ferrites, ICF-6*, 1992, Japan Soc. Powder and Powder Met., Tokyo, 317.
20. F. Mahloojchi and F. R. Sale: *Ceram. Int.*, 1989, **15**, 51.
21. Y-T. Chien and F. R. Sale: *Brit. Ceram. Proc.*, 1994, **4**, 53.
22. T. Nomura and M. Takaya: *Hybrids*, 1987, **3**, 15.
23. M. Satoh, A. Ono, T. Maruno and N. Kaihara: *Proc. 6th Int. Conf. Ferrites, ICF-6*, 1992, Japan Soc. Powder and Powder Met., Tokyo, p. 1210.
24. A. Nakano, H. Momoi and T. Nomura: *ibid*, 1225.

25. S. Sugihara and K. Okazaki: *ibid*, 370.

26. T. Araki, H. Morinaga, K-I. Kobayashi, T. Oomura and K. Sato: *ibid*, 1185.

27. A. Goldman: *Amer. Ceram. Soc. Trans.*, 1995, **47**, 105.

28. M. J. Ruthner: *Proc. 6th Int. Conf. Ferrites, ICF-6*, 1992, Japan Soc. Powder and Powder Met., Tokyo, 40.

29. M. J. Ruthner: Eu Patent, 0 186 042 B1.

30. F. R. Sale, J. Fan and Y-T. Chien: *Amer. Ceram. Soc. Trans.*, 1995, **47**, 155.

31. Y-T Chien and F. R. Sale: *Proc. 6th Int. Conf. Ferrites, ICF-6*, 1992, Japan Soc. Powder and Powder Met., Tokyo, 301.

Microwave Dielectric Ceramics – Current Materials and Future Needs

R. FREER

Manchester Materials Science Centre, University of Manchester and UMIST, Grosvenor Street, Manchester, M1 7HS, UK

ABSTRACT

A brief review is presented of current materials suitable for microwave dielectric resonators; relative permittivities are in the range of 20–90 and Q values up to 40000 at 10 GHz are attainable. Materials in the systems $(Zr,Sn)TiO_4$ and $BaO–R_2O_3–TiO_2$ (where R = rare earth) are used as examples to identify some of the factors controlling Q values. Future needs in terms of materials properties and a fundamental understanding of intrinsic behaviour (and thus the origin of dielectric loss) are briefly examined.

1. INTRODUCTION

Microwave dielectric ceramics are used extensively for electronic packaging and as resonators. This paper will consider only the latter application. Since solid dielectric resonators offer the possibility of reducing component and system size, according to $1/\sqrt{\varepsilon_r}$ dependence (where ε_r is relative permittivity), they have replaced the traditional air-filled resonators.[1] The first microwave filters using TiO_2 ceramic resonators were realised in the 1960's but the poor temperature dependence of dielectric properties precluded practical exploitation of the devices.[2]

Since the 1970's there has been a worldwide effort to develop high quality temperature-stable dielectric ceramics suitable for resonator and filter applications.[3,4] The primary characteristics of any candidate material are as follows:

(i)   high ε_r, to minimise component size
(ii) small dielectric loss (tan δ), more conveniently described in terms of the Q (1/tan δ), value
(iii) small coefficient of linear expansion (α)
(iv) small, preferably zero, temperature coefficient of resonant frequency (τ_f), which is given by

$$\tau_f = -\frac{1}{2}\tau_k - \alpha \tag{1}$$

where τ_k is the temperature coefficient of dielectric constant. From classical dispersion theory it is predicted that at microwave frequencies the relative permittivity is approximately constant, whilst tan δ increases, and therefore Q decreases, with increasing frequency (f). The product $Q \times f$ provides a useful 'figure of merit' to compare different materials.

This paper seeks to provide a brief overview of currrent materials, focusing on two systems to give insight into factors that control important dielectric properties, and considers the prospects for improvements in materials and our theoretical understanding.

2. CURRENT MATERIALS

There are at present five main families of ceramics suitable for applications at microwave frequencies. Table 1 summarises the properties of representative members of these families. Details of individual materials are given in a number of recent reviews.[2,5,6,7] Typical applications are described by Wakino,[2] but the major growth area in recent years has been cellular radio, with a growing demand for materials for components in base stations and mobile telephones.

2.1 $(Zr,Sn)TiO_4$ Ceramics

Zirconium titanate-based materials have been recognised as temperature stable dielectrics for over fifty years,[8] but incorporation of Sn is required to provide both high Q and near zero τ_f. Within the ZrO_2–TiO_2–SnO_2 system single phase materials only exist over a limited range of compositions;[9] commercial products are often close to $Zr_{0.8}Sn_{0.2}TiO_4$.

Newnham[10] demonstrated that the structure of $ZrTiO_4$ was orthorhombic; having the form of α-PbO_2. McHale and Roth[9,11] found that the length of the c-axis was highly sensitive to processing conditions, changing from approximately 5.50Å, in materials sintered and annealed at high temperature (~ 1400°C), to approximately 5.35Å, in materials sintered at high temperature and annealed at low temperature (≤ 1100°C). The critical annealing range appears to be around 1130–1170°C. The difference between materials with a long c-axis and those with a short c-axis appears to be the state of cation ordering; with shortening of the c-axis there is an increase in the degree of cation ordering.[11] Doping with Sn stabilises the high temperature, longer c-axis form of zirconium titanate.

Table 1.   Dielectric properties of resonator materials

	ε_r	Q	f_o (GHz)	$Qf \times 10^3$
$MgTiO_3$–$CaTiO_3$	21	8000	7	56
$Ba(Mg,Ta)O_3$–$BaSnO_3$	24	43000	10	430
$Ba(Zn,Ta)O_3$	30	14000	12	168
$(Zr,Sn)TiO_4$	38	10300	5	51
$Ba_2Ti_9O_{20}$	36	10700	4.5	48
$BaO.Sm_2O_3.TiO_2$	80	3700	2.9	11
$BaO.Nd_2O_3.TiO_2$	86	3000	3.5	10.5

Azough[12] prepared ceramics of $Zr_{0.8}Sn_{0.2}TiO_4$ having a range of states of cation ordering by the use of controlled cooling rates (air quenched to 1°C min^{-1} after sintering). The state of ordering was assessed by both TEM electron diffraction (Fig. 1) and X-ray diffraction. It was found that the dielectric Q value increased as the degree of ordering increased (Fig. 2). Whilst ordering *appears* to have a profound effect on the loss behaviour of zirconium titanate-based materials, it is not the only factor. Reduction of titanium, e.g.

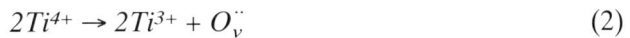

$$2Ti^{4+} \rightarrow 2Ti^{3+} + O_v^{\cdot\cdot} \tag{2}$$

with the generation of oxygen vacancies, leads to a substantial reduction in the Q value.[3] Similarly, the presence of trivalent impurities, such as Fe, Cr or La, which are located in the primary grains, cause a catastrophic reduction in the Q value,[13,14,15] by a mechanism involving the generation of electrons, 14 e.g.

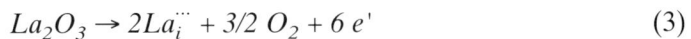

$$La_2O_3 \rightarrow 2La_i^{\cdot\cdot\cdot} + 3/2\ O_2 + 6\ e' \tag{3}$$

Clearly, cation ordering, impurities and lattice vacancies affect the loss properties of $ZnTiO_4$-based materials. For the complex perovskites, such as $Ba(Mg,Ta)O_3$, cation ordering can enable Q values up to 40,000 to be achieved.[16] There is also a suggestion that ordering may be important in $MgTiO_3$ ceramics.[17]

2.2 BaO–R$_2$O$_3$–TiO$_2$ Ceramics

Materials in the systems $BaO–R_2O_3–TiO_2$, where R is a rare-earth species (Nd, La, Sm, Pr) have high ε_r (typically 80–100) and Q values of a few thousand (Table 1). The higher relative permittivities offer the possibility of smaller resonators than those based on zirconium titanate or the complex perovskites. During the 1980s the system $BaO–Nd_2O_3–TiO_2$ was explored in some detail.[13,18] Kolar and co workers [18] demonstrated that the highest ε_r values and highest Q values were achieved for ceramics having the three components in the ratio 1:1:3, 1:1:4 or 1:1:5. Subsequently, interest focussed on materials around 1:1:4 and 1:1:5. Negas et al[5] reviewed the effect of processing and composition in $BaO–Nd_2O_3–TiO_2$ ceramics. Alternative systems based on Sm_2O_3, Pr_2O_3 and La_2O_3 have also been investigated.[6] Compositions involving $BaO–Nd_2O_3–TiO_2$ (1:1:4) with PbO and/or Bi_2O_3 additions offer good, temperature-stable dielectric properties. A full understanding of structure-property relationships in these $BaO–R_2O_3–TiO_2$ materials has been hampered by problems of refining the structures.

Matveeva et al[19] suggested that the space group of $Ba_{3.75}Pr_{9.5}Ti_{18}O_{54}$ was either Pbam or Pba2. Subsequent investigations of $BaO.Nd_2O_3.5TiO_2$ and $BaO.Nd_2O_3.4TiO_2$ suggested Pbam and Pba2 respectively.[20,21] More recently Azough et al[22] used TEM electron diffraction techniques and proposed that the space group of $Ba_{3.75}Pr_{9.5}Ti_{18}O_{54}$ and $Ba_{3.75}Nd_{9.5}Ti_{18}O_{54}$ was Pnam. This space group enabled the structure of the Nd analogue to be refined from high resolution X-ray diffraction data;[23] three of the large cation sites were found to be disordered

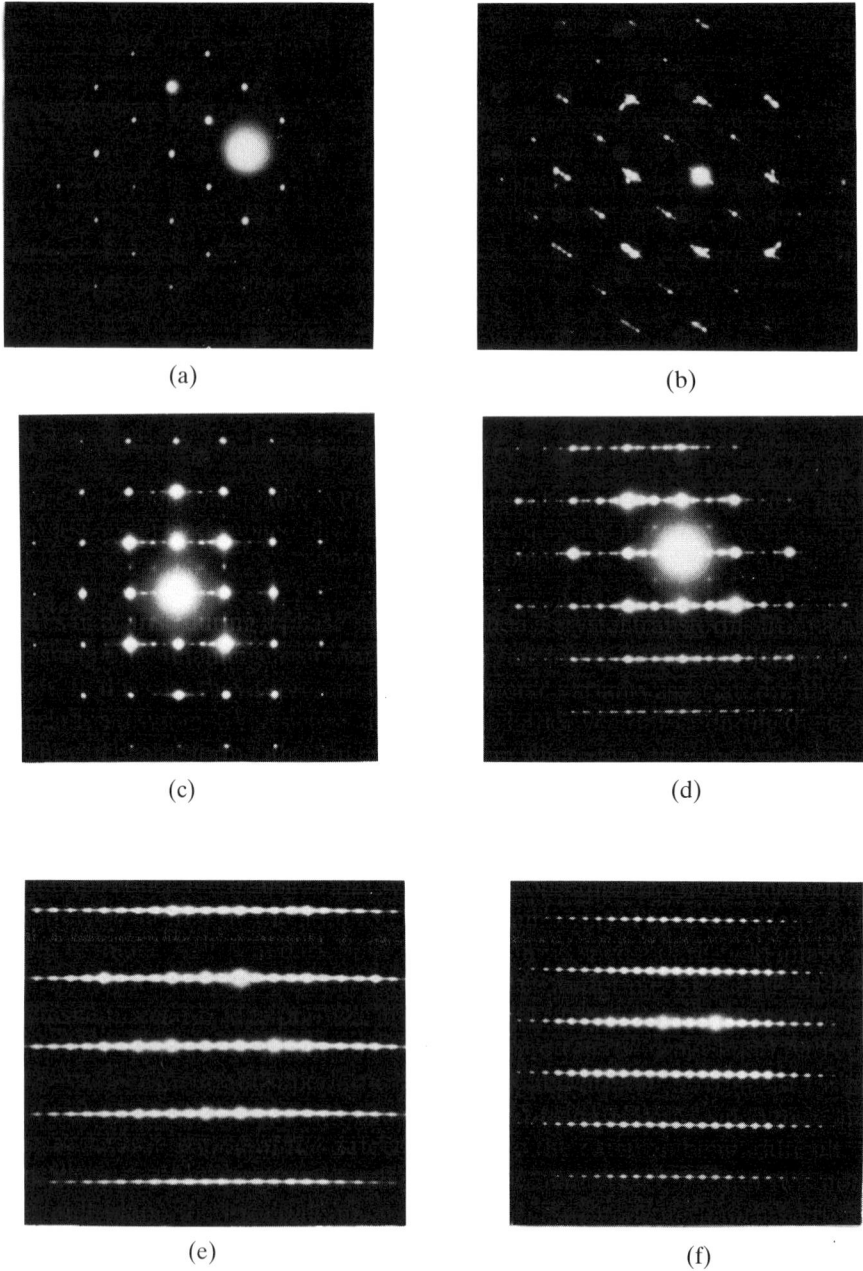

Figure 1.   Transmission electron microscope electron diffraction images of
$(Zr_{0.8}Sn_{0.2})TiO_4$ ceramic cooled at different rates after sintering: (a) air cooled,
(b) cooled at 360°C/hour, (c) 120°C/hour, (d) 6°C/hour, (e) 1°C/hour, (f) 1°C/hour,
and annealed at 1000°C.

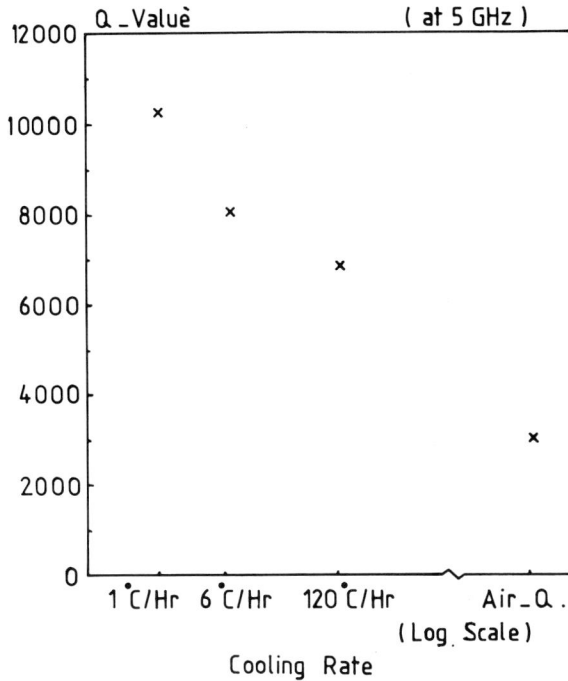

Figure 2.   Dielectric Q value of $(Zr_{0.8}Sn_{0.2})TiO_4$ ceramic as a function of cooling rate after sintering.

in Ba/Nd, whilst the fourth was fully occupied by Nd. The cell parameters are summarised in Table 2.

Improvements in ceramic processing techniques, removal of impurities and the inclusion of selective additives has enabled the Qxf product for $BaO–R_2O_3–TiO_2$ ceramics to be increased to 11000 during the last decade.[5,7]

3. FUTURE NEEDS

As designers of mobile telephones and mobile telephone systems demand further miniaturization, there will be a greater need for high quality materials of high ε_r, around 100. Once relative permittivities increase significantly above this value the tolerances on the size and shape of individual components become very small,

Table 2.   Lattice parameters of $Ba_{3.75}Nd_{9.5}Ti_{18}O_{54}$ (after [23]).

a	22.3476 (2) Å
b	12.1938 (1) Å
c	7.6771 (1) Å
vol	2092.01 (4) Å^3

placing considerable strain on final machining operations. However, there is likely to be increased need for materials with moderately high ε_r values (25–40) with higher Q values to utilise the new frequency bands that become available for the so-called Personal Communication Network (PCN).

If current and new materials are to be fully exploited a greater fundamental understanding will be required. This embodies both the effect of compositional variations, impurities, structural distortion and defects etc on dielectric properties and the ability to predict the properties of perfect and 'real' materials.

4. INTRINSIC DIELECTRIC PROPERTIES

4.1 Classical Theory

Since the 1960s far infrared spectra of perovskite structure materials have been analysed by the Kramers-Kronig relations and by classical dispersion theory.[24] Using group theory, the allowed lattice vibrations, and normal modes of vibration accessible in far IR and Raman spectra may be predicted. The relative permittivity and tan δ at microwave frequencies can be calculated via:

$$\varepsilon = \varepsilon_\infty + \sum_j 4\pi\rho_j \tag{4}$$

$$tan\delta = \frac{\sum_j 4\pi\rho_j(\gamma_j.\omega)/\omega_j^2}{\varepsilon_\infty + \sum_j 4S\pi\rho_j} \tag{5}$$

where the strength $4\pi\rho_j$, width γ_j, and resonant frequency ω_j of each oscillator are the dispersion parameters and ε_∞ is the high frequency dielectric constant caused by electronic polarisation. Summation over all oscillators and adjustment of the dispersion parameters normally gives good agreement between calculated and measured IR spectra. Table 3 lists the experimentally-determined and calculated dielectric properties of Ba(Zn,Ta)O$_3$ ceramic.[24] There is good agreement between relative permittivity data, but the experimental Q value is less than half the predicted value. The differences between the Q values in part reflect the trial and adjustment approach used to fit the dispersion parameters, but also the problem of trying to predict intrinsic properties from 'real' materials.

Table 3.   Experimental and calculated dielectric characteristics of Ba(Zn,Ta)O$_3$ ceramics (after [24]).

	ε_r	Q at 7 GHz
Experimental	29.1	9800
Calculated	30.3	20200

4.2 Simulation Studies

An alternative approach to predicting structural and dielectric properties is simulation techniques.[25] The basis of the approach is that the energy of the system is described as a function of atomic co-ordination. Potential models include both Coulombic energies (fully ionic model) and short range energies (repulsive and attractive terms). All interactions between the core ions in the central region (amounting to some 100–300 ions) are treated explicitly, and the region beyond is treated as a continuum. This approach has proved very successful for the simulation of a wide range of oxides and silicates,[25] but the dielectric properties of titanates have proved a problem. Calculated dielectric properties of materials involving TiO_2 are very sensitive to Ti potentials; predicted ε_r values may change by ± 100%, after only small perturbations in input data. However, newly developed potentials for Ti and other relevant ions suggest that reliable simulation studies should be possible within a few years.

5. CONCLUSIONS

Five main families of ceramic dielectrics, with relative permittivities in the range 20–90, are currently available for microwave resonator applications. Dielectric Q values of up to 40,000 are possible at 10 GHz. The primary factors controlling Q values include impurities, cleanliness of the grain boundaries, structural defects, lattice defects and state of cation ordering. For development of mobile telephone systems in the future, dielectrics of both higher Q and high ε_r will be needed. Complete exploitation will require deeper understanding of current materials and improved theory to enable reliable prediction of the properties of new candidate materials. There is a good prospect that simulation studies will be sufficiently developed within a few years to predict the dielectric properties of perfect and defective materials.

REFERENCES

1. R. D. Richtmyer: *R. Appl. Phys.*, 1939, **10**, 391–398.
2. K. Wakino: *ISAF '86, Proc. 6th Int. Symp. Appl. Ferroelectrics*, 1986, 97–106.
3. K. Wakino and H. Tamura: *Ceramic Trans.*, **8**, Amer. Ceram. Soc. Westerville, Ohio, 1990, 305–314.
4. K. Wakino, M. Murata and H. Tamura: *J. Amer. Ceram. Soc.*, 1986, **69**, 34–37.
5. T. Negas, G. Yeager, S. Bell and R. Amren: NIST Special Public, 804, *Chemistry of Electronic Ceramic Materials*, Proc. of conf. held at Jackson, WY, Aug 1990, 1991, 21–34.
6. R. Freer: *Silicate Industriels*, 1993, **59**, 191–197.
7. T. Negas and H. C. Ling (eds): 'Materials and Processes for Wireless Communication', *Ceramic Trans.*, **53**, Amer. Ceram. Soc. Westerville, Ohio, 1995, 228.
8. W. Rath: *Keram. Radsch*, 1941, **49**, 137–139.
9. A. E. McHale and R. S. Roth: *J. Amer. Ceram. Soc.*, 1983, C-18–C-20.
10. R. E. Newnham: *J. Amer. Ceram. Soc.*, 1967, **50**, 216.
11. A. E. McHale and R. S. Roth: *J. Amer. Ceram. Soc.*, 1986, **69**, 827–832.

12. F. Azough: unpublished Ph.D. Thesis, University of Manchester, 1991.
13. K. Wakino, K. Minai and H. Tamura: *J. Amer. Ceram. Soc.*, 1984, **67**, 278–281.
14. D. M. Iddles, A. J. Bell and A. J. Moulson: *J. Mater. Sci.*, 1992, **27**, 6303–6310.
15. F. Azough and R. Freer: *Br. Ceram. Proc.*, 1989, **42**, 225–232.
16. H. Matsumoto, H. Tamura and K. Wakino: *Japan J. Appl. Phys.*, 1991, **30**, 2347–2349.
17. V. M. Ferreira, F. Azough, J. L. Baptista and R. Freer: Proc. ECAPD-2, London, *Ferroelectrics*, 1992, **133**, 127–132.
18. D. Kolar, Z. Stadler, S. Gaberscek and D. Suvorov: *Ber. Deut. Keram. Ges.*, 1978, **55**, 346–348.
19. R. G. Matveeva, M. B. Varfolomeev and L. S. Il'yusheenko: Trans. frm *Zh. Neorg. Khimi*, 1984, **29**, 31–34.
20. R. S. Roth, F. Beach, A. Santoro and K. Davis: Abstract 07.9-9, Fourteenth Int. Congress of Crystallography, Perth, Australia, August 1987.
21. D. Kolar, S. Gabrsek and D. Suvorov: *Proceedings Third Euro. Ceramics Conf.*, P. Duran and J. F. Fernandoz eds, *Faenza Iberica S.L.*, 1993, 229–234.
22. F. Azough, P. Champness and R. Freer: *J. Appl. Crystallog.* (accepted).
23. F. Azough, P. Setasuwon and R. Freer: *Ceram. Trans.*, **53**, Am. Ceram. Soc., Westerville, Ohio, 1995, 215–228.
24. H. Tamura, D. A. Sagala and K. Wakino: Japan. *J. Appl. Phys.*, 1986, **25**, 787–791.
25. C. R. A. Catlow: in A. V. Chadwick and M. Terenzi (eds), 'Defects in Solids', *Modern Techniques*, Plenum, New York, 1986, 269–302.

Characterisation of PZT Thin Films Prepared from a Diol Sol-Gel Route using Different Precursors

Y. L. TU and S. J. MILNE

School of Materials, University of Leeds, Leeds, LS2 9JT, UK

ABSTRACT

A sol-gel route using 1,3 propanediol as the solvent has been used to synthesise single-layer PZT films. Two zirconium precursors, namely acetylacetonate stabilised zirconium n-propoxide and zirconium acetylacetonate were employed to make PZT sols for the preparation of thin films. The thermal decomposition behaviour was similar between the sols made from these two precursors, but preferred orientation, microstructure, and electrical properties were dependent on precursor type. The employment of zirconium acetylacetonate precursors generally led to stronger (111) preferred orientation, smaller grain size, lower ε_r and higher P_r. PZT films were fired using two heating schedules: the heating rate effects on the orientation, microstructure and electrical properties are also presented. The electrical properties for films prepared from either precursor and fired using a 'direct' insertion method exhibited a P_r in the range of 30–33 μCcm^{-2}, ε_r in the range of 1100–1260 and an E_c of 46 $kVcm^{-1}$.

1. INTRODUCTION

Various thin film fabrication processes have been devised for the fabrication of PZT thin films including physical vapour deposition, MOCVD and sol-gel routes.[1] Over the past decade, sol-gel processing has received growing interest for the fabrication of PZT films. The processing of sol-gel thin films generally employs spin-coating or dip-coating to form gel coatings, and a two step post-deposition firing procedure to convert the organic bearing gel layers to ceramic films. In the literature, the maximum single-layer thickness for most of the sol-gel routes is ~0.1 μm, and multiple deposition is necessary to build up thicker films.

The first reports of solution deposition methods for preparing PT-PZT films appeared in 1984; butanol was used as the solvent medium for lead 2-ethylhexanoate, titanium tetra-butoxide and zirconium acetylacetonate.[2] Others around the same time developed a solution system employing lead acetate, titanium tetra-propoxide and zirconium tetra-propoxide in a methoxyethanol solvent.[3] Subsequently, there were several modifications using either different precursors such as lead acetylacetonate,[4] titanium and zirconium butoxide[5] or different solvents such as simple alcohol[6] or carboxylic acid.[7] Because of the moisture sensitive nature of titanium and zirconium alkoxide, chemical modifying agents such as acetic acid,[7] acetylacetonate[8] and diethanolamine[9] were also introduced by some groups. The reasons for changing precursors and solvents or adding chemical modifying agents can be to increase the limiting thickness of crack-free films,[7] to improve solution stability[4,7,8] and to simplify preparation procedures.[7]

Our group was the first to report a novel sol-gel route based on diol solvents.[10–12] In this paper, two zirconium precursors were used: one is moisture-sensitive zirconium n-propoxide and the other is zirconium acetylacetonate. We report the precursors effects on the thermal decomposition, XRD patterns, microstructures and electrical properties of single-layer PZT films made from this diol sol-gel route. The effects of heating rate are also compared.

2. EXPERIMENTAL

The details of the diol PZT sol synthesis procedure are described elsewhere.[12] The starting reagents for PZT (53/47) solution synthesis were lead acetate trihydrate $[Pb(OOCCH_3)_2.3H_2O]$ (Aldrich Co.) and titanium diispropoxide bisacetylacetonate $[Ti(OC_3H_7)_2(CH_3COCHCOCH_3)_2]$ (75% in isopropanol, Alfa Co.,), abbreviated TIAA whereas two zirconium precursors, namely zirconium acetylacetonate $[Zr(CH_3COCHCOCH_3)_4]$ (Aldrich Co.), abbreviated ZAA, and zirconium n-propoxide $[Zr(OCH_2CH_2CH_3)_4]$ (70% in n-propanol, Alfa Co.,) were used to investigate the effect of precursors. Lead acetate trihydrate was mixed with 1,3 propanediol (Aldrich Co.,) in a ratio of 1:1; the mixture was refluxed for 2 hrs to form a lead-diol precursor. Zirconium n-propoxide was first stabilised by reaction with acetylacetone in a molar ratio of 1:2. The stabilised zirconium n-propoxide is abbreviated ZPA. TIAA was then also reacted with diol and either ZPA or ZAA under refluxing conditions for 2 hrs. These two solutions were then combined. In total 5 moles of 1,3 propanediol were used per mole of lead. Further reflexing for 5 hrs with one distillation after 2 hrs resulted in a stock solution concentration of 1.1–1.2M. All starting reagents were assayed gravimetrically; final sols contained the equivalent of 10 mol% excess $Pb(OOCCH_3)_2)$ to compensate for PbO losses during firing.

Films were fabricated by spin coating sols on to platinised silicon substrates with Pt(111)/Ti/SiO_2/Si configuration. The films prepared from 1M sols were spin coated at 1500 rpm for 1 min. The coated wet substrates were then transferred to a hot plate set at 350°C for 1 min. These prefired films were inserted directly into a furnace set at 700°C for 15 min.

Phase analysis of thin films was examined using an X-ray diffractometer (Philips APD 1700) at room temperature. Scanning electron microscopy (Hitachi S700) was used to examine the surface microstructure.

For electrical characterisation, gold top electrodes were applied on to the surface of the films using a shadow masking method. The diameter of the gold electrodes was ~300 μm. The exact diameters on each film were examined using an optical microscope fitted with a calibrated graticule eyepiece. The relative permittivity, ε_r, and dissipation factor, D, were examined using a Hewlett Packard 4192A impedance analyser at a frequency of 1 kHz. P-E characteristics were studied using a Radiant Technology RT66A Ferroelectric Tester at a frequency of ~60 Hz and 300 kV cm^{-1}. At least thirty permittivity measurements and six P-E measurements were taken on different dot capacitors for each sample in order to ensure that values recorded were representative. All the electrical properties were measured 10 days after firing to avoid variations due to any possible ageing effects.

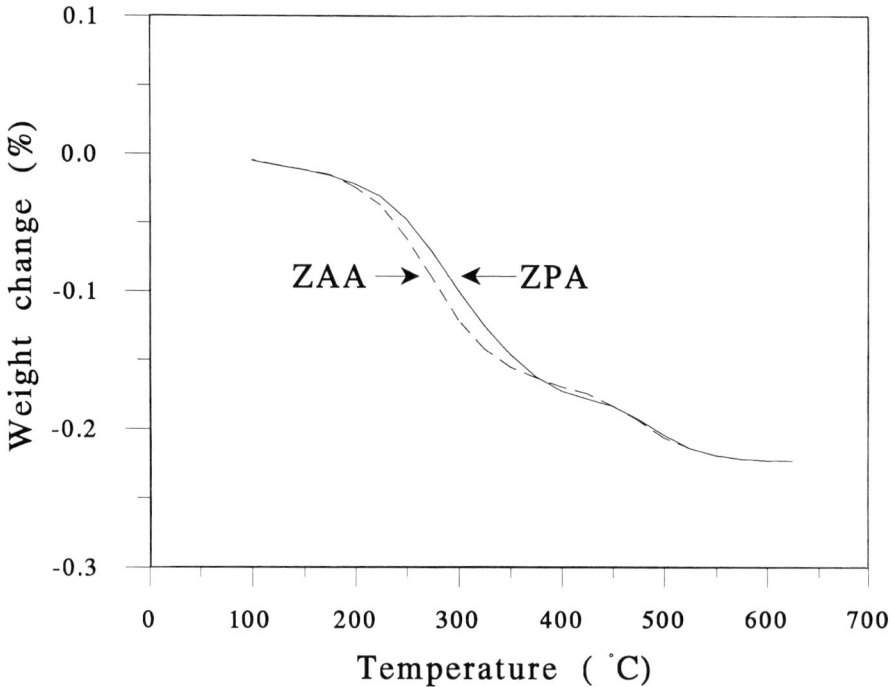

Figure 1(a).   TGA curves of PZT gels made from different precursors.

3. RESULTS AND DISCUSSION

TGA data for PZT precursors gels prepared using different precursors are compared in Fig. 1a. Gels were prepared by drying the stock solution at 120°C for 24 hrs. The organic decomposition steps in those gels prepared using different precursors were similar, but some subtle variations maybe existed due to experimental variations. The decomposition under these heating conditions is completed by ~550°C. The corresponding DTA curves are presented in Fig. 1b. The DTA trace derived from gels made using ZPA showed several exothermic peaks: the peaks at 255°C, 280°C, 405°C, and 480°C corresponded to steps in the TGA trace indicating them to be associated with the elimination of decomposition products from the gel, Fig. 1a, while the use of ZAA gave rise to peaks at 240°C, 290°C, 400°C, 475°C and 490°C which can be also correlated to decomposition of the gel. However, the precise mechanisms of the decomposition are not yet understood.

The concentrations of stock solutions made from ZPA and ZAA were 1.1M and 1.2M, respectively. Single-layer films obtained by depositing 1 M sols diluted with n-propanol exhibited a thickness of 0.5 μm for ZPA precursor and 0.4 μm for ZAA precursor.

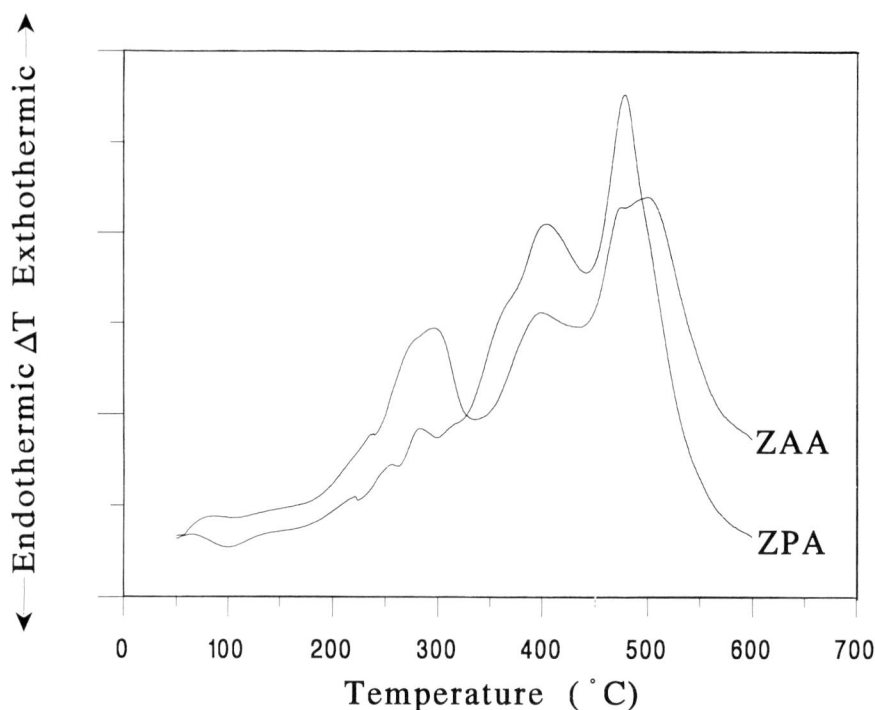

Figure 1(b).   DTA curves of PZT gels made from different precursors.

The XRD patterns of these two films fired at 700°C for 15 min using the direct firing method are compared in Fig. 2. The 0.5 μm single-layer film prepared from ZPA precursor exhibited weaker intensity of 111 peak compared to the 0.4 μm single-layer film prepared from ZAA precursor. ZPA precursor derived film fired at 700°C using slow heating rate 10°Cmin⁻¹ also gave rise to weaker intensity of 111 peak compared to ZAA precursor derived film, Fig. 3. No pyrochlore phase was present in those films. The reason for the precursor effect on slight change in orientation is not yet understood. Comparisons between Fig. 2 and Fig. 3 indicated that slow heating resulted in films with stronger (111) orientation. This could be attributed to prolonged firing time at low firing temperature promoting heterogeneous nucleation and growth from Pt(111) and film interface. It has been suggested that heterogeneous nucleation at the Pt(111) bottom electrode is the reason for (111) orientation in the deposited films.[13]

The microstructure of 0.5 μm single-layer films prepared from ZPA precursor is presented in Fig. 4a. The film was fired directly at 700°C for 15 min. Rosette-like grains up to ~0.5 μm in size were present, together with a fine grain matrix exhibiting dark contrast, Fig. 4a. Within the rosettes it was possible to identify a sub-structure of smaller grains. It has been suggested that the fine grain fraction is a lead deficient

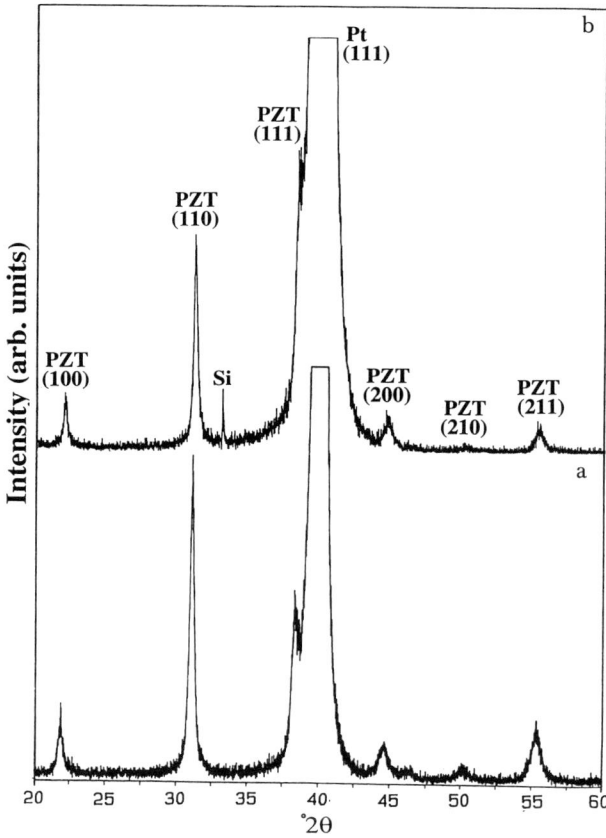

Figure 2.   XRD patterns of PZT films prepared from (a) ZPA and (b) ZAA precursors. Films were fired at 700°C for 15 min using a 'direct' insertion method.

pyrochlore phase.[14] The surface microstructure of 0.4 μm single-layer films prepared from ZAA precursor was composed of fine grains ~0.1 μm in size, Fig. 4b. The differences in microstructure most possibly resulted from solution chemistry induced during reaction and concentration of different precursors. Other workers have reported that the changes in solution chemistry caused by altering the mixing order of precursors[15] or distillation of reaction by-product[16] affected the microstructure of PZT films.

Slow heating to 700°C yielded a quite different microstructure for each precursor, as shown in Fig. 5. A 0.5 μm film prepared from ZPA precursor exhibited an indistinct microstructure but individual grains ~0.07 μm in size could be resolved in some regions, Fig. 5a. The 0.4 μm film prepared from zirconium acetylacetonate possessed grains ~0.05 μm in size throughout the surface, Fig. 5b. It also exhibited regions of light and dark contrast. This contrast variations may imply topographic contrast due to an unsmooth surface.

The comparisons of SEM micrographs between Fig. 4 and Fig. 5 indicated that slow heating led to finer grain size. The differences in grain size also correlate with

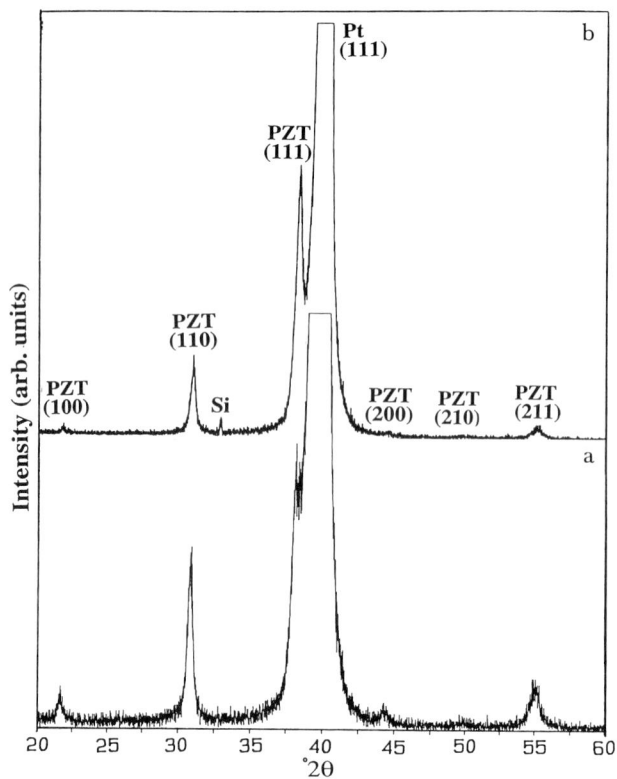

Figure 3.   XRD patterns of PZT films prepared from (a) ZPA and (b) ZAA precursors.
Films were fired at 700°C for 15 min using slow heating 10°Cmin⁻¹.

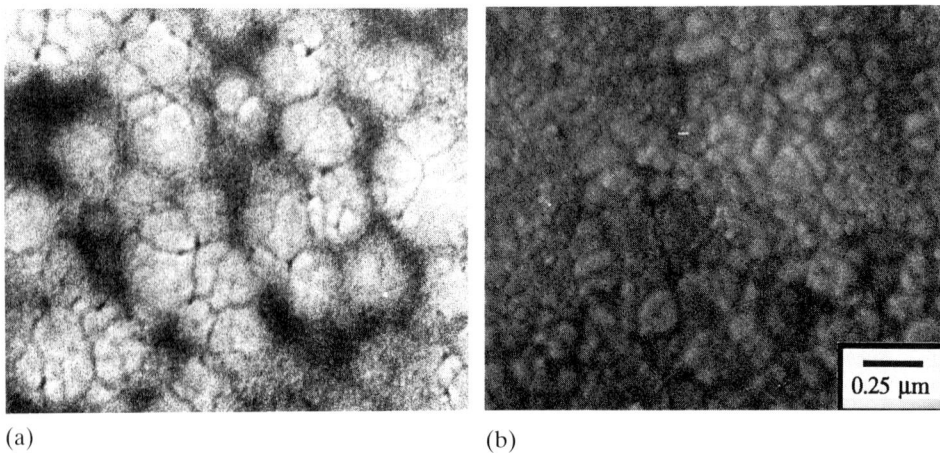

(a)   (b)

Figure 4.   Surface microstructures of PZT films prepared from (a) ZPA and
(b) ZAA precursors. Films were fired at 700°C for 15 min using a 'direct'
insertion method.

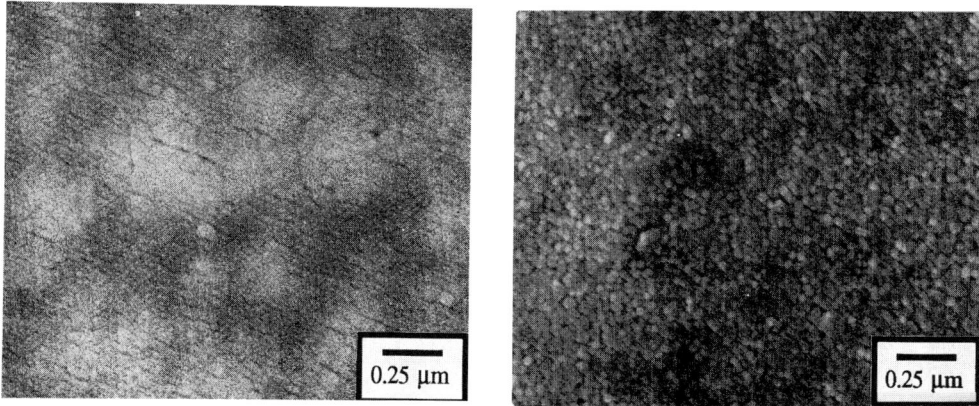

Figure 5.   Surface microstructures of PZT films prepared from (a) ZPA and
(b) ZAA precursors. Films were fired at 700°C for 15 min using slow heating rates
(10 °Cmin^{-1}).

the degree of (111) preferred orientation, i.e. the finer the grains the stronger the
(111) orientation. Thus, the heating rate dependence of grain size is likely to be the
result of changes to the crystallisation of the gel brought about by different heating
conditions.

The electrical properties of both types of films are summarised in Table 1a and 1b
for films fired using different heating rate. Final firing was at 700°C for 15 min. The
corresponding P-E hysteresis loops for both films measured at an applied field of 300
kV cm^{-1} are illustrated in Fig. 6. The remanent polarisation, P_r, and relative permittivity, ε_r, values of films made from zirconium n-propoxide using the direct insertion
method, were 30 μCcm^{-2} and 1260, whereas films prepared from ZAA had a higher
P_r of 33 μCcm^{-2} and lower ε_r of 1100; E_c was 46 kV cm^{-1} for both films, Table 1a.
Similarly, a slowly heated film prepared from ZAA precursor possessed a higher P_r
of 25 μCcm^{-2} and lower ε_r of 840 compared to the slowly heated film prepared from
ZPA precursor which had a P_r of 20 μCcm^{-2} and ε_r of 970, Table 1b; E_c was

**Table 1(a).   Electrical properties of PZT films made from sols using two different
precursors. Films were fired at 700°C for 15 min using a 'direct' insertion method.**

Precursor	P_r(μCcm^{-2})	E_c(kVcm^{-1})	ε_r	D
ZPA	30 ± 1.8	46 ± 2.1	1260 ± 90	0.07
ZAA	33 ± 2.1	46 ± 1.4	1100 ± 85	0.05

**Table 1(b).   Electrical properties of PZT films made from sols using two different
precursors. Films were fired at 700°C for 15 min using slow heating 10°Cmin^{-1}.**

Precursor	P_r(μCcm^{-2})	E_c(kVcm^{-1})	ε_r	D
ZPA	20 ± 0.3	51 ± 0.8	970 ± 70	0.04
ZAA	25 ± 0.5	51 ± 2.0	840 ± 75	0.05

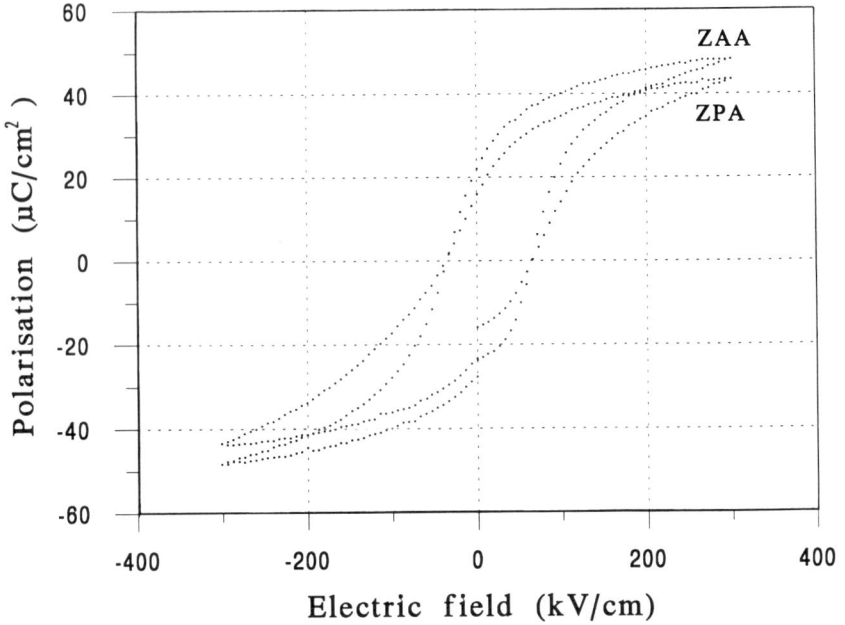

Figure 6(a).   Comparisons of P-E hysteresis loops of films prepared from sols made using
the two different precursors and fired at 700°C for 15 min using 'direct' insertion method.

51 kVcm[-1] for both films, i.e. independent of the precursors used. The discrepancies in microstructures and orientation using different precursors may account for the observed electrical properties. However, the differences in film thickness between films prepared from both precursors may also affect the electrical properties. It has been reported that P_r and ε_r increased and E_c decreased with increasing film thickness.[17]

The comparisons of electrical properties in Table 1a and Table 1b for films prepared from both precursors showed that slow heating to 700°C resulted in films with a lower P_r and ε_r and a higher E_c. The reasons for this and other observed trends are under investigation.

4. CONCLUSIONS

A diol sol-gel route has been demonstrated for depositing single-layer PZT films on platinised Si substrates using sols made from two different precursors. The employment of the ZPA precursor led to weaker (111) orientation compared to the orientation of the films prepared from ZAA precursor. Using the direct firing schedule, the film prepared from ZPA precursor was composed of rosette-like grains ~0.5 μm in size and a fine grain matrix, while ZAA precursor exhibited individual grains ~0.1 μm in size. P_r and ε_r were 30 μCcm[-2] and 1260 using ZPA precursor and 33 μCcm[-2] and 1100 using ZAA precursor. E_c was 46 kVcm[-1] for both precursors. Slow heating led

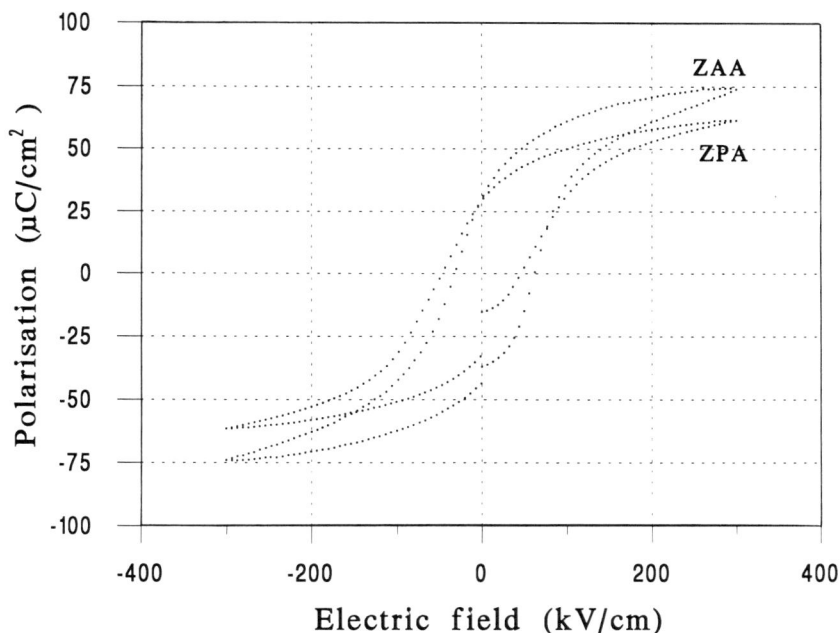

Figure 6(b).   Comparisons of P-E hysteresis loops of films prepared from sols made using the two different precursors and fired at 700°C for 15 min using slow heating rates (10°Cmin^{-1}).

to finer grain size and stronger (111) orientation for films prepared from both precursors. Subsequently, lower P_r and ε_r and higher E_c were observed.

REFERENCES

1. L. M. Sheppard: *Am. Ceram. Soc. Bull.*, 1992, **71**, 85–95.
2. J. Fukushima, K. Kodaira and T. Matsushita: *J. Mater. Sci.*, 1984, **19**, 595–598.
3. K. D. Budd, S. K. Dey and D. A. Payne: *Brit. Ceram. Soc. Proc.*, 1985, **36**, 107–121.
4. U. Selvaraj, K. Brooks, A. V. Prasadarao, S. Komarneni, R. Roy, L. E. Cross: *J. Am. Ceram. Soc.*, 1993, **76**, 1441–1444.
5. N. Tohge, S. Takahashi and T. Minami: *J. Am. Ceram. Soc.*, 1991, **74**, 67–71.
6. K. C. Chen, A. Janah and J. D. Mackenzie: *Mater. Res. Soc. Symp. Proc.*, 1986, **73**, 731–736.
7. G. Yi, Z. Wu and M. Sayer: *J. Appl. Phys.*, 1988, **64**, 2713–23.
8. S. Hirano, T. Yogo, K. Kikuta, Y. Arak, M. Saitoh and S. Ogasahara: *J. Am. Ceram. Soc.*, 1992, **75**, 2785–89.
9. Y. Takahashi, Y. Matsuoka, K. Yamaguchi, M. Matsuki and K. Kobayashi: *J. Mater. Sci.*, 1990, **25**, 3960–64.
10. N. J. Phillips and S. J. Milne: *J. Mater. Chem. Lett.*, 1991, **1**, 893–894.
11. N. J. Phillips, M. L. Calzada and S. J. Milne: *J. Non-Cryst. Solids*, 1992, **147 & 148**, 285–90.
12. Y. L. Tu and S. J. Milne: *J. Mater. Sci.*, 1995, **30**, 2507–2516.

13. K. G. Brooks, I. M. Reaney, R. Klissurska, Y. Huang, L. Bursill and N. Setter: *J. Mater. Res.,* 1994, **9**, 2540–53.
14. A. H. Carim, B. A. Tuttle, D. H. Doughty and S. L. Martinez: *J. Am. Ceram. Soc.,* 1991, **74**, 1455–58.
15. R. W. Schwartz, B. C. Bunker, D. B. Dimo, R. A. Assink, B. A. Tuttle, D. R. Tallant and I. A. Weinstock: *Proc. of 3rd ISIF*, 1991, 535–546.
16. C. D. E. Lakeman and D. A. Payne: *J. Am. Ceram. Soc.,* 1992, **75**, 3091–96.
17. K. Amanuma, T. Mori, T. Hase, T. Sakuma, A. Ocui and Y. Miyasaka: *Jpn. J. Appl. Phys.,* 1993, **32**, 1150–52.

PMW–PT–PZ Low Firing Multilayer Actuators

S. SIRISOONTHORN* and D. HIND

School of Materials, University of Leeds, Leeds, LS2 9JT, UK

ABSTRACT

The $Pb(Mg_{1/2}W_{1/2})O_3$–$PbTiO_3$–$PbZrO_3$ (PMW–PT–PZ) ternary system was selected for study because of its promise in offering useful strain, in micropositioning applications, accompanying a transition from the antiferroelectric to ferroelectric states. An investigation of the antiferroelectric (AFE) ceramic composition 0.4PMW–0.4PT–0.2PZ using hot pressing followed by annealing studies shows that an ordering of the structure takes place. Microstructural characterisation illustrates the development of ferroelectric domains. Tape casting of an organic-based PMW–PT–PZ slip plus the fabrication and sintering at <1000°C of multilayer actuators with dielectric layers of thickness 60 μm are described. Characterisation of prototype actuators shows they are capable of producing strains of approximately 0.7% for an applied electric field of ~ 1.5kV mm⁻¹.

1. INTRODUCTION

Over recent years there has been a rapid growth of interest in the development of micromovement (~1 μm) actuators. Such devices are in demand for applications as diverse as conformable optics, semiconductor microchip processing machinery and automotive control systems.[1] With conventional actuators there is difficulty in achieving precise movements below 10 μm but multilayer actuators, based on the piezoelectric and electrostriction effects, can achieve this objective.[2,3] By fabricating devices using largely well-established multilayer (ML) technology, drive voltages have been brought down to ≤150 V, for which high stability power sources with good frequency response and dynamic range are available. Most importantly it enables unit costs to be cut dramatically by the use of mass-production techniques.

The PMW–PT–PZ ternary system has attracted attention because of its promise in offering useful strain accompanying a transition from the AFE to ferroelectric states.[4,5] Measurement of the electromechanical strains in the system PMW–PT–PZ has been reported elsewhere.[6,7,8]

The present paper is concerned with the composition 0.4PMW–0.4PT–0.2PZ which has been reported to show optimum properties for room temperature micropositioning applications.[8] The objective of the study was to characterise the relevant physical properties of the ceramic and to demonstrate the feasibility of fabricating prototype multilayer structures.

---

* Now at National Metal and Materials Technology Center (MTEC), Bangkok, Thailand

2. EXPERIMENTAL METHOD

2.1 Powder Synthesis & Materials Characterisation

The 0.4PMW–0.4PT–0.2PZ powder was prepared by the mixed powder route using appropriate amounts of reagent grade lead(II) oxide, magnesium carbonate hydroxide pentahydrate, tungsten(VI) oxide, titanium(IV) oxide and zirconium(IV) oxide. The mixtures were calcined in an alumina crucible at 800°C for 4 h. Discs were pressed and sintered at 1050°C for 2 h in a closed system designed to ensure control over the sintering atmosphere.[8]

Dielectric properties were measured on specimens in the form of parallel-sided discs ~0.5 mm thick, carrying sputtered gold electrodes. The capacitance was measured using a Hewlett Packard 4192A Impedance Analyser. Polarisation-electric field data were obtained at room temperature using a modified Sawyer-Tower circuit. Using the same samples as described above, electric field-induced strains were measured directly at room temperature, using a capacitative displacement transducer.

2.2 Hot Pressing and Annealing Studies

To study the effects of annealing time on structural order, pellets were hot-pressed in an alumina die at approximately 9 MPa in an O_2 atmosphere at 1050°C for 1 h; the pellet was surrounded by an 'atmosphere' powder consisting of 0.4PMW–0.4PT–0.2PZ mixed with MgO. After hot-pressing the specimens were annealed with atmosphere control (PMW powder) at 1050°C for periods between 2 h and 32 h. It was not possible to achieve density values higher than 98% theoretical because of limitations set by the hot-pressing die-set.

Materials were characterised by X-ray diffraction analysis and by SEM of polished and etched sections.

2.3 Fabrication of a Multilayer Actuator

From a review of the literature[9,10] an organic binder/solvent based tape casting system was selected based on polyvinyl butyral (PVB) binder. The components of the slip system consisted of ceramic powder, organic solvents, dispersant, binder and plasticiser. The slip was formulated using a proprietary copolymer, closely related to PVB, a fatty acid ester as dispersant and a three-solvent mixture in order to provide a range of drying rates. The required amount of slip components were blended and the slip prepared by ball-milling (ZrO_2 media) for 24 h prior to tape-casting.

The slip was cast on to a polymer film using a laboratory tape-casting unit which has been described elsewhere.[10] The doctor blade was replaced with a paint film applicator which has gap heights of 25–200 μm in 25 μm steps. Using a 200 μm gap gave a tape thickness after drying of ~100 μm. The initial tapes were assessed subjectively for general handling qualities – relative strength, plasticity, flexibility or brittleness and adjustments made as required to slip formulations.

MODEL ACTUATOR DESIGN

SQUARE TYPE (Dimensions : mm)

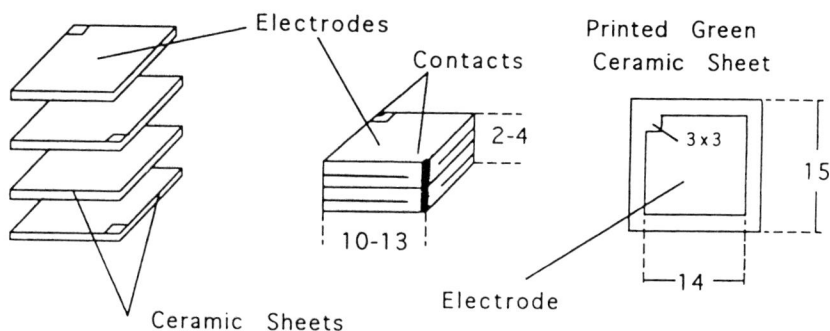

Figure 1.   Schematic diagram of model actuator design (square type).

The square type model actuator design shown in Fig. 1, which incorporates electrodes 14 mm square with a 3 mm square re-entrant at one corner, was adapted from earlier work.[10] Electrodes were screen-printed with Pt-based ink using a bench top manual machine equipped with 495 mesh polyester 45° screens, emulsion thickness 10 μm. This was predicted to give a dry electrode thickness of ~5 μm. Pieces of tape were held in place for printing by means of a vacuum bed. The problem of tape deformation due to suction into the vacuum apertures was eliminated by screening the pattern before the tape was stripped from the polymer carrier.

Ceramic tapes printed with the electrode pattern were cut out as 15 mm square sections and stacked in 180° registration in a 15 mm square die, fitted with low temperature resistance heaters. Warm-pressing was carried out by pressing a 20 layers stack at 14 MPa at 55°C for 3 minutes in order to form a laminate, with the electrodes buried within the structure. The laminate was carefully cut to expose the electrodes at the edges.

Binder burnout is all-important in the fabrication of good quality ML structures. Organic components have to be eliminated slowly in order to avoid delamination and were removed by a carefully controlled heat-treatment up to 500°C. Afterwards, the laminate was sintered at 975°C for 2 h. After the sintering stage the terminations were made using a silver based conductor fired at 700°C for 5 minutes prior to solding on connecting leads.

The sintered multilayer structures were examined in polished section.

3. RESULTS

3.1 Bulk Ceramics

The ceramic bulk densities achieved were >95% of the theoretical value. XRD confirmed the formation of the perovskite structure. Relative permittivity and loss tangent as a function of frequency and temperature are shown in Fig. 2.

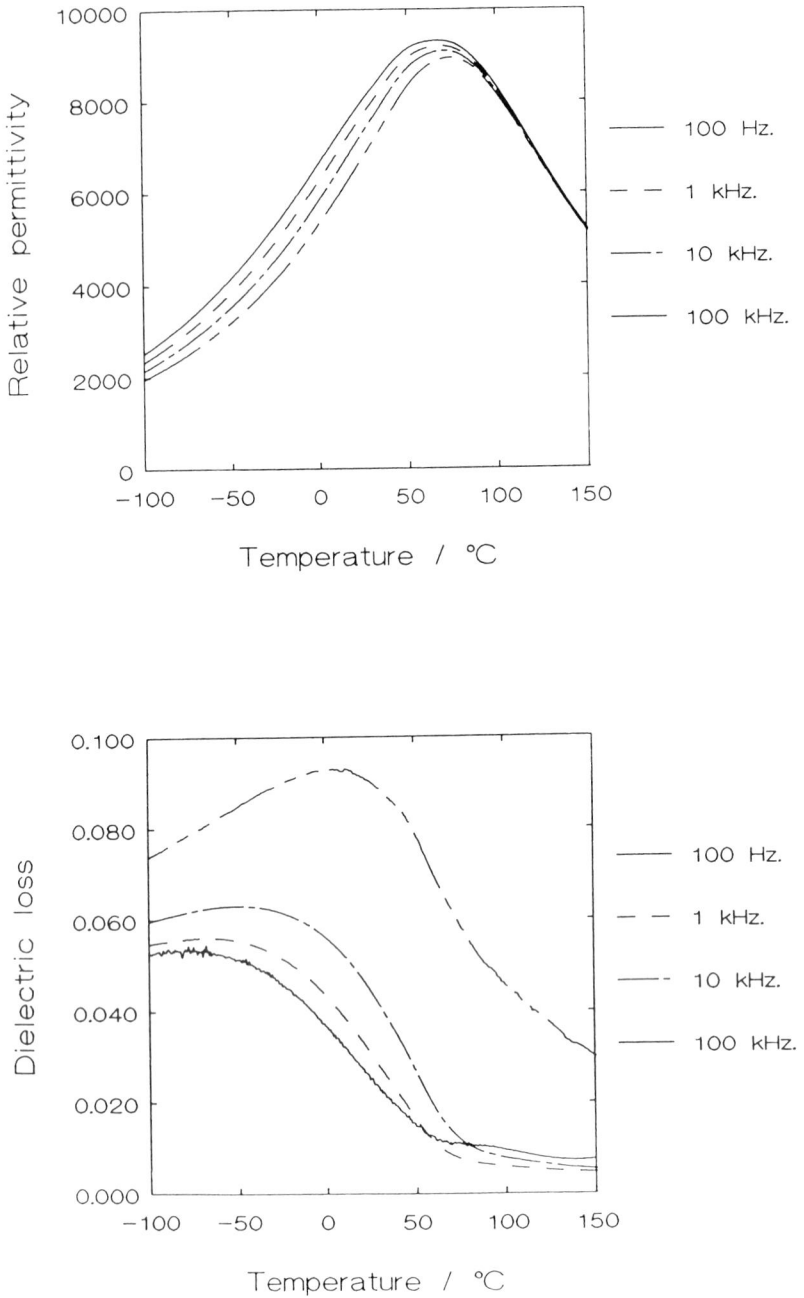

Figure 2. Relative permittivity and loss tangent as a function of frequency and temperature for composition 0.4PMW–0.4PT–0.2PZ.

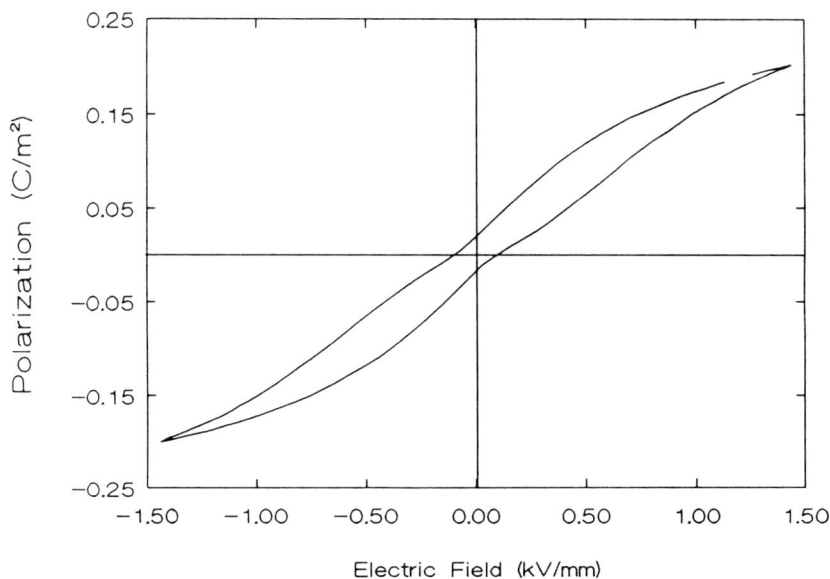

Figure 3.   Polarisation as a function of electric field for composition
0.4PMW–0.4PT–0.2PZ.

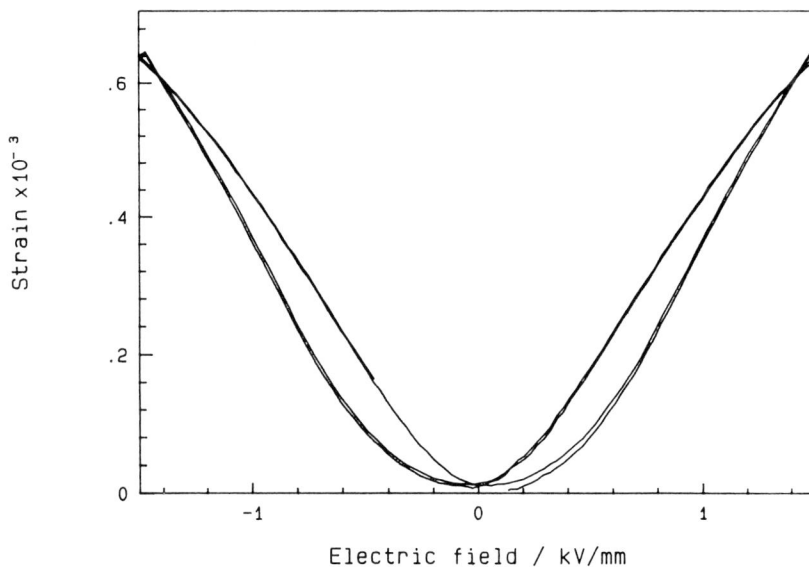

Figure 4.   Electrostrictive strain as a function of electric field for composition
0.4PMW–0.4PT–0.2PZ.

(a)

(b)

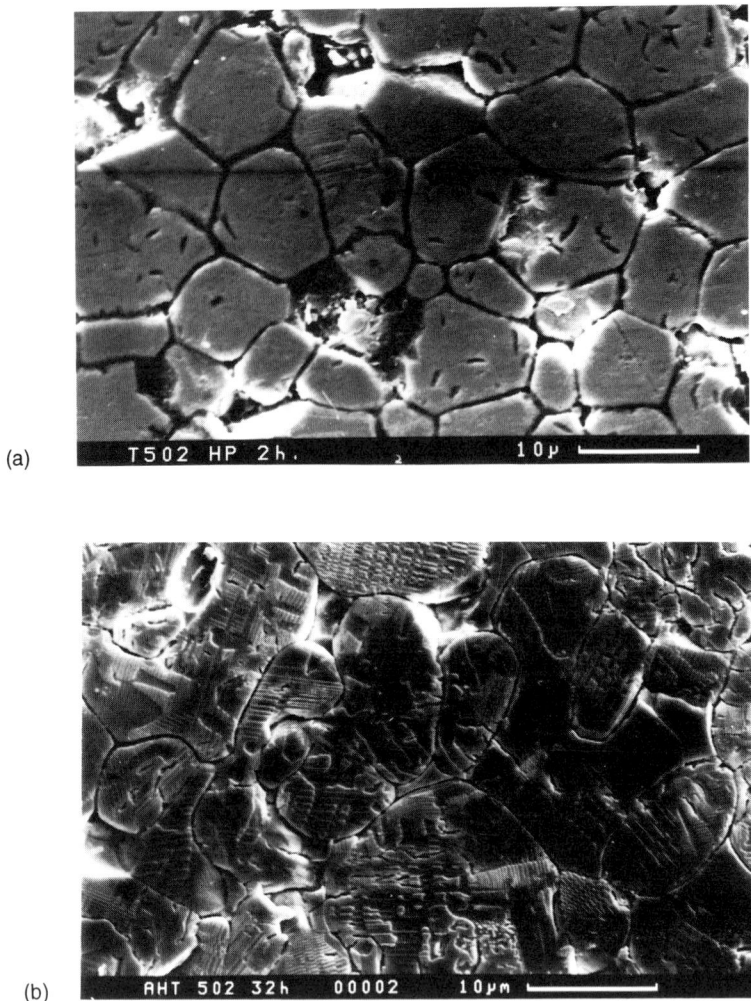

Figure 5.   Microstructures of hot pressed composition 0.4PMW–0.4PT–0.2PZ annealed for
a) 2h b) 32h.

The polarisation as a function of electric field, Fig. 3, shows a small tendency towards a double hysteresis loop suggesting some antiferroelectric character in this material. The electrostrictive strain-field relationships are shown in Fig. 4.

Figure 5 shows SEM micrographs of specimens annealed for 2 hr and 32 hr respectively. The polished sections have been etched with a hydrochloric/hydrofluoric acid mixture in order to reveal the grain structures, which show no evidence of grain-growth occurring with increasing annealing time. Figure 5b, after a 32 hr anneal, shows a very similar grain size to that in Fig. 5a but the domain structure is quite evident, which is most likely related to order/disorder effects as during the long anneal the structure would be expected to order, changing from relaxor to ordered ferroelectric character.

Table 1.   Slip formulation for tape-casting

Materials	PVB Slip (Volume%)
PMW-PT-PZ powder	15
Mixed organic solvents	69
Plasticiser	4.0
Dispersant	3.0
Binder	8.6

The development of ferroelectric domains has presumably been accompanied by a structural ordering process. The X-ray diffraction analysis, shown in Fig. 6, gives clear evidence of an ordering process with annealing time. For the disordered structure the lattice constant is ~4Å and the ordered structure the lattice constant is ~8Å.

3.2 Slips and Tapes

The precursor ceramic powder was calcined at 800°C for 4 hours giving a particle size (d_{50}) of ~1.5 µm. Table 1 gives a typical formulation for the tape casting slip which was developed in order to achieve a working viscosity of ~1–2 Pa.s and near Newtonian behaviour up to shear rates of 500 s^{-1}. Tape thickness uniformity was excellent, being about ±1 µm variation over a cast area of 430mm × 76mm. The specific Pt electrode ink was procured on the advice of the respective manufacturer in order to achieve compatibility with the PVB-based tape. Lamination was satisfactorily achieved as illustrated in Fig. 7, which shows the interdigitated electrode structure, uniform electrode layer separation and good green ceramic-electrode bonding.

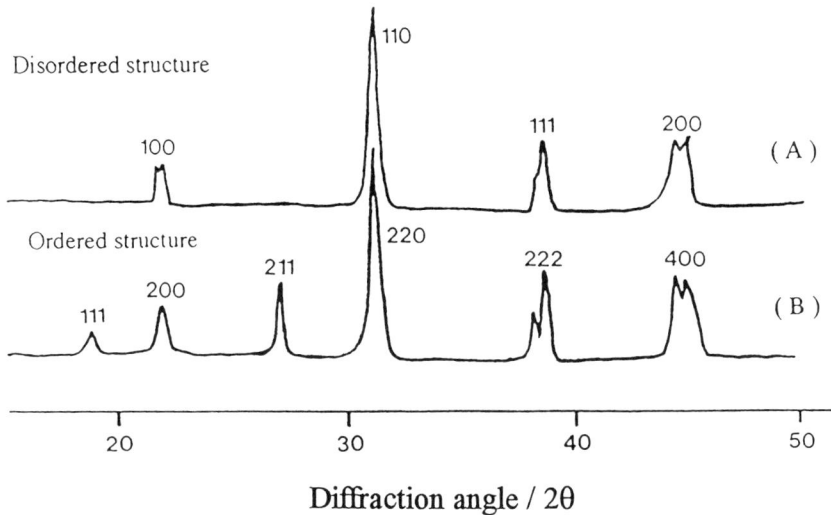

Figure 6.   XRD data for hot pressed 0.4PMW–0.4PT–0.2PZ (a) Annealed at 1050°C for 2h (b) annealed at 1050°C for 32h.

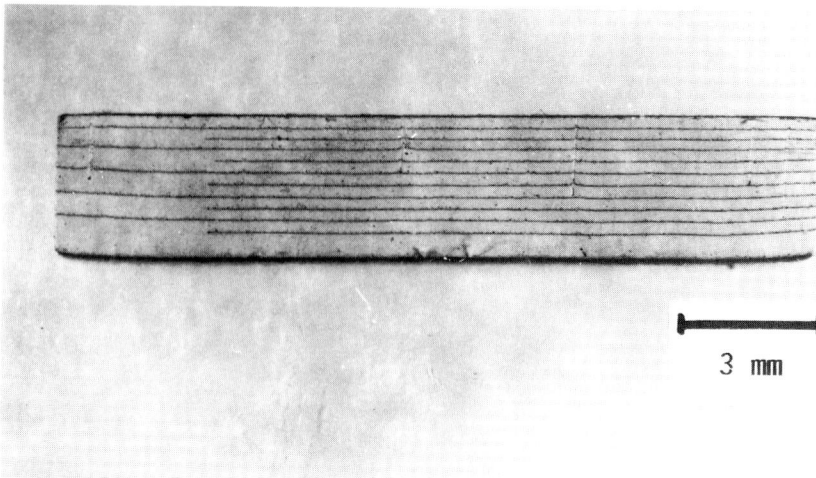

Figure 7.   Cross section of green 0.4PMW–0.4PT–0.2PZ ML structure.

After initial studies to determine the binder burnout characteristics good results were finally achieved for burnout and cofiring of the multilayer structures. Figure 8 shows a polished section through a prototype actuator, orthogonal to the electrode layers. The uniformity of the green macrostructure in Fig. 7 has been maintained, giving ~60 μm PMW–PT–PZ layers separated by ~5 μm thick Pt electrodes. The ceramic-electrode bonding appears to be good.

The displacement of the actuator as a function of electric field is shown in Fig. 9, the maximum displacement at 100 V being ~0.7 μm. The capacitance of the actuators,

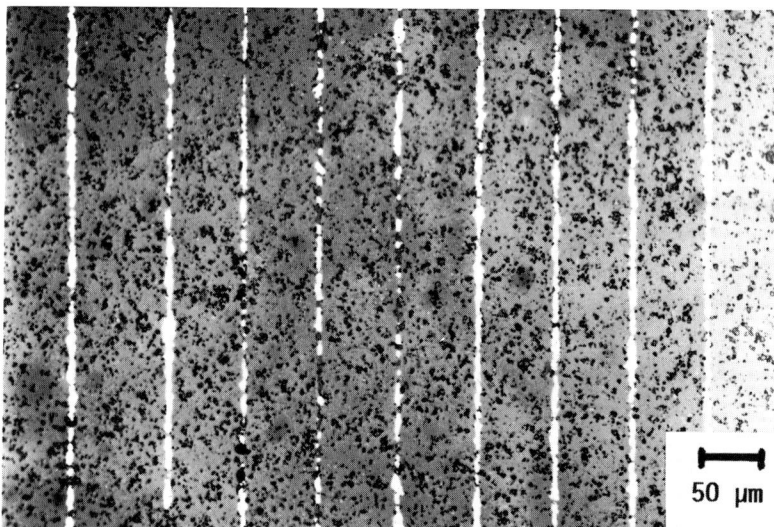

Figure 8.   Microstructure of prototype PMW–PT–PZ actuator.

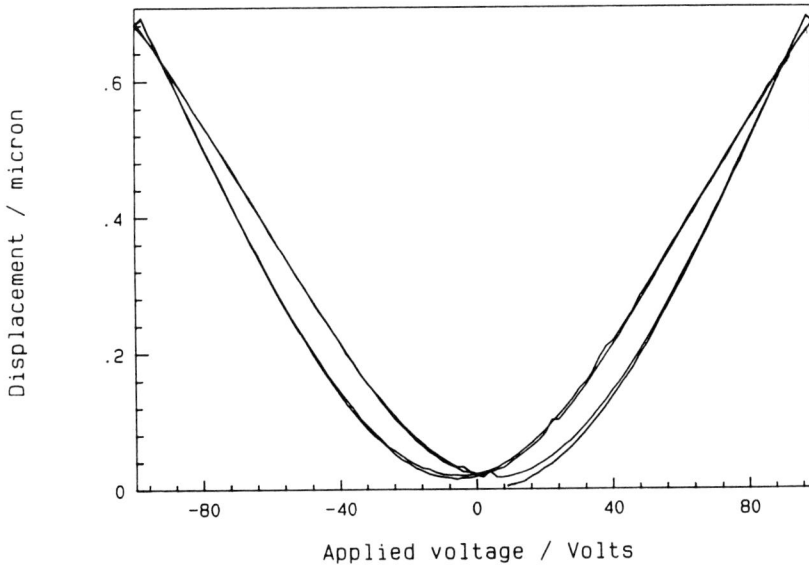

Figure 9. Displacement as a function of electric field for prototype PMW–PT–PZ actuator.

which had 20 layers of ceramic of thickness ~60 μm, was approximately 2 μF, and the dielectric loss approximately 7×10^{-4}.

4. CONCLUSIONS

The antiferroelectric (AFE) ceramic composition 0.4PMW–0.4PT–0.2PZ has been shown to be capable of producing useful strains of approximately 0.7% for an applied electric field of ~1.5 kV mm^{-1}. Hot pressing and annealing studies of this composition have demonstrated that the development of ferroelectric domains is accompanied by a structural ordering process and change of lattice constant. This work has confirmed that the PMW–PT–PZ system especially the composition 0.4PMW–0.4PT–0.2PZ is a candidate material for the fabrication of good quality mutilayer actuators which can be sintered at < 1000°C. Optimisation of processing could lead to the development of ML structures incorporating cheaper electrode materials and thinner active ceramic layers so that drive voltages could be achieved using low-cost microelectronic circuitry.

ACKNOWLEDGEMENTS

The authors would like to thank Dr. A.J. Bell for suggesting the study and Dr. A.J Moulson and Prof. R. Stevens for their helpful discussions in the preparation of the paper. One of the authors (S.S.) wishes to express his gratitude to the Thai Government for financial support.

REFERENCES

1. K. Uchino: *Am. Ceram. Soc. Bull.*, 1986, **65**(4), 647.
2. K. Abe, K. Uchino and S. Nomura: *Ferroelectrics*, 1986, **68**, 215.
3. D. Berlincourt: *Ferroelectrics,* 1976, **10**, 111.
4. S. Nomura and K. Uchino: *Ferroelectrics,* 1982, **41**, 117.
5. M. Yonezawa, M. Miyauchi, K. Utsumi and S. Saito: *High Tech Ceramics*, Vol. 2, P. Vincenzini ed., Elsevier Science Publishers, 1987, 1493.
6. N. N. Kranik and A. I. Agranovskaya: *Soviet Phys-Solid State*, 1960, **2**(1), 63.
7. S. Sirisoonthorn, A. J. Moulson and R. Stevens: *Third Euro-Ceramic, Vol. 2, Properties of Ceramics*, P. Duran and J.F. Fernandez eds, Faenza Editrice Iberica, Castellón de la Plana, Spain, 1993, 77.
8. S. Sirisoonthorn: '*Electric field-induced mechanical strains in the system* $Pb(Mg_{1/2}W_{1/2})O_3$–$PbTiO_3$–$PbZrO_3$', Ph.D. Thesis, Leeds University, 1993.
9. J. C. Williams: '*Treatise on Materials Science and Technology*', Vol. 9, F. F. Y. Wang, ed., Academic Press, 1976, 173.
10. D. Hind and P. R. Knott: '*Electroceramics: Production, Properties and Microstructures*', W. E. Lee and A. Bell, eds, The Institute of Materials, 1994, 107.

Planar Defects in Electroceramics

W. E. LEE, I. M. REANEY and M. A. McCOY

Department of Engineering Materials, University of Sheffield, Mappin Street,
Sheffield, S1 3JD, UK

ABSTRACT

The various forms of planar defects found in electroceramics are reviewed along with their image characteristics in the transmission electron microscope. Illustrative examples include stacking faults in $BaTiO_3$, antiphase domain boundaries in $Pb(Sc_{1/2}Ta_{1/2})O_3$, crystallographic shear planes in $SrTiO_3$, ferroelectric domain boundaries in $BaTiO_3$ and PLZT, and inversion domain boundaries in ZnO.

1. INTRODUCTION

Present understanding of the characteristic (extrinsic) properties of electroceramics governed by, for example, grain size, porosity and grain boundary structure, while by no means complete, is at a reasonably high level.[1] The effects of these relatively coarse microstructural features have been examined extensively in many systems using optical and scanning electron microscopes. However, studies of the influence of intragranular, planar (extended two-dimensional) defects on electroceramic properties are much more limited. The reasons for this are associated with the often secondary nature of the defects on properties and the need for complex transmission electron microscope techniques to fully characterise them. Nonetheless, in certain systems where planar defects have a critical effect on properties, such as the domain boundaries in ferro-electrics and piezoelectrics, they have also been examined in some detail.[2]

In the twenty first century it is hoped that processing techniques will improve so that coarse microstructural defects such as pores and inclusions will not occur. It is also likely that future developments in electroceramic devices will involve the use of ever thinner films and multilayers. In this case the influence of grain boundaries, particularly through the film/layer thickness, will be minimised. Planar defects will assume greater importance in the quest for even better properties since they are ubiquitous in electroceramic microstructures. In this paper we review the common types of planar defect and illustrate them with examples in electroceramic microstructures.

2. BACKGROUND

2.1 Types of Planar Defect

Three types of planar interface can be defined by considering the macroscopic crystallographic symmetry:[3]

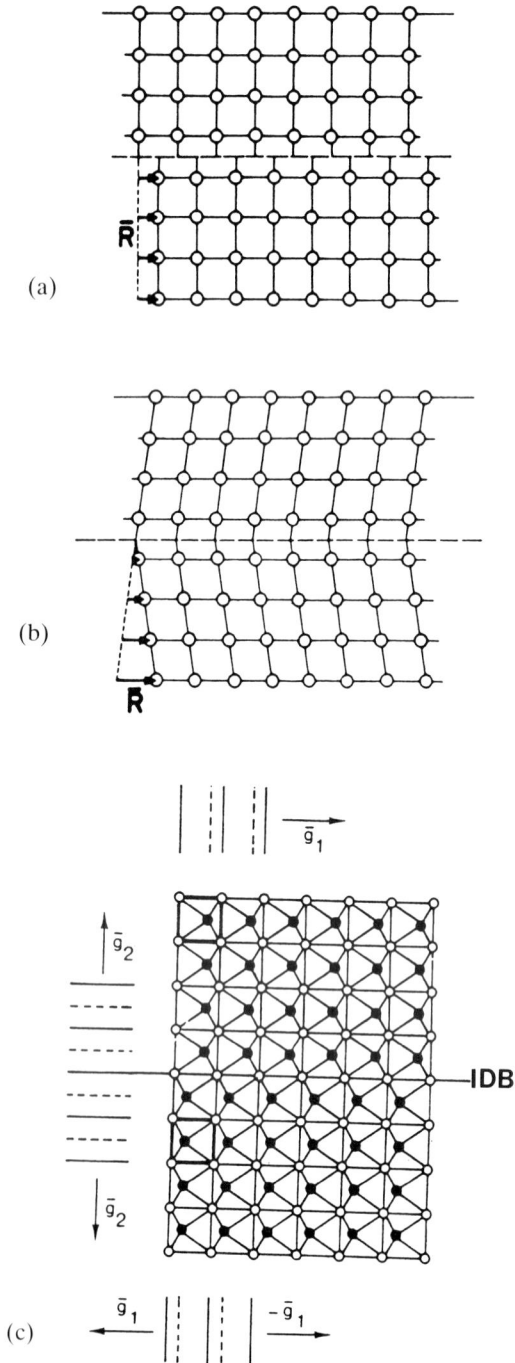

Figure 1. The three main types of planar interface, a) translation (e.g. SF's, APBs and CS planes), b) orientation (e.g. twins) and c) inversion (e.g. IDBs) interfaces.

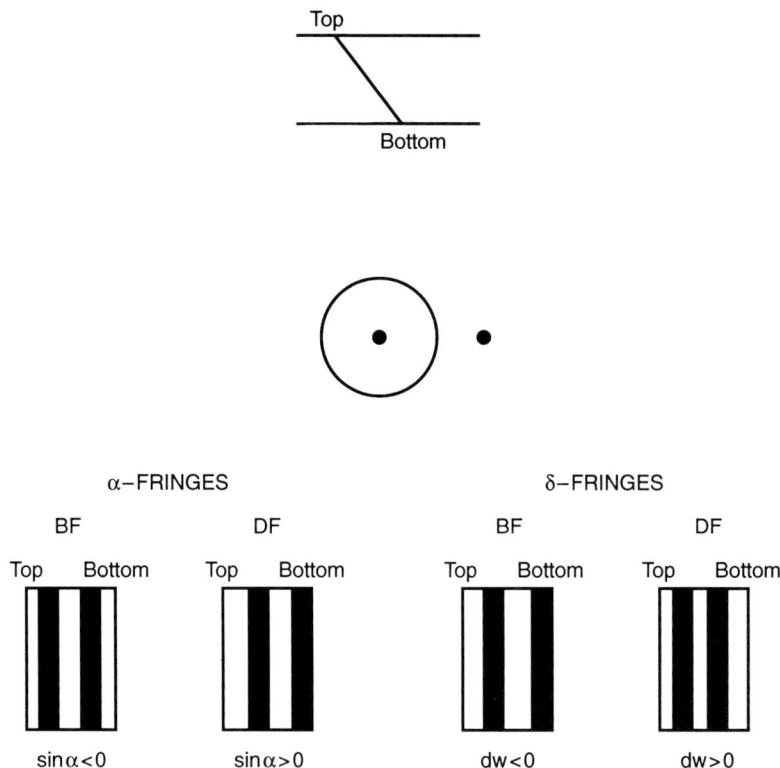

Figure 2.   Conditions for diffraction contrast imaging of fringes. The fault must be inclined in the foil and imaged at a two beam condition (shown in bright field). The resulting fringe contrast is shown for α and δ boundaries. BF = bright field, DF = dark field, w is a measure of the deviation from the Bragg condition.

2.1.1 Translation interfaces
(Fig. 1a) which separate two parts of a crystal (variants or domains) that are related to each other by a constant displacement vector $\underline{R}$, independent of the distance from the interface. Examples include stacking faults (SFs), antiphase domain boundaries (APB's) and crystallographic shear (CS) planes.

2.1.2 Orientation interfaces
(Fig. 1b) which separate variants which possess a displacement field in which $\underline{R}$ increases linearly with distance from the interface. These separate parts of a crystal at a different orientation e.g. in twin variants related by a point symmetry element such as a mirror plane or rotation axis giving reflection and rotation twins respectively.

2.1.3 Inversion interfaces
(180° boundaries or inversion twins) which separate two domains related to one another by an inversion operation (Fig. 1c), e.g. inversion domain boundaries (IDBs).

2.2 TEM Analysis of Planar Defects

While there are many TEM techniques available for examination of planar defects in a perfect matrix (see e.g. ref. 4) only diffraction (fringe) and phase contrast techniques will be considered here.

To examine a defect which is inclined in the thin TEM specimen (Fig. 2) two-beam conditions are used with one set of planes at the Bragg condition. Interference between the electrons diffracted above and below the plane of the defect gives rise to fringes whose characteristics define the nature of the defect.[5] A pure translation or inversion across the boundary results in α fringes due to the extra phase factor ($\alpha = 2\pi g.R$, where g is the vector defining the plane at the Bragg condition) introduced by the defect. In the general case in which $\alpha \neq \pi$ the fringes are symmetric (i.e. the outer fringes are either both black or both white) in bright-field images but asymmetric (one black and one white outer fringe) in dark field. Examples of planar defects exhibiting α fringes include SF's, APB's and IDBs. For the special case $\alpha = \pi$ fringes in both BF and DF images are symmetric and pseudo-complementary. A misorientation (translation plus rotation, rotation or mirror) across the boundary results in δ fringes due to interference arising from a difference in g or s, the deviation from the Bragg condition. In this case the fringes are asymmetric in bright-field images but symmetric in dark field. Additionally, δ-fringe defects may give rise to spot splitting in diffraction patterns, a feature not

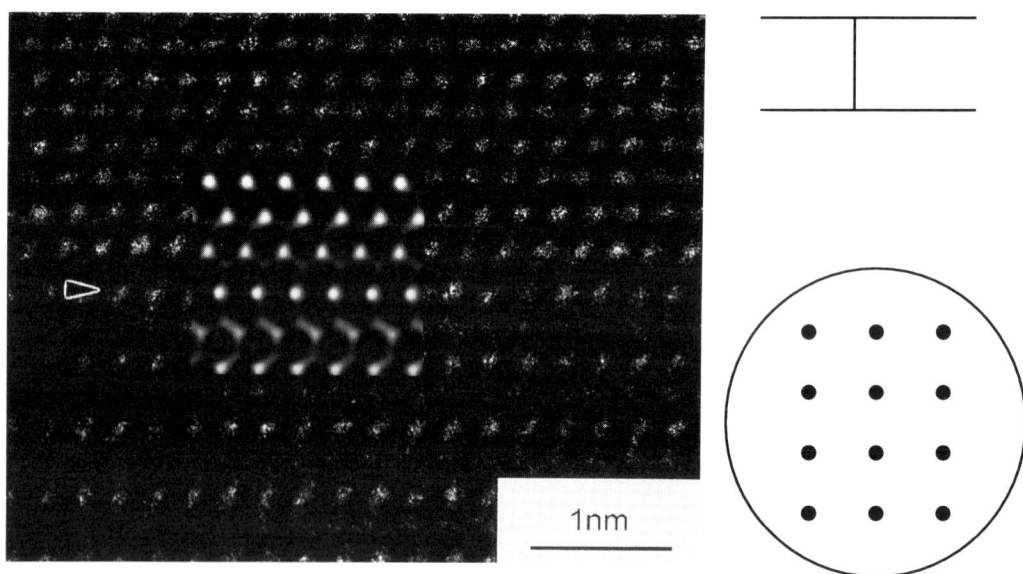

Figure 3.   Conditions for phase contrast imaging in the HREM. The fault must be parallel to the incident electron beam. A large objective aperture is used to allow the beams to interfere with each other to produce a structure image of rows of atoms in the crystal projection. The resulting image shows an IDB (arrowed) in Sb-doped ZnO with the calculated image as an insert.

caused by defects giving rise to α-fringes. Planar defects giving rise to δ fringes are orientation domain boundaries (ODBs) including twins.

To examine a defect which is parallel to the incoming electrons (Fig. 3) many Bragg diffracted beams are allowed to interfere with each other and the transmitted beam to give a phase contrast or structure image of rows of atoms in the crystal. Interpretation of these HREM (High Resolution Electron Microscopy) images is complex requiring comparison of images calculated from image contrast theories to those obtained in the microscope.[6]

The three general types of planar defect will now be considered in more detail and illustrated with electroceramic examples.

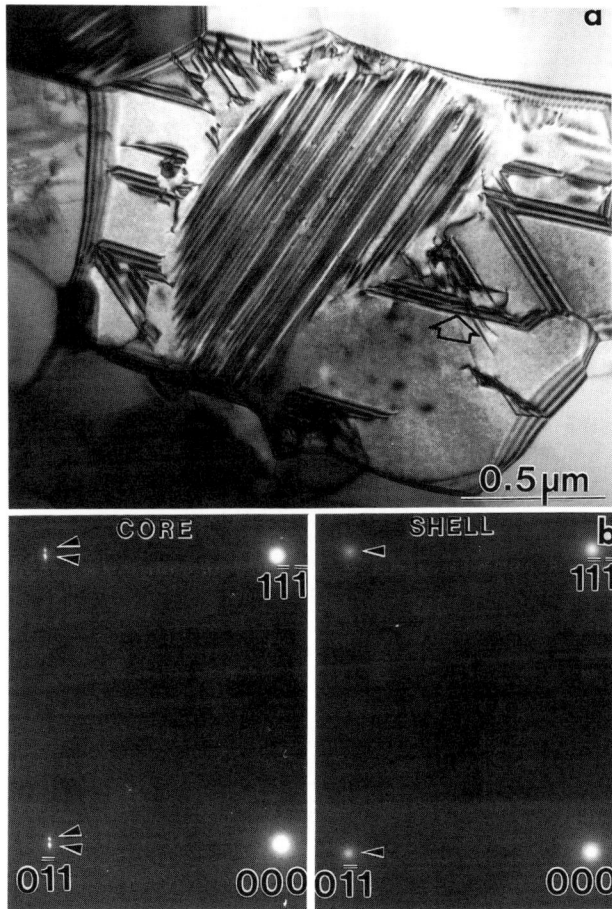

Figure 4a). Bright-field TEM image showing α fringe contrast from SFs (e.g. arrowed) in the Nb_2O_5/Bi_2O_3 containing cubic shell of a single $BaTiO_3$ grain. The pure $BaTiO_3$ tetragonal core region contains 90° ferroelectric domains. b) [211] zone axis diffraction patterns from the core and shell revealing the spot splitting characteristic of ODB's (δ boundaries).

3. TRANSLATIONAL DEFECTS

The main translational defects are stacking faults, antiphase domain boundaries and crystallographic shear planes.

3.1 Stacking Faults

A SF is a planar defect where the regular stacking sequence of the lattice of a single grain has been interrupted. They are generally straight and can form when a partial dislocation (with a Burgers vector $\underline{b}$ which is some fraction of a lattice repeat) traverses a crystal (glides), during crystal growth, or on a phase transformation from one layer structure to a different structure made up of similar layers (a polytypic transition). The simplest type of SF involves a displacement of the structure by a displacement vector $\underline{R}$ which lies in the fault plane. SF's are most common in structures which can be described in terms of layers (SiC, mica, pyroxenes, ZnS (wurtzite or zinc blende structures)). Often in a ceramic the $\underline{R}$ may leave some aspect of the structure unfaulted so that e.g. the SF may refer only to the cation sublattice.

When SF's occur singly they are clearly describable as planar defects. However, when they are periodic they may become part of the structural principle describing some layering sequence, leading to *polytypes* i.e. compounds with the same chemical composition but different stacking sequences of the same layers e.g. SiC.[7] Figure 4a shows a stacking fault in the shell region of a single grain in a Nb_2O_5/Bi_2O_3-doped $BaTiO_3$ temperature stable dielectric. The temperature stability is conferred by the heterogeneous core-shell microstructure in which the shell region is cubic paraelectric containing the dopants while the core is pure, tetragonal ferroelectric $BaTiO_3$[8].

3.2 Antiphase Domain Boundaries

In general, the term SF is applied to crystal defects which can be constructed from stacked planes and so possess specific values of $\underline{R}$; an APB is a translation interface

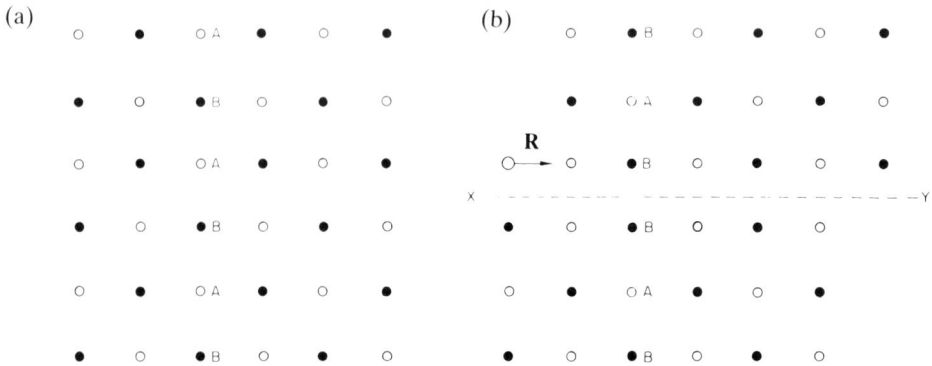

Figure 5a). Perfect crystal with ABAB stacking and b) containing a conservative APB at XY with displacement vector R in the boundary plane. Note the ABAB stacking sequence is now broken with like atoms occurring as neighbours so changing the chemistry at the APB.

with an R equal to a vector connecting different atom species in the unit cell. In APBs the R is usually parallel to the fault plane and the defect is generally curved. SF's occur only on close-packed planes; APBs can, in principle, occur on any plane. They can also accommodate nonstoichiometry within the boundary region and can form between ordered and disordered structures as well as between two ordered structures. They occur in ordered ceramics when there is a change in the identity of the atom at a given lattice point (Fig. 5) but there is no atomic stacking change as in a SF. More commonly in perovskite electroceramics APB's arise from the impingement of two regions of octahedral tilting which have nucleated out of phase.[9]

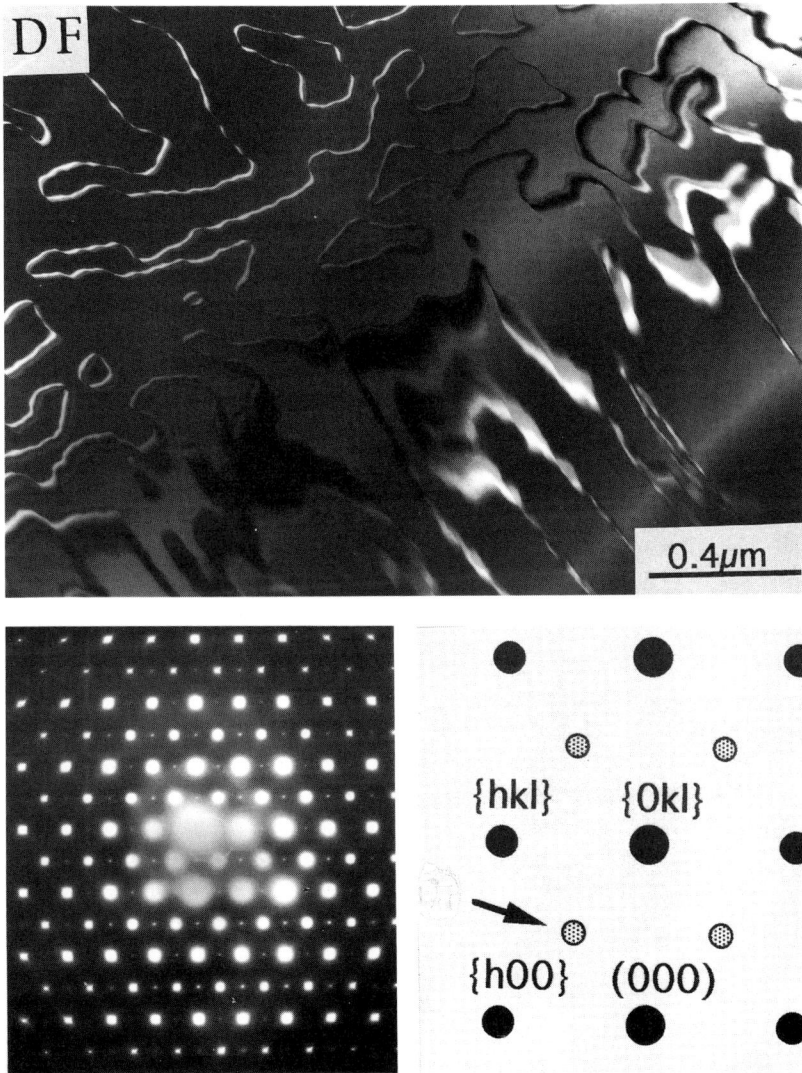

Figure 6.   Dark-field TEM image of APBs in annealed $Pb(Sc_{1/2}Ta_{1/2})O_3$ imaged using the arrowed superlattice reflexion. The zone axis diffraction pattern is [110] and the superlattice reflexion $h\pm1/2$ $k\pm1/2$ $1\pm1/2$.

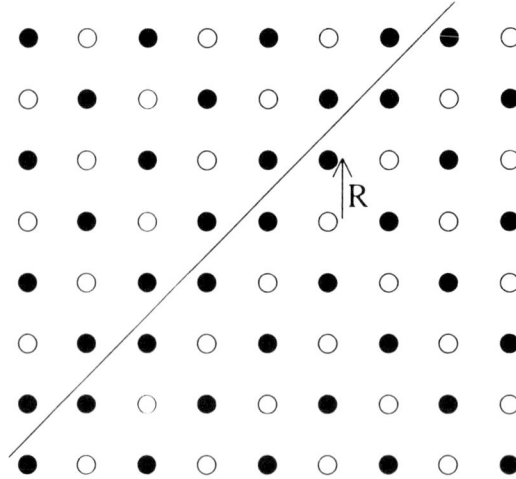

Figure 7.   A non-conservative APB with $\underline{R}$ not parallel to the boundary plane. This defect leaves a row of black atoms either side of the boundary and is termed a CS plane in many nonstoichiometric compounds.

APB's may arise during a phase transition when a crystal transforms from a higher to a lower symmetry retaining a similar structure but losing some translational symmetry element. In this case they represent the translational variants associated with a reduction in translational symmetry, $\underline{R}$ across the APB is a lattice vector lost during the transformation. They may also form either on crystallisation where two growing crystals meet at the APB, during ordering (either substitutional ordering or ordering of vacancies) or by glide.

When APB's are present in a grain while the basic structure remains the same it is modified to reduce the translational symmetry. Consequently, the fundamental diffraction pattern is altered by the addition of extra reflexions. These superlattice reflexions are used to image APB's in dark field. Figure 6 shows such a dark-field image revealing APB's in $Pb(Sc_{1/2}Ta_{1/2})O_3$ associated with Sc/Ta ordering. SF's and APB's can be distinguished because SF's produce fringe contrast for certain fundamental reflexions but are out of contrast for all superlattice reflexions. On the other hand APB's produce fringe patterns for certain superlattice reflexions and are out of contrast for all fundamental reflexions. APB's are not usually imaged using HREM since they are curved making exact orientation of the defect parallel to the electron beam difficult.

APB's are *conservative* if $\underline{R}$ is in the boundary plane (Fig. 5) but if $\underline{R}$ is not parallel to the fault plane the local chemical composition around the fault is different from that in the bulk structure. In this case the APB is termed *non-conservative* (Fig. 7). The name crystallographic shear (CS) plane is given to such interfaces in many non-stoichiometric ceramic oxides.

3.3 Crystallographic Shear Planes

CS planes change the stoichiometry of the crystal and correspond to regions of a second phase intergrown with the parent phase. They may be as small as half a unit

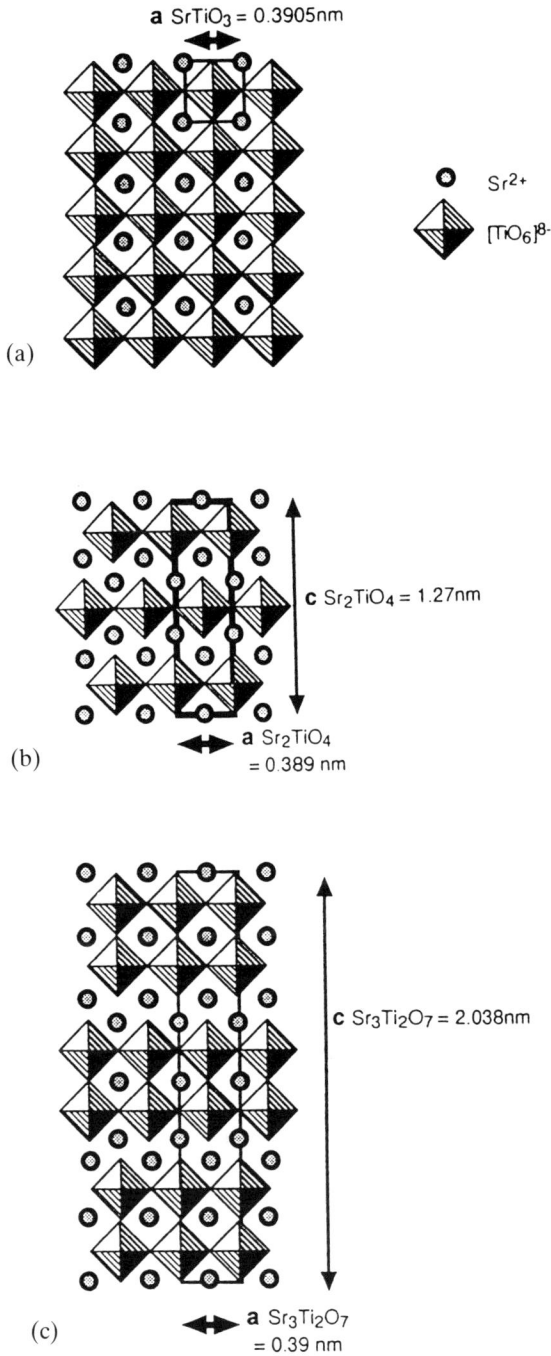

Figure 8a) Perovskite $SrTiO_3$, b) Sr_2TiO_4 and c) $Sr_3Ti_2O_7$. Ruddlesden-Popper defect structures of b) and c) in SrO-rich $SrTiO_3$ are generated by removing an (001) sheet of TiO_6 octahedra from $SrTiO_3$ and shearing the layer above by 1/2 [111]. The general formula of the phases is $Sr_{1+n}Ti_nO_{3n+1}$, where n is the number of perovskite layers between defects.

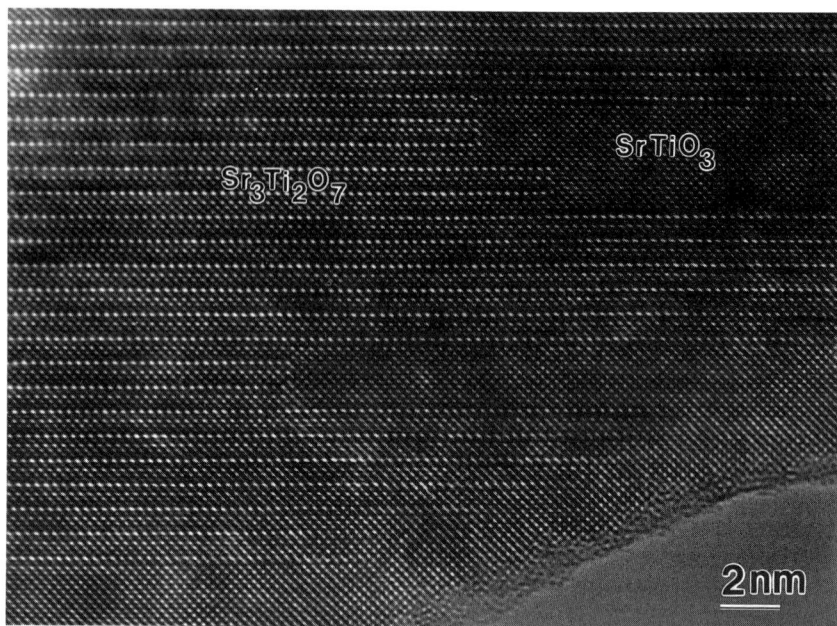

Figure 9.   HREM image at Scherzer defocus in a 6nm thick sample imaged down [010].
The repeat unit along [001] $Sr_3Ti_2O_7$ (2.038nm) is twice the width of the bright white dots
(Sr atom positions).

cell wide for a single fault although they often occur in regular sequences. The
stoichiometry varies over a range governed by the spacing between the CS planes i.e.
the proportion of parent phase. They have been studied extensively by solid state
chemists in accommodating oxygen deficiency in tungsten, niobium and titanium
oxide systems.[10–12] Depending on details of the defect and system these may also be
termed Wadsley defects, Magneli phases or Ruddlesden-Popper defects. Wadsley
defects are randomly spaced CS planes.

CS planes also occur in electroceramics such as $SrTiO_3$.[13] $SrTiO_3$ is a cubic perovskite (Fig. 8a) while Sr_2TiO_4 is tetragonal and made up of interleaved SrO (cubic
rock salt) and $SrTiO_3$ (Fig. 8b). Similarly, $Sr_3Ti_2O_7$ is comprised of double $SrTiO_3$
slabs and SrO (Fig. 8c). These can be regarded as members of a homologous series of
oxides of general formula $Sr_{n+1}Ti_nO_{3n+1}$. Such phases in this system are termed
Ruddlesden-Popper defects. Similarly, a homologous series of oxides exists between
Ti_2O_3 (corundum type) and TiO_2 (rutile) producing Magneli phases of composition
Ti_nO_{2n-1}. In SrO-rich $SrTiO_3$ nonstoichiometry is accommodated by formation of
Ruddlesden-Popper slabs whereas TiO_2-rich $SrTiO_3$ forms Ti_nO_{2n-1} Magneli phases
at grain junctions.[14] Figure 9 shows a HREM image of slabs of $Sr_3Ti_2O_7$ in a $SrTiO_3$
parent phase in a SrO-rich $SrTiO_3$ composition.[15]

4. ORIENTATION INTERFACES

4.1 Orientation Domain Boundaries (ODB's)

Where the two parts of the crystal separated by the defect are at different orientations δ fringes arise in bright-field images under two beam conditions along with spot splitting in zone axis diffraction patterns. Twin boundaries, a subgroup of ODB's, separate domains which are related by a point symmetry element i.e. a mirror or a rotation axis. The two general types of twin are *reflection or type I* twins and *rotation or type II* twins. Twin boundaries form during phase transitions, growth or deformation. Twin domains arising from phase transitions represent the *orientational variants* associated with a reduction in the point group symmetry, and the geometric relationships between the domains are defined by the symmetry element lost during the transformation.

A number of ceramics have polar crystal structures and may form regions within a single grain in which the polarisation vectors are in different orientations (domains). Such domains typically arise during ferroelectric phase transitions. If an order parameter Q is defined relating the free energy of the high temperature phase to that of the low temperature phase, then Q may be a function of polarisation (P) and strain (e). A purely ferroelectric phase transition is only associated with the loss of a centre of symmetry (e.g. ($R\bar{3}c$ to R3c in $LiNbO_3$) and the presence of P. This may only lead to formation of 180° domains which are related by an inversion of the polar axis (i.e. IDB's). In this case domain formation lowers the overall free energy by reducing the net electric field of the grain to zero.

Displacive ferroelectric phase transitions (e.g. cubic $Pm\bar{3}m$ to tetragonal P4mm $BaTiO_3$) not only involve the loss of centrosymmetry but also a change in the point group symmetry, thus permitting formation of ODBs. Unlike 180° domains they are present to relieve the stress associated with clamping of the spontaneous strain (e) that appears at the phase transition (i.e. the cubic to tetragonal distortion in $BaTiO_3$ when a,b ≠ c). However, because e and P are coupled, there is still a change in the polarisation vector across the ODB and so they are referred to as ferroelectric rather than ferroelastic domains. In ferroelastic domains there is only a change in the direction of strain (e) across the ODB. Depending on the crystal structure of the low temperature phase, the change in e and P across the ODBs may be through 90° (tetragonal and orthorhombic) or 71/109° (rhombohedral). When the c/a ratio, in e.g. a tetragonal crystal, is large, spot splitting may be observed in a diffraction pattern taken from at least two 90° domains. In $BaTiO_3$ such 90° domains (also termed deformation twins) lead to clear spot splitting in diffraction patterns. Figure 4b shows diffraction patterns from the core and shell regions of the $BaTiO_3$ grain in Fig. 4a. The spot splitting associated with the 90° domains in the grain core is evident.

The effect of composition on the domain structure in PLZT's heat treated under identical conditions is illustrated in Fig. 10 which shows microstructures from an x/65/35 PLZT where x is the mol % La[16]. With x = 8 the typical relaxor ferroelectric microstructure of nanometre scale regions of long range order is visible (Fig. 10a). With x = 7 much coarser domains occur; the microstructure is transforming to that of a typical ferroelectric ceramic (Fig. 10b). With x = 5 coarse ferroelectric domains are visible with the 109° boundaries in this rhombohedral phase exhibiting δ fringes (Fig. 10c).

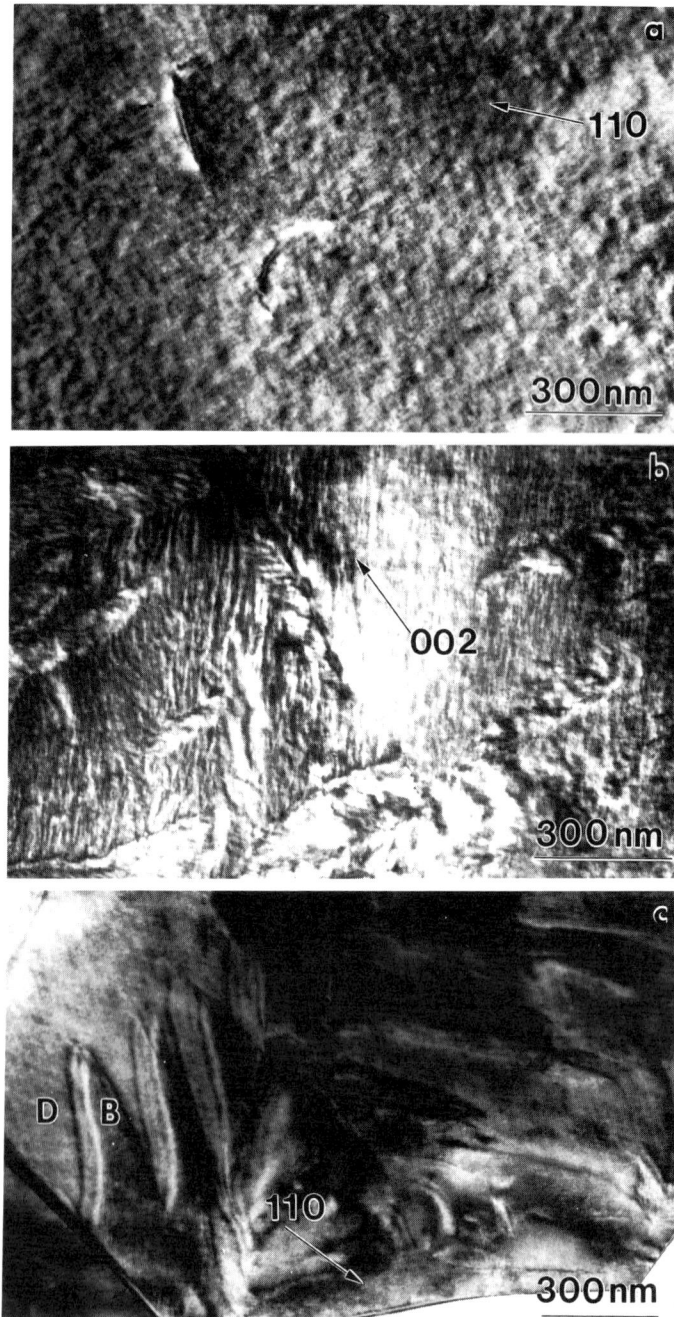

Figure 10.   Bright-field TEM images of domains in x/65/35 PLZT with a) 8 mol% La, b) 7 mol % La and c) 5 mol % La. The gradual transition from a relaxor microstructure (a) to ferroelectric with δ boundaries (c) is apparent.

5. INVERSION INTERFACES

5.1 Inversion Domain Boundaries

In this case the lattice is the same on both sides of the boundary but the structure is inverted. IDB's occur when there is a *polarity* reversal across the interface i.e. *an inversion*. IDB's are sometimes called inversion twin or 180° boundaries and can only occur in noncentrosymmetric crystal structures. Although they may in many cases form spontaneously (as in quartz and $BaTiO_3$) they have been studied most extensively in AlN where they are associated with the presence of O impurity[17] and in ZnO where they are associated with Sb doping.[18] In fact IDB formation may be a simple method of removing nonstoichiometry from the lattice as a result of the presence of such impurities. The left hand part of Fig. 3 is an HREM image of an IDB in Sb-doped ZnO. The calculated HREM image (insert) confirms the presence of a Zn/Sb spinel monolayer at the boundary. Since most grains in ZnO varistor microstructures contain an IDB and Bi is known to segregate to them this may have significant commercial implications since the varistor response is governed by the presence of Bi at grain boundaries.

6. SUMMARY

The various forms of planar defect commonly found in electrical ceramics and their contrast mechanisms in the transmission electron microscope have been reviewed. It is likely that planar defects will assume greater importance in the future, in particular in thin film and multilayer electroceramic devices.

REFERENCES

1. L. M. Levinson and S. Hirano (eds): *Grain Boundaries and Interfacial Phenomena in Electronic Ceramics*, ACerS, Columbus, Ohio, USA, 1993.
2. L. A. Bursill and P. Ju Lin: *Chemistry of Electronic Ceramic Materials*, P. K. Davies and R. S. Roth, eds., Technomic, Pennsylvania, USA, 1990, 67–77.
3. S. Amelinckx and J. Van Landuyt: *Diffraction and Imaging Techniques in Materials Science*, S. Amelinckx, R. Gevers and J. van Landuyt, eds., N. Holland, Amsterdam, Holland, 1978, 107–151.
4. D. B. Williams: *Practical Analytical Electron Microscopy in Materials Science*, Philips, Eindhoven, Holland, 1984.
5. S. Amelinckx and J. van Landuyt: *Electron Microscopy in Mineralogy*, H-R Wenk, ed., Springer-Verlag, Berlin, Germany 1976, 68–112.
6. P. Busseck, J. Cowley and L. Eyring (eds.): *High Resolution Transmission Electron Microscopy and Associated Techniques*, OUP, Oxford, UK, 1992.
7. A. Putnis: *Introduction to Mineral Sciences*, CUP, Cambridge, UK, 1992.
8. S. Pathumarak and W. E. Lee: 1994, *Brit. Cer. Trans.*, **93** (3), 114–118.
9. E. Colla, I. M. Reaney and N. Setter: 1993, *J. Appl. Phys.* **74** (5), 3414–3425.
10. A. D. Wadsley and S. Andersson: *Perspectives in Structural Chemistry Vol. III*, J. D. Dunitz and J. A. Ibers, eds., Wiley, London, UK, 1970, 1–59.

11. B. G. Hyde and L. A. Bursill: *The Chemistry of Extended Defects in Non Metallic Solids*, L. Eyring and M. O'Keefe, eds, N. Holland, Amsterdam, Holland, 1970, 347–374.
12. S. Iijima: 1975, *J. Solid State Chem.*, **14**, 52–65.
13. R. J. D. Tilley: 1977, *J. Solid State Chem.*, **21**, 293–301.
14. M. Fujimoto and M. Watanabe: 1985, *J. Mat. Sci.*, **20**, 3683–3690.
15. M. A. McCoy, R. W. Grimes and W. E. Lee: 1995, *Proc. 4th Euro. Ceramics*, **5**, 451–458.
16. M. A. Akbas, I. M. Reaney and W. E. Lee: submitted to *J. Mat. Res.* 1995.
17. A. D. Westwood, R. A. Youngman, M. R. Mecartney, A. N. Cormack and M. R. Notis: 1995, *J. Mat. Res.* **10** [5] 1270–1286.
18. M. A. McCoy, R. W. Grimes and W. E. Lee: Submitted to *J. Mat. Res.*, 1995.

Non-Linear Piezoelectric Metrology:
Coefficients and Fatigue under Impact

J. A. CLOSE and R. STEVENS

School of Materials, University of Leeds, Leeds, LS2 9JT, UK

ABSTRACT

Samples of poled PZT piezoceramics were subject to longitudinal, compressive impact loading by a free falling striker. Compositions under test were: PC4*, PC4A*, PC5*, PC5H*, Pz-26** and Pz-27**. Samples were sandwiched in-between a Split Hopkinson Pressure Bar instrumented with foil strain gauges linked to a high frequency amplifier. The apparatus produced planar strain waves, ranging from 32 to 82 μs in duration and varying in magnitude between 2 and 40 MPa. The piezoceramic was discharged under short-circuit conditions and from these data, values of the piezoelectric charge coefficient, d_{33} were derived and compared with those obtained via the conventional low frequency (100 Hz), low stress (10 kPa) resonance method. The results obtained from the impact rig gave values for the d_{33} coefficient that are up to twice those gained from the resonance rig.[1] It is apparent that there exists a relationship between the impact stress and the piezoelectric charge coefficient of the material i.e., the larger the stress, the larger the d_{33} value. The degree of d_{33} variation with stress is related to the ferroelectric hardness/softness i.e., coercivity of the material.

Ageing of the piezoelectric device under stress was studied by means of plotting d_{33} as a function of the number of impacts suffered. The compositions displayed a drop in the values of d_{33} of between 18 and 49% when subject to 5000 impacts of approximately 40 MPa in magnitude. These results are related to the ferroelectric coercivity of the material.

As a control and reference exercise, a sample of X-cut α quartz was subjected to the same tests as the PZT compositions. An examination using scanning electron microscopy was carried out to determine if there was any microcracking present following impact.

It is evident that the mobility of ferroelectric domains and therefore the electric current they produce can be correlated with the stress to which it is subjected and that repeated impacting of these materials leads to rapid depolarisation and a deterioration of the piezoelectric properties.

1. INTRODUCTION

Recent years have seen many new applications for piezoelectric devices; impact detonation devices requiring very rapid, large voltage pulses, high speed sensors and actuators such as those utilised in active automobile suspension systems and ignition devices for gas appliances and flash bulbs. In many of these applications, the piezoelectric device is subject to relatively large stresses of short duration, conditions to which conventional metrology techniques do not lend themselves. Currently, standard characterisation methods subject the piezoelectric device to cyclic stresses

* Supplied by Morgan Matroc, Unilator Division, Ruabon, Wrexham, Clwyd. LL14 6HY.
** Supplied by FerroPerm, Piezoelectric Ceramics Division, Hejreskovvej 6, DK-3490 Kvistgård, Denmark.

of very low magnitude at low frequencies or specific resonance conditions. Manufacturers data relating the performance of their materials is becoming less relevant as they are employed under conditions far removed from those under which they were characterised. It has therefore become necessary to develop a technique whereby piezoelectric device characteristation can take place under conditions relevant to today's demanding applications.

2. EXPERIMENTAL TECHNIQUE

The 'drop weight' impact testing rig, illustrated below comprises a free falling striker bar, running on an air bearing that provides a compressive impact that can be varied between 32 and 82 µs in duration and of magnitude 2 to 40 MPA. The piezoelectric device under test (DUT) is situated in a vertically mounted, EN24 steel Split Hopkinson Pressure Bar (SHPB) that consists of a top 'impact' bar and a lower 'receiver'

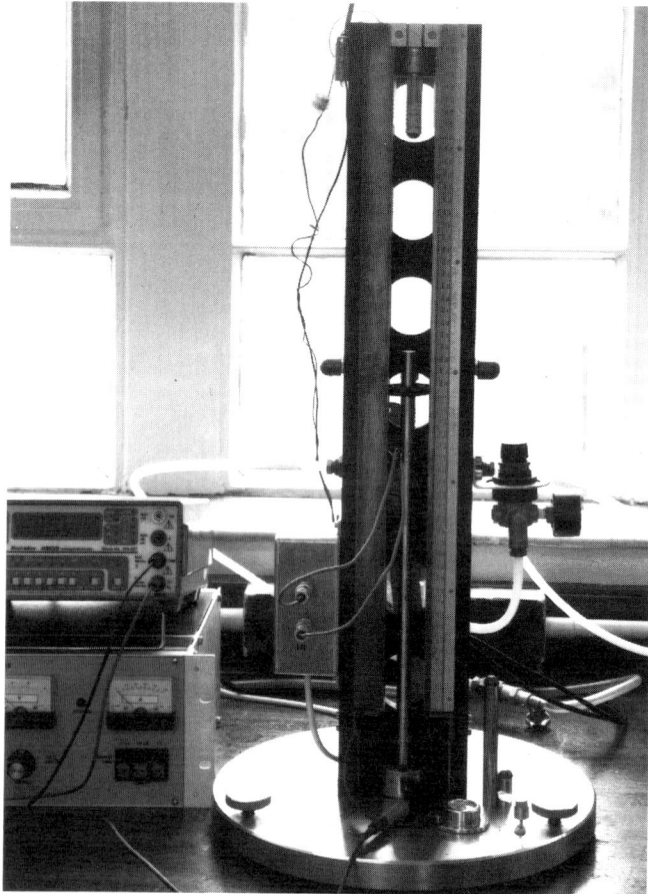

Figure 1.   'Drop weight' impact test rig.

bar. The SHPB arrangement is instrumented with four miniature foil strain gauges that monitor the passage of the impact induced stress wave through the receiver bar following the impact of the striker bar on the impact bar and its passage through the DUT. The strain gauge signal received from the SHPB is sent to a custom made high frequency (2 MHz bandwidth) virtual-earth pulse amplifier*. The amplifier output is then appropriated for analysis by a computer controlled digital storage adapter (DSA).

The DUT was mounted with the poling direction anti-parallel to the direction of wave propagation on a 12 mm diameter EN24 steel stub by means of silver loaded, conducting epoxy resin serving as the negative electrode of the sample. The positive electrode of the DUT was a fillet of aluminium foil, 20 μm thick on the top of the device. The piezoceramic was discharged under short circuit conditions by applying a large capacitance in parallel to the DUT. The charge held on the load capacitor was then read into the DSA, simultaneously with the signal from the strain gauge amplifier.

As a control and reference experiment, results taken from the PZT compositions were compared with those from samples of X-cut α quartz.

3. INTERPRETATION OF RESULTS

From the captured data, a calculation is made of the piezoelectric charge coefficient (d_{33}/pC N^{-1});

$$d_{33} = \left(\frac{dD}{dX}\right)_{E,T} \tag{1}$$

where D is the dielectric displacement, X is the stress, the subscripts E and T denote constant (zero) field and constant temperature respectively; whilst '33' indicates the direction of the applied stress and that of the current flow from the DUT ('3' being the polarisation direction in the ceramic device).

4. UNDERLYING THEORIES

The theories presented below are based on the assumption that the system obeys the laws of linear elasticity i.e., that initial and final body configurations are identical. The duration of the impact is dependent on the time taken for an (acoustic) stress wave, travelling at a speed C_0 (where C_0 is the *longitudinal* acoustic wave velocity in the material equal to the square root of the material's Young modulus, E, divided by its mass density, ρ), to traverse the striker bar twice[2] at which time the tensile release wave extricates the striker bar from the top of the impact bar. The rise time of the stress wave is governed by the geometry of the striker bar facing. When the impact is served by a hemispherical striker bar, the stress wave originates from point source,

---

* Designed and built by the author and Mr. Rodney Holt, University of Leeds.

producing a radial wave, but as this propagates down an impact bar of specific aspect ratio, the stress wave assumes a planar, uniaxial profile.[3]

From the signal picked up by the strain gauges on the receiver bar, a measure of the stress put upon the DUT can be derived. The axial stress on the DUT, X_s is directly related to the engineering strain, x_r measured in the receiver bar,[4] i.e.,

$$X_s = \frac{A_r}{A_s} E_r x_r \qquad (2)$$

where A_s is the area of the cross section of the sample and A_r is that of the receiver bar whose Young modulus is E_r. The value of x_r is derived from a function involving the peak voltage of the recorded strain wave, the current in the strain gauges, the strain gauge factor and the degree of amplification provided.

5. RESULTS

Measurements were made of the d_{33} charge coefficient on samples of each material 2 mm and 5 mm thick and 10 mm in diameter using the conventional low frequency (100 Hz), low stress (10 kPa) method i.e., a Berlincourt meter. These data appear in the table below, together with the value quoted in the manufacturer's data tables.

Material	Pz26	Pz27	PC4	PC4A	PC5	PC5H	Quartz
Quoted d_{33}/pC N^{-1}	275	375	287	296	409	529	2.27
Measured d_{33}/pC N^{-1}	249	398	217	229	340	486	2.52

In view of the limited space available here, the results of the coefficients taken under impact that appear below are taken from only two of the six PZT compositions along with those of the quartz reference. The two materials selected, Pz-26 and PC5H give a good representation of the results that can be expected from a 'hard' and a 'soft' ferroelectric material respectively. As can be seen from the plots below of d_{33} as a function of X_s, the PZT piezoceramics exhibit charge coefficients substantially higher than those taken under low frequency, low stress conditions and they show an essentially linear increase in the value of d_{33} with increasing impact stress. The 'hard' Pz-26 material exhibits a much lower variation in the value of the charge coefficient with impact velocity than does the 'soft' PC5H material. It should be noted that when the Pz-27 piezoceramic was impacted with the largest and most massive striker bar, there was a substantial drop in the recorded readings of d_{33}. Subsequent readings re-taken for the striker bars of lower mass all displayed a similar drop in d_{33} values of approximately 20%. This result is indicative of the fact that the onset of depolarisation had begun in the DUT at this relatively low stress magnitude. In contrast to the PZT materials, the quartz proved to have a relatively static response under varying impact stresses.

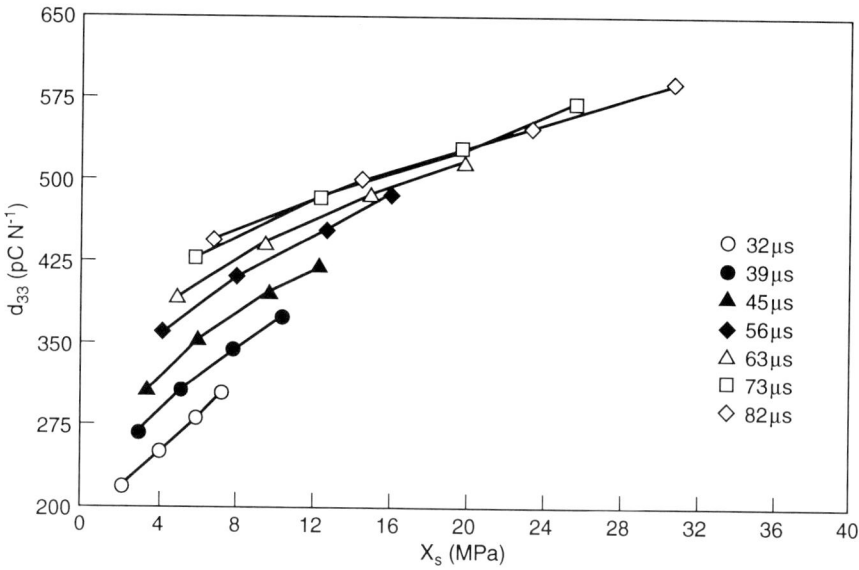

Figure 2. d_{33} as a function of impact stress: Pz-26[5]

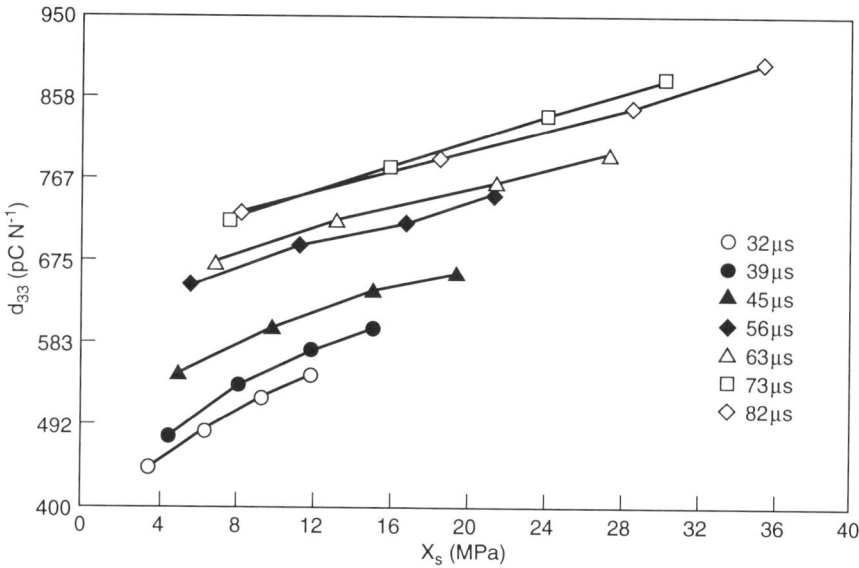

Figure 3. d_{33} as a function of impact stress: PC5H[5]

The results obtained from the impact rig gave values of the d_{33} coefficient that are up to twice those provided by the conventional test method.[5] It is apparent that there exists a relationship between the impact stress magnitude and the piezoelectric charge coefficient (and also the electromechanical coupling coefficient), i.e., the larger the stress, the greater the d_{33} and k values, provided that depolarisation has not occurred. It is also apparent that, the shorter the duration of the impact stress wave, the lower the coefficient values, i.e., there is a frequency dependence for ferroelectric domain reorientation. The degree of variation of d_{33} and k with stress and impact duration is related to the ferroelectric softness/hardness of the ceramic; the softer materials exhibiting the greatest dependence on these experimental parameters. Evidently, it is possible to produce a wide array of piezoelectric coefficient values by applying specific stressing conditions. Examination of the d_{33} versus impact stress curves reveals that the periphery of the plotted lines traces a curve not dissimilar to the cusp of a hysteresis curve of polarisation under applied field. Since the hysteresis loop of polarisation under applied field is associated with the characteristic (composition dependent) domain reorientation behaviour, it can be seen that there is a direct analogy with the domain behaviour and therefore charge flow from the material under stimulation by an applied stress. As the magnitude and period of the stress field applied increases, the degree of domain realignment increases in a manner according to the stability of each type present. The non-linear domain reorientation arises as each individual domain is subject to a different component of the stress field according to its relative crystallographic orientation.

6. FATIGUE BEHAVIOUR UNDER IMPACT

Samples were subject to the maximum possible stress magnitude (40 MPa) and duration provided by the impact rig (82 µs). The following table shows the percentage drop in the value of the piezoelectric charge coefficient, d_{33} for all six PZT compositions and the quartz sample following 5000 impacts.

Material	Pz26	Pz27	PC4	PC4A	PC5	PC5H	Quartz
% d_{33} drop	19	45	30	36	18	22	<2

It is evident from these results and the plots shown below, that all the ferroelectric materials show a distinct drop in their piezoelectric properties under these conditions though there is a tendency for the initially large decrease to flatten off. The degree of fluctuation in the d_{33} value for quartz at <2% is tending to the limit of the resolution of the apparatus. The average recorded value of 2.54 pC N^{-1} is less than 1% greater than the value recorded by the conventional technique. This result is of significance in that it demonstrates the accuracy and reproducibility of results taken on the impact rig.

The initially abrupt drop in the d_{33} value is superseded by a flattening off of the decay in d_{33} as the number of impacts progresses. The current drop is a manifestation of the domain switching that has occurred in the ceramic as a means of stress relief, into the more favourable permitted orientations perpendicular to the stressing axis. The initially rapid and subsequent decay can be represented as an exponential drop in the charge coefficient under repetitive impact stressing in the poled ferroelectric materials and good fits to the data are obtained as illustrated by the linearity of the experimental points when plotted on a logarithmic axis. The gradient of the plot, K_i,

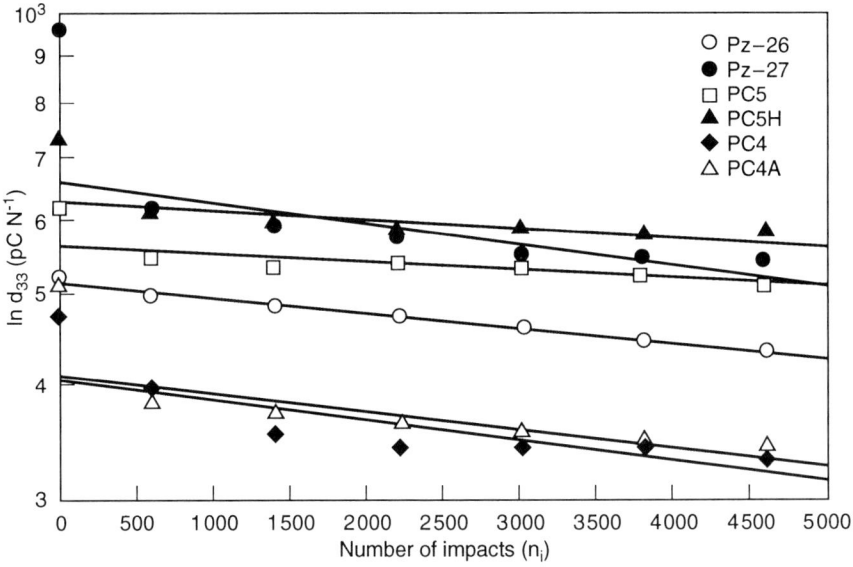

Figure 4. d_{33} as a function of impact repetition: all PZT compositions

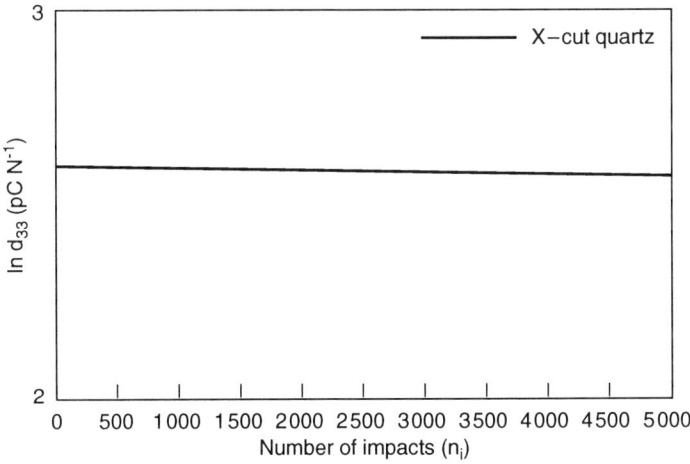

Figure 5. d_{33} as a function of impact repetition: X-cut α quartz

the rate of change of d_{33} with n_i, the number of impacts, may be construed as an ageing coefficient for the material under impact, where d_{33}^0 is the initial value of the charge coefficient prior to any impacts:

$$d_{33} = d^0_{33}\, e^{-K_i n_i} \qquad (3)$$

Poled ferroelectric devices are most prone to fatigue behaviour and a loss of piezoelectric properties when i) compressive stress is applied parallel to the poling axis and they are under short circuit conditions since there is no electric field between the electrodes to sustain the domain configurations in their poled positions and ii) under tensile stress parallel to the poling axis as there is a depolarising field produced if open circuit conditions prevail. Also, in view of the frequency dependence of non-linear behaviour, the fatigue behaviour of a ferroelectric device can be expected to be directly related to the duration of loading, be it mechanical or electrical. Extended application of compression parallel to the poling axis, even at relatively low levels will lead to fatigue of the device though mechanical creep as a means of stress relief. The manner of deployment of the piezoelectric device is important i.e., mode of operation; the electrical and mechanical loading (including its mountings, the con-straints it imposes upon the device and its direction with respect to the poling axis) put upon it will dictate its fatigue behaviour. Thus, a poled ferroelectric device cannot be exposed to different sets of operating conditions and be expected to behave similarly in each instance.

7. MICROSTRUCTURAL ANALYSIS

Microcracking induced in the test specimens by the impacting process will adversely affect the electrical output from the device as dielectric breakdown occurs across the fissures. High localised stresses arising from direct impact by mis-aligned or uneven projectiles are a major cause of microcracking. Below are two scanning electron micrographs showing the degree of microcracking present in a polished sample of PC5. In this particular extreme case, it is evident that the majority of cracking was initiated in or adjacent to areas of porosity or contamination, already present in the sample.

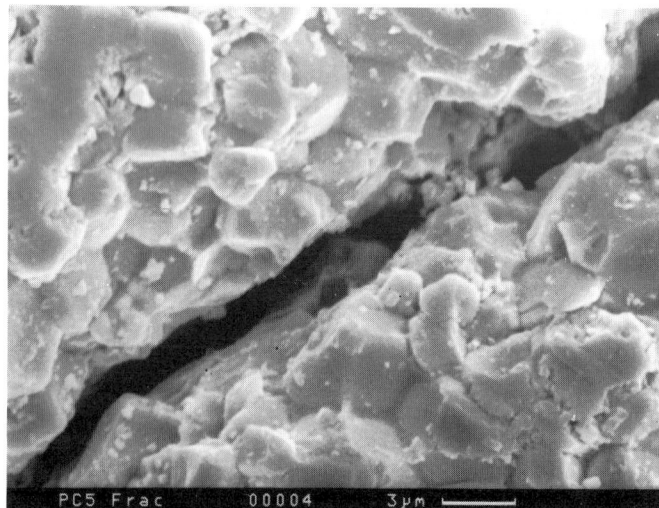

Figure 6.   Fissure induced by the impacting process

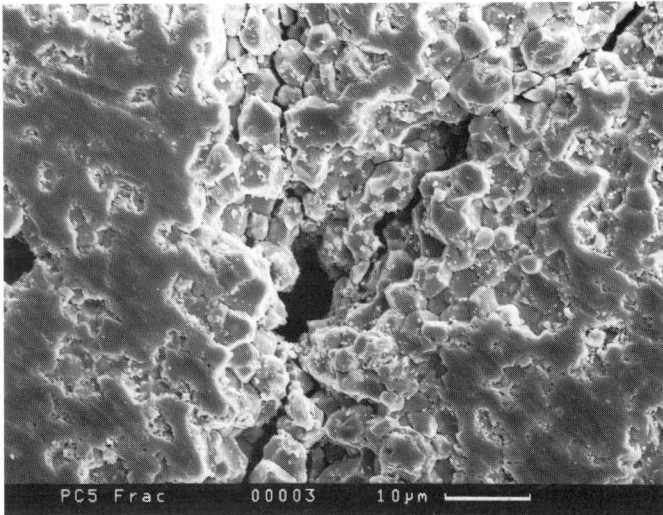

Figure 7.   Poor sample integrity leads to microcracking

8. CONCLUSIONS

From the results presented above, the following observations can be made:

1) The piezoelectric response i.e., the charge and electromechanical coupling co-efficients of the ferroelectric materials does not show a static value when subject to variations in stress, as one might conclude from conventional metrology methods. The magnitude of variation in d_{33} with stress magnitude and duration is dependent on the mobility of the ferroelectric domains in the material i.e., its coercivity, this being dependent on the chemical composition and degree of doping in the material. The almost linear response of the X-cut α quartz sample, though not immune to changes in its charge coefficient under stress, proved to be an excellent reference with which to calibrate the impact testing rig and compare the ferroelectric materials.

2) It is possible to obtain extremely high values of d_{33} and therefore output voltages from the ferroelectric materials, particularly those of a 'soft' nature, though under repetitive impacting there is a rapid diminution of this phenomena and the charge coefficient tends towards a value more in accordance with that taken under conditions of low stress and low frequency.

3) It is imperative that when piezoceramic devices are to be used under demanding conditions such as those exercised in the preceding experiments that the material has high microstructural integrity and is free from manufacturing defects if catastrophic failure is to be avoided. Flaws in the DUT are liable to cause internal stress wave reflections, in consequence, inhomogeneous stress states occur and there is a proclivity to crack initiation.

ACKNOWLEDGEMENTS

This research was supported by the National Physical Laboratories, Teddington, Middlesex, England, TW11 0LW.

REFERENCES

1. J. A. Close and R. Stevens: *Ferroelectrics*, 1992, **134**, 181–187.
2. B. Hopkinson: *Phil. Trans. Royal Soc.* London, **A213**, 1913–1914, 437–456.
3. R. M. Davies, *Phil. Trans. Royal Soc.* London, **A240**, 1946–1948, 375–457.
4. S. Nemat-Nasser, J. B. Isaacs and J. E. Starrett: *Proc. Royal Soc.* London, 8th Nov. 1991, **A435**, 371–391.
5. J. A. Close: Ph.D. Thesis, University of Leeds, UK., 1993.

Structural Ceramics

Alumina/SiC Nancomposites: Big Things from Small Packages

R. I. TODD

Manchester Materials Science Centre, University of Manchester and UMIST,
Grosvenor Street, Manchester, M1 7HS, UK

ABSTRACT

The literature on alumina/SiC nanocomposites is reviewed. These are structural ceramics consisting of alumina grains of conventional size (~ 1–4μm) containing a dispersion of SiC 'nanoparticles' <300nm in diameter. At high temperatures the properties are dominated by the difficulty in removing and depositing material at the SiC/alumina interface. This hinders pressureless sintering, but improves the creep resistance and microstructural stability. At room temperature the SiC particles cause an apparent strengthening of the alumina grain boundaries. This improves the quality of machined surfaces, increases the wear resistance, and is also associated with strength improvements. There is no universally acknowledged explanation for this grain boundary strengthening, although the residual stresses caused by the thermal expansion mismatch between the alumina and the SiC are thought to play an important role.

1. INTRODUCTION

'Ceramic nanocomposites' are a class of structural ceramics which first came to prominence between 1988 and 1991 with the publication of a series of papers by Niihara and co-workers[1–4] describing their processing, microstructure and properties. Excellent properties were reported for a number of systems, but the combination which has attracted most attention consists of alumina reinforced with SiC 'nanoparticles' less than 300nm in diameter. The claim responsible for this interest is that the strength of alumina, already the most widely used structural ceramic because of its excellent mechanical properties, can be increased by a factor of three simply by adding 5vol% SiC nanoparticles. The strength can be increased by a further 50% to 1500MPa on annealing at 1300°C.[4]

The high strength of Niihara's material has since been verified by independent testing, but despite several attempts to fabricate alumina/SiC nanocomposites by other researchers[5–8] strength increases comparable to those of Niihara have not been reproduced. This has led to a degree of scepticism regarding the value of these materials, and even an aura of mysticism. This paper is a review of the available literature regarding alumina/SiC nanocomposites, and demonstrates (i) that these materials do offer useful and reproducible improvements over pure alumina in a number of properties, (ii) that there is substantial agreement regarding most aspects of behaviour and that there are no irreconcilable contradictions between the various results which have been reported, and (iii) that most of the observed effects of the SiC additions can be rationalised in terms of two simple ideas, one for high temperatures and one for low temperatures.

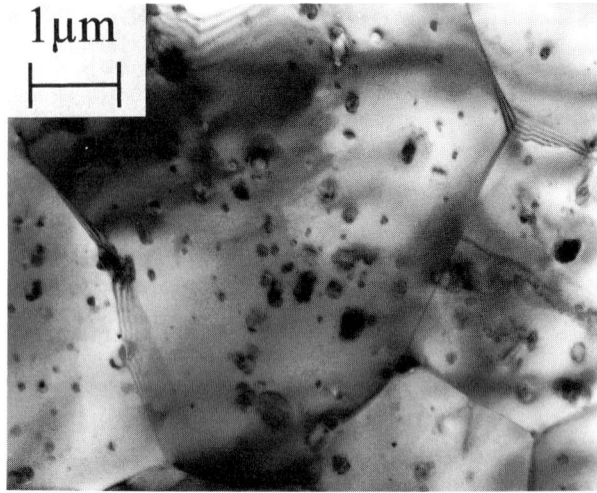

Figure 1. Typical microstructure of alumina/SiC nanocomposite, showing alumina grains and SiC 'nanoparticles' situated both within the grains and on the alumina grain boundaries.[9]

2. DEFINITION OF SCOPE AND MICROSTRUCTURAL OVERVIEW

All processing of alumina/SiC nanocomposites to date has been by ball milling mixtures of fine alumina and SiC powders followed by densification at temperatures of 1600–1800°C. The resulting microstructure depends largely on the relative proportions of the constituents. The present paper is confined to the consideration of composites made from powders with roughly equal sizes of less than 300nm, containing less than 30vol% SiC. An example of the microstructure produced under these conditions is shown in Fig. 1. The alumina grains are of conventional size, typically

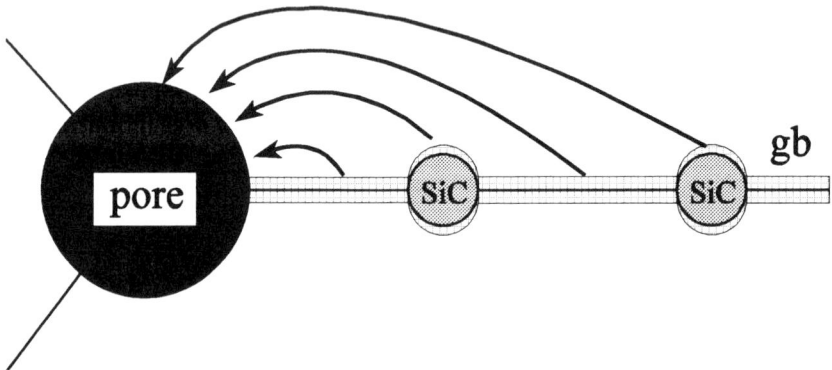

Figure 2. During sintering, compatibility requires material to be removed both from the alumina grain boundaries and from the periphary of grain boundary SiC particles.

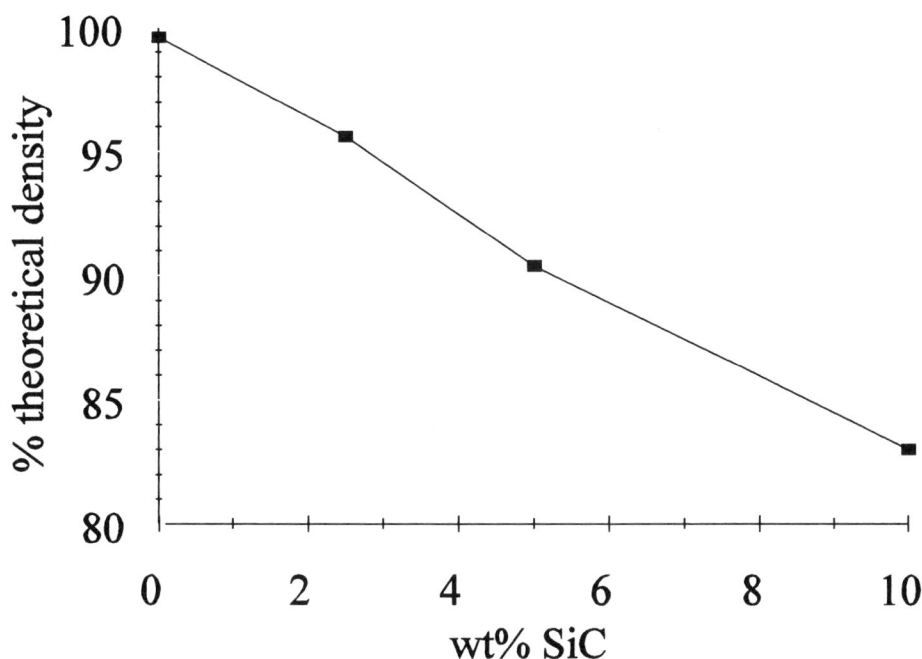

Figure 3.   Sintered density versus SiC content, for sintering at 1700°C, 2 hours.[7]

1–4µm. The SiC particles are uniformly dispersed, with some situated within the alumina grains and others at the alumina grain boundaries. The intragranular particles are thought to be important to the improvements in room temperature properties,[10] whilst the grain boundary particles are significant at high temperatures.[11] TEM under suitable diffraction conditions has also revealed the presence of dislocations in the alumina.[4] The dislocations tend to form networks which develop further during annealing at 1300°C.

3. HIGH TEMPERATURE BEHAVIOUR

3.1 Background

High resolution microscopy has shown that the SiC is directly bonded to the alumina at the particle/matrix interfaces with no separating impurity phase. Simple considerations suggest that it should be difficult to remove or deposit alumina at such an interface, because for this to be achieved by diffusion requires that *both* phases which meet at the interface be mobile if bonding is retained across the interface.[12] SiC is *not* very mobile at temperatures below the melting point of alumina because it has a very high melting point (~2700°C) and has very strong directional covalent bonding. The difficulty in removing/depositing alumina at the particle/matrix interfaces dominates the behaviour of nanocomposites at high temperatures.

3.2 Sintering

An essential step in sintering nanocomposites is for alumina to be removed from the grain boundaries and deposited in the pores. Material must be removed evenly over a grain boundary, otherwise incompatibility stresses build up and oppose continuation of the process. This implies that alumina must be removed from the interfaces between the alumina matrix and grain boundary SiC particles (Fig. 2), and since this is difficult, sintering is inhibited. Figure 3 shows the density attained during pressureless sintering at 1700°C in Argon–1% Hydrogen.[7] The addition of only 2.5 wt% SiC causes a marked reduction in density, and the downward trend continues with further additions. With different powders and processing details Stearns et al[5] were able to achieve a sintered density of 99% at 1700°C with 5vol% SiC, but lower temperatures yielded much lower densities, despite the fact that the unreinforced alumina sintered to near theoretical density at only 1400°C. There are no reports of the production of high densities by pressureless sintering for SiC contents of more than 5vol%.

The scope for improving pressureless sintering by raising the temperature is limited by the melting point of alumina (2040°C), but dense material can be produced routinely by hot pressing,[4,6,7] which increases the driving force for densification sufficiently to overcome the opposing action of the SiC/alumina interfaces.

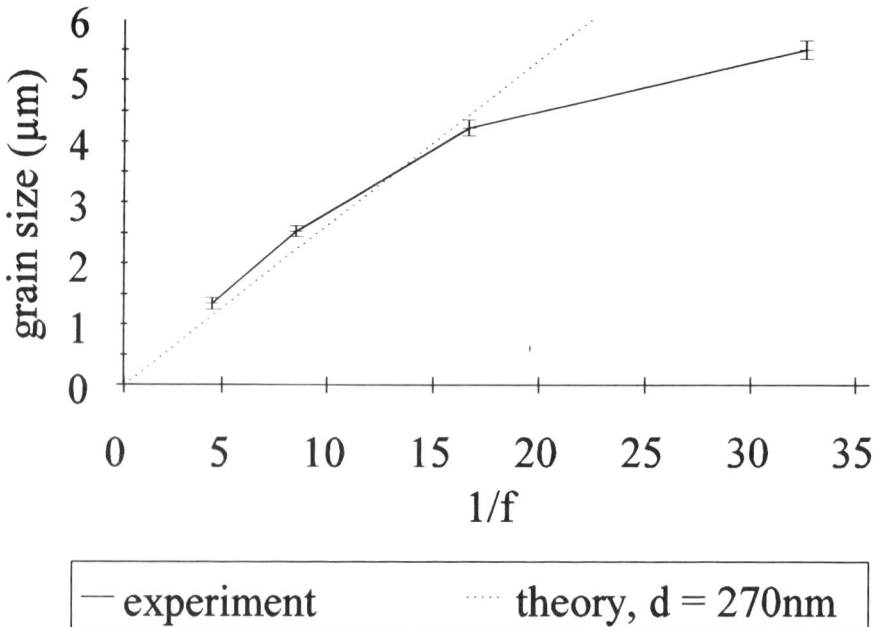

Figure 4.   Grain size versus 1/volume fraction SiC for nanocomposites hot pressed at 1700°C,[7] and the predicted line according to Zener's theory for 270nm SiC particles.

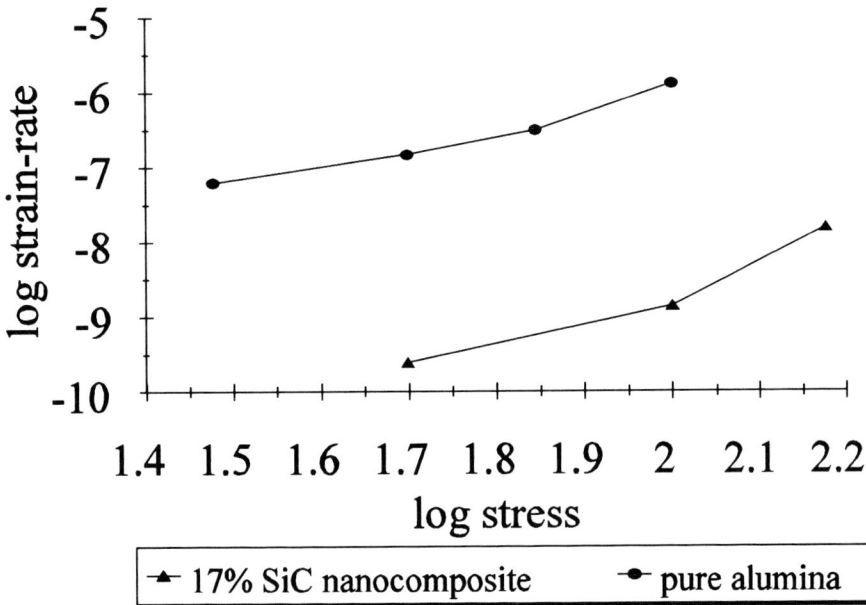

Figure 5. Logarithmic plot of strain-rate against stress at 1200°C. Data from.[11]

3.3 Grain Growth

Particles and grain boundaries are mutually attracted because the presence of a particle effectively removes a section of the grain boundary along with the associated interfacial energy. There are three general possibilities during grain growth. (i) If the driving force for grain growth is sufficiently high and the particles are relatively immobile, the grain boundary can simply break away from the pinning effect of the particles, and grain growth ensues. (ii) If the particles are mobile they can be dragged along with the migrating grain boundary allowing grain growth to continue. (iii) If the driving force for grain growth is low and the particles are immobile, the grain boundaries are pinned by the particles and grain growth ceases. For the particles to be mobile, matrix material must be removed from the particle/matrix interface in front of the particle and deposited behind it. This is difficult in the alumina/SiC system so the particles are virtually immobile. Microstructural development is therefore simple. The SiC particles remain stationary throughout, retaining their initial uniform distribution. When the grains are small the driving force for grain growth is high and grain growth occurs by grain boundary breakaway until the driving force falls below the pinning force exerted by the particles, at which point grain growth ceases.

The stagnation grain size was predicted by Zener[13] to be $\sim d/f$, where d is the particle size and f is the particle volume fraction. Borsa et al[7] reported their results to be in good agreement with this relationship for SiC volume fractions of more than 3vol% (Fig. 4). The straight line on the graph corresponds to a particle size of 270nm, which compares well with the nominal size of 200nm. The relationship breaks down

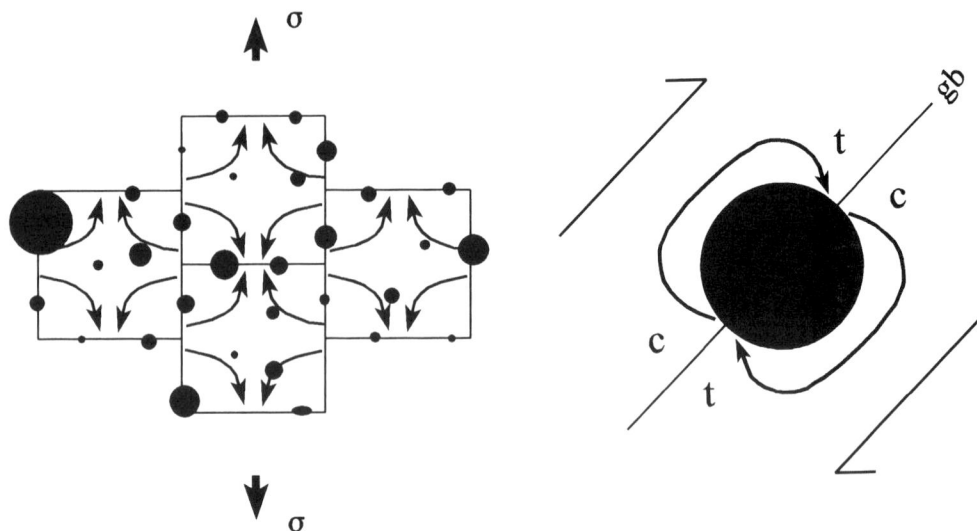

Figure 6.   Schematic illustration of material transfer during (a) diffusion creep and
(b) grain boundary sliding.

at low volume fractions because there is not enough time for the grains to grow to the
very large equilibrium sizes predicted. Stearns *et al*[5] reported a grain size of 4.8µm for
5vol% SiC material, which gives a particle size of 240nm according to the Zener
relationship. The nominal particle size was 150nm, so the level of agreement is
comparable to that of Fig. 4, although these authors point out that the Zener grain
size represents an upper limit and the sense of the discrepancy therefore indicates
poor agreement. However, there is evidence that the presence of grain boundary
glassy phases in grain boundaries can help the boundary to escape from the pinning
particles,[14] and it is argued here that given the uncertainties in the particle size and
the approximate nature of the Zener theory, the reported agreements to within a
factor of 2 are remarkably good.

3.4 Creep

The presence of SiC reduces the creep rate by two to four orders of magnitude
compared to pure alumina (Fig. 5), depending on the SiC content, stress and loading
method.[11] The creep life is increased by one order of magnitude,[11] and the maximum
operating temperature from 800°C to 1200°C.[4] The creep strain at failure is reduced,
however, because the grain boundary SiC particles act as nucleation sites for cavita-
tion. Nevertheless, these particles are necessary to obtain the improvement described
because they are expected to reduce the rate of diffusion creep and the ease of grain
boundary sliding. Diffusion creep requires the uniform removal of material from
compressive grain boundaries and its redeposition on tensile boundaries (Fig. 6a). As
with sintering, this necessitates removal (and deposition) of material at the interfaces

of grain boundary particles, and diffusion creep is inhibited because this process is slow. Similarly, grain boundary sliding is impeded because it requires material to be removed from some sections of the particle/matrix interfaces and redeposited at others (Fig. 6b).

4. ROOM TEMPERATURE BEHAVIOUR

4.1 Fracture Mode

All researchers agree that the fracture mode is predominantly transgranular in the nanocomposites, in sharp contrast to the intergranular fracture exhibited by pure alumina.[4,6,7,8,15] SiC additions as low as 4vol% are capable of producing this change (Fig. 7). The change in fracture mode is a manifestation of a general idea which can be used to rationalise much of the room temperature behaviour, *viz* that the SiC nanoparticles cause an apparent strengthening of the grain boundaries.

4.2 Tribological Properties

The apparent strengthening of the alumina grain boundaries also dominates the tribological properties which have been reported. The grain pullout by grain boundary fracture during diamond polishing, which is well known as a cause of poor surface finish in alumina, is eliminated in the nanocomposites.[6,15] Similarly, the appearance

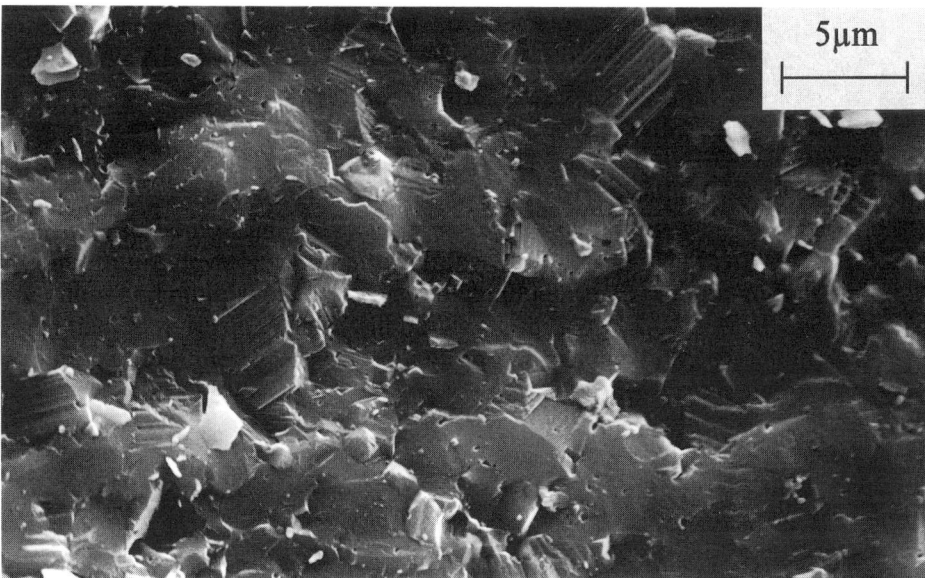

Figure 7.   Fracture surface of a 4 vol% SiC nanocomposite showing predominantly transgranular fracture.

of diamond ground alumina surfaces resembles the facetted appearance of its fracture surfaces, because whole grains are removed by fracture along their grain boundaries. Similar areas can also be found on ground nanocomposites, but the majority of
the surface exhibits clear grinding marks, suggesting that material is removed predominantly by a plasticity based mechanism.[15]

The same effect increases the wear resistance of the nanocomposites. In one
example, the wet erosive wear rate of a 5% SiC nanocomposite was reduced to a
third that of pure alumina with the same grain size, presumably because each grain
has to be worn away rather than being removed whole by grain pullout.[15]

4.3 Strength and Toughness

Comparisons of strengths for hot-pressed nanocomposites and pure alumina are
shown in Fig. 8, and further details from each study are listed in Table 1. Comparisons are given only where there is some evidence that the average grain size was
similar for both materials.

Only studies 1 and 6 report the effect of annealing on the strength of the nanocomposite, and in both cases a strength increase resulted. Observation of the radial cracks
surrounding hardness indentations showed that annealing causes crack healing,
which offers a convincing explanation for this phenomenon.[16] The effect may be
heightened relative to alumina by the ability of the nanocomposite to retain any
compressive surface stresses caused by machining because of its better creep resistance,[6] although since no study has reported strengths for both annealed and unannealed pure alumina it remains to be seen whether this is borne out by experiment.
That the two studies found different strength increases is not surprising, since the
degree of healing which is possible presumably depends on the size and nature of the
original flaw, and the healing of surface cracks may cause a different type of flaw

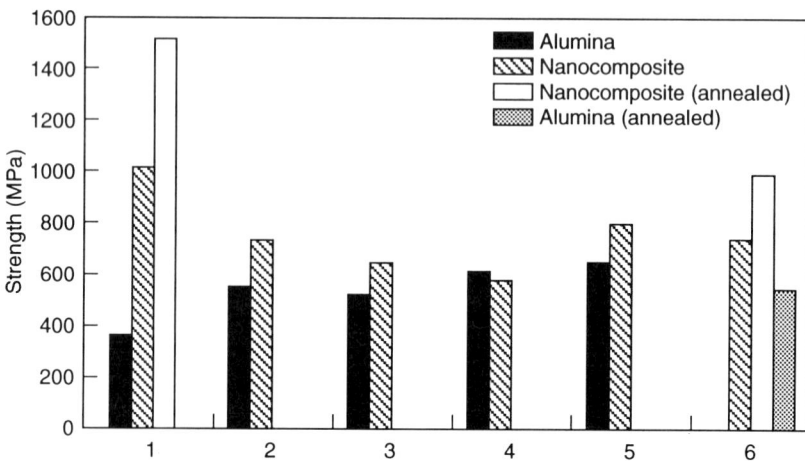

Figure 8.   Reported strengths for nanocomposites and comparable pure alumina. Further
details are given in Table 1 for each study number.

Table 1. Details of reported strength results for hot pressed SiC/alumina nanocomposites. Study numbers refer to Figure 8.

Study number	Reference	NC grain size (μm)	SiC content (vol%)	Bend strengths (MPa) unannealed NC	PA	annealed NC	PA	NC/PA strength ratio (unannealed)
1	[4]	2	5	1017	335	1520	–	2.86
2	[7]	1.4	24	730 ± 20	560 ± 77	–	–	1.30
3	[7]	2.4	12	648 ± 27	520 ± 10	–	–	1.25
4	[8]	1.1	5	592 ± 3	618 ± 54	–	–	0.96
5	[8]	1.1	5	803 ± 60	660 ± 27	–	–	1.21
6	[6]	4.2	5	760 ± 28	–	1001 ± 102	559 ± 51	–

NC = nanocomposite, PA = pure alumina.

which is more difficult to heal (e.g. a large grain, porosity) to take over the role of critical defect. In addition, similar observations on a SiC whisker reinforced alumina showed that the strength could be increased or decreased simply by varying the oxygen content of the atmosphere during annealing,[17] thereby altering the nature of the oxidation occurring at the surface. Since this seems to be an established phenomenon in alumina/SiC systems, the remainder of this section will be restricted to the unannealed strength.

Results 1–5 allow comparison of strengths of unannealed materials. The nanocomposite is stronger than the comparable pure alumina in all cases except no. 4, where there is no significant difference. The strengths reported for both materials vary from one study to another, as is usual for ceramics, where processing details can greatly influence the strength by altering the nature and size of the critical flaw. The details of strength testing also have a part to play. Higher strengths should be expected from studies 1, 4 and 5, because they used three point bending, which subjects a smaller volume of material to high stress than the four point bending method used in 2 and 3. Surface finish can also influence the strength level.

The disparity between the results in Fig. 8 cannot be explained purely in these terms however. Differences in processing and testing might be expected to have at least qualitatively similar effects on both the alumina and the nanocomposite, but it is evident that the attainment of the biggest nanocomposite/alumina strength ratio in study 1 is as much attributable to the pure alumina being the weakest in this case as it is to the nanocomposite being the strongest. The reason for the low strength of the pure alumina relative to studies 2–5 cannot be determined with certainty since no details of its preparation or microstructure are given. Nevertheless, assuming that the processing was of the same high quality which must have been used to obtain such high strengths in the nanocomposite, a strength of only 355MPa with an average grain size of 2μm may indicate abnormal grain growth. Indeed, Niihara remarks in ref. 4 that the presence of the nanoparticles reduces the abnormal grain growth. It is certainly pertinent to state the obvious points (i) that for a true assessment of the direct relevance of the nanoparticles to fracture, the microstructure of the alumina grains should be identical in both the reference alumina and the nanocomposite, and (ii) that this may be very difficult to achieve.

Microstructural details may play a part in some of the results of Table 1, but they are certainly not the whole story. In study no. 5 the nanocomposite was 22% stronger than the alumina, and Poorteman *et al* inspected the grain size distribution as well as the average grain size. The higher strength of the nanocompsotie is not attributable to a reduction in abnormal grain growth because the nanocomposite had a greater spread of grain sizes, and therefore a *bigger* maximum grain size (7μm), than the pure alumina (3μm). This shows that the SiC nanoparticles *do* have a direct influence on fracture.

Only study 1 reports a comparison of toughness for unannealed pure alumina (3.5MPa m$^{1/2}$) and the nanocomposite (4.8MPa m$^{1/2}$ for 5% SiC). This toughness increase of 37% is sufficient to explain the strength improvements in studies 2, 3, (4), and 5, but not the large increase of study 1 itself. If some of the increase in study 1 is attributable to a reduction in the critical defect size through the microstructural benefits of the SiC nanoparticles, as suggested above, this is understandable. Nevertheless, comparison of the data of 4 and 5 reveals a more subtle effect. These two sets of data are from the *same* pure alumina and nanocomposite, the only difference between the measurements being that the bend bars in 5 were machined to a higher quality finish than those in 4. The improved machining made no significant difference to the strength of the alumina, but increased the nanocomposite strength by 36% to 803±60MPa. These results strongly suggest that there is a difference in the crack *initiation* behaviour between the alumina and the nanocomposite, since the improvement in machining reduced the critical flaw size of the nanocomposite, but did not change the alumina flaw size significantly. There is clearly a connection between this conclusion and the superior tribological properties of nanocomposites described in section 4.2 and this links at least part of the strength increase produced by the nanoparticles to the apparent strengthening of the alumina grain boundaries which unifies the other room temperature observations. The toughness measurements of study 1 are undoubtedly relevant to the crack initiation mechanism, since they were obtained using the indentation crack method, which samples the toughness at the short crack lengths of interest. However, a simple single valued toughness approach cannot provide a complete explanation of the observations, and a more detailed investigation of short crack *R*-curve behaviour and the mechanics of polishing and grinding is required. It is also noted here that any improvements in strength obtained through an increased resistance to crack initiation may be limited or even obscured by the presence of pre-existing processing defects whose size is *not* determined by this factor.

5. EXPLANATIONS FOR ROOM TEMPERATURE BEHAVIOUR

Section 4 shows that the key to understanding the room temperature behaviour of nanocomposites is to explain the apparent strengthening of the alumina grain boundaries caused by the SiC nanoparticles. Most known strengthening mechanisms related to second phase additions, such as crack interface bridging and microcrack toughening, become less effective as the particle size diminishes, and can therefore be ruled out. A number of alternative mechanisms have been proposed, and all rely on the residual stresses which arise during cooling from processing temperatures as a result of the thermal expansion mismatch between the SiC and the alumina matrix. The residual stresses are the most obvious effect of the SiC additions and there is experimental support for their importance from a study of alumina/TiN nanocomposites.[15] The thermal expansion mismatch between the matrix and the reinforcement is very small in this system, and it does not exhibit the distinctive change in fracture mode, improved surface finish and increased strength characteristics of alumina/SiC nanocomposites. There are three main ideas for the way in which these residual stresses can bring about the apparent strengthening of the grain boundaries. (i) Subgrain formation.[4] The dislocations which are observed in the grains are attributed to deformation under the action of the residual stresses. The dislocations form low angle boundaries, and fracture takes place along these instead of along the full grain boundaries.

This is an attractive concept at first sight, but it is difficult to see why low angle boundaries should be very much weaker than defect free alumina, because only a very small fraction of their area is comprised of 'bad' material associated with the dislocation cores. The TEM evidence shows that the subgrain boundaries are not well developed in unannealed material, and only one or two per grain can be seen. In addition, there is no provision in the model for an apparent *strengthening* of the grain boundaries, simply an alternative weak path. In consequence, the mechanism does not provide a convincing explanation for the observed reduction in crack initiation and grain pullout.

(ii) Crack Deflection.[4,6] The residual stress state is such that the SiC particles are in approximately hydrostatic compression, and there are tensile hoop stresses in the surrounding matrix. The hoop stresses around an intragranular particle may be sufficient to attract an intergranular crack out of the grain boundary and into the bulk of the grain (Fig. 9). Once the crack tip has propagated along its new transgranular path beyond the tensile hoop stresses of the SiC particle, it experiences the high toughness of the grain interior, making continued advance more difficult. Qualitatively, this mechanism provides a plausible explanation for many of the experimental results. Quantitatively, however, there is reason to be cautious. Although a detailed solution of the geometry of Fig. 9 would be difficult to develop, an estimate of the contribution of the hoop stresses to the stress intensity, K_h, can be made by assuming a long straight crack with the average hoop stress, $\bar{\sigma}_h$, acting over a distance from the crack tip equal to the interparticle spacing, d:[18]

$$K_h = 2\bar{\sigma}_h \sqrt{\frac{2d}{\pi}}$$

With $\bar{\sigma}_h = 90\text{MPa}$[19] and $d = 400\text{nm}$, K_h is $0.09\text{MPa m}^{1/2}$, which is only 2.5% of the fracture toughness of alumina ($\sim 3.5\text{MPa m}^{1/2}$), i.e. the contribution is *small*, and may not be sufficient for the mechanism to work. This conclusion is reinforced when it is

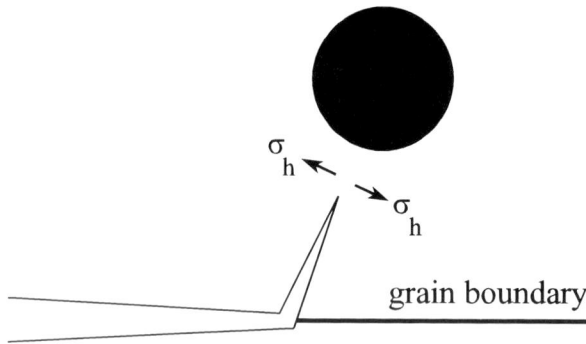

Figure 9.   Tensile hoop stresses around intragranular particles can in principle attract grain
boundary cracks into the grain interior.

considered that the equation applies to a case in which the hoop stresses act over the whole length of the crack front. In reality only small sections would be close to intragranular particles.

(iii) Reduction of Anisotropy Stresses.[19] Recent measurements of the residual stresses[19] have shown that the thermal expansion anisotropy of alumina is effectively reduced by the presence of the SiC. This is attributable to anisotropic relaxation of the interphase thermal expansion mismatch at high temperature by the removal of matrix material from the particle matrix interface followed by preferential redeposition on the {006} alumina planes by climb of rhombohedral dislocations. (Although removal of material from the interface is difficult, it is possible under the high compressive residual stresses approaching 1GPa in question). A similar process has been observed in star sapphire,[20] and explains the formation of dislocation networks. The thermal expansion anisotropy of alumina also leads to residual stresses. The magnitude of these 'anisotropy stresses' is comparable to that of the interphase stresses, but they act over much larger distances (~the grain size cf. the interparticle spacing) and so can have a greater influence on fracture. In particular, the anisotropy stresses may favour intergranular fracture,[19] and help crack initiation on grain boundaries with a tensile residual stress. Any reduction in their size would favour transgranular fracture and impede crack initiation, as found experimentally. Like (ii), however, this effect is small. For a 4vol% SiC composite, which exhibited the characteristic apparent grain boundary strengthening described, the reduction in the anisotropy stresses was estimated to be only 7%.[19]

It is not clear which, if any, of the above mechanisms is responsible for the experimental observations. Although the mechanisms reviewed are weak effects, the balance between transgranular and intergranular fracture may be sufficiently fine to be tipped one way or the other by small changes, particularly if more than one mechanism act in concert. Progress will require both more sophisticated modelling and carefully conducted experiments designed to establish the details of crack initiation and short-crack R-curve behaviour in alumina/SiC nanocomposites.

6. CONCLUSIONS

1. Alumina/SiC nanocomposites offer significant and reproducible improvements over unreinforced alumina in a number of useful properties.
2. At high temperatures the properties are dominated by the difficulty in removing or depositing alumina at the SiC/alumina interface. This leads to increased creep resistance and microstructural stability.
3. At room temperature the properties are dominated by an apparent strengthening of the alumina grain boundaries. This leads to improved polishing and machining behaviour, increased wear resistance and higher strength.
4. There is substantial agreement regarding the existing experimental results, but further work is required to ascertain the mechanisms responsible for the room temperature observations and to optimise the benefits to be gained. Experiments should concentrate on short crack fracture behaviour and initiation, and must take into account microstructural details such as the grain size distribution and the nature of the critical flaw.

ACKNOWLEDGEMENTS

Numerous discussions with C. E. Borsa, C. N. Walker, R. W. Davidge and R. J. Brook are gratefully acknowledged. I am indebted to C. E. Borsa for providing Fig. 1, and to M. Poorteman for supplying a preprint of reference 8.

REFERENCES

1. K. Niihara and A. Nakahira: 'Strengthening of Oxide Ceramics by SiC and Si_3N_4 Dispersions', *Proceedings of the Third International Symposium on Ceramic Materials and Components for Engines*, The American Ceramic Society, 1988, 919–926.
2. K. Niihara, A. Nakahira, G. Sasaki and M. Hiabayashi: 'Development of Strong Al_2O_3/SiC Composites', *Proceedings of MRS International Meeting on Advanced Materials, Volume 4*, MRS, Japan, 1989, 129–134.
3. K. Niihara and A. Nakahira: 'Particulate Strengthened Oxide Nanocomposites', *Advanced Structural Inorganic Composites*, P. Vincenzini (ed.), Elsevier, 1990.
4. K. Niihara: 'New Design Concept of Structural Ceramics-Ceramic Nanocomposites', *The Centennial Memorial Issue of the Ceramic Society of Japan*, 1991, **99**, 974–982.
5. L. C. Stearns, J. Zhao and M. P. Harmer: 'Processing and Microstructural Development in Al_2O_3-SiC "Nanocomposites" ', *J. Eur. Ceram. Soc.*, 1992, **10**, 473–477.
6. J. Zhao, L. C. Stearns, M. P. Harmer, H. M. Chan, G. A. Miller and R. F. Cook: 'Mechanical Behavior of Alumina-Silicon Carbide "Nanocomposites" ', *J. Am. Ceram. Soc.*, 1993, **76**, 503–510.
7. C. E. Borsa, S. Jiao, R. I. Todd and R. J. Brook: 'Processing and Properties of Al_2O_3/SiC Nanocomposites', *Journal of Microscopy*, 1995, **177**, 305–312.
8. M. Poorteman, P. Descamps, F. Cambier, D. O'Sullivan, B. Thierry and A. Leriche: 'Optimisation of Dispersion of Nanosize SiC Particles into Alumina Matrix and Mechanical Properties of Corresponding Nanocomposite', to be published in proceedings of 8th Cimtec, Firenze, 1994.

9. C. E. Borsa: DPhil Thesis, University of Oxford, 1995.
10. K. Niihara: 'SiC-Al$_2$O$_3$ Composite Sintered Bodies and Method of Producing the Same' European Patent Specification no. 0 311 289 B1, 1993.
11. T. Ohji, A. Nakahira, T. Hirano and K. Niihara: 'Tensile Creep Behaviour of Alumina/Silicon Carbide Nanocomposite', *J. Am. Ceram. Soc.*, 1994, **77**, 3259–62.
12. M. F. Ashby and M. A. Centamore, 'The Dragging of Small Oxide Particles by Migrating Grain Boundaries in Copper', *Acta Metall*, 1968, **16**, 1081–1092.
13. C. S. Smith: 'Grains, Phases and Interphases: an Interpretation of the Microstructure', *Trans. Met. Soc. AIME*, 1948, **175**, 15–51.
14. L. A. Xue, K. Meyer and I.-W. Chen: 'Control of Grain-Boundary Pinning in Al$_2$O$_3$/ZrO$_2$ Composites with Ce^{3+}/Ce^{4+} Doping', *J. Am. Ceram. Soc.*, 1992, **75**, 822–29.
15. C. N. Walker, C. E. Borsa, R. I. Todd, R. W. Davidge and R. J. Brook: 'Fabrication, Characterisation and Properties of Alumina Matrix Nanocomposites', *British Ceramic Proc.*, 1994, **53**, 249–264.
16. A. M. Thompson, H. M. Chan, M. P. Harmer and R. F. Cook: 'Crack Healing and Stress Relaxation in Al$_2$O$_3$-SiC "Nanocomposites" ', *J. Am. Ceram. Soc.*, 1995, **78**, 567–571.
17. H.-E. Kim and A. J. Moorhead: 'Oxidation Behaviour and Effects of Oxidation on the Strength of SiC-Whisker Reinforced Alumina', *J. Mat. Sci.*, 1994, **29**, 1656–1661.
18. A. G. Evans, A. H. Heuer and D. L. Porter: 'The Fracture Toughness of Ceramics', *Proc. 4th Int. Conf. Frac.*, 1977, vol. 1, 529–56.
19. R. I. Todd, M. A. M. Bourke, C. E. Borsa and R. J. Brook: 'Measurement and Role of Residual Stresses in Alumina/SiC Nanocomposites', *Fourth Euro-Ceramics vol 4*, A. Bellosi ed., Gruppo Editoriale Faenza Editrice, Faenza, Italy, 1995, 217–224.
20. D. S. Phillips, T. E. Mitchell and A. H. Heuer: 'Precipitation in Star Sapphire III. Chemical Effects Accompanying Precipitation', *Phil. Mag. A*, 1980, **42**, 417–432.

Optimised High-Temperature Sialon Ceramics Containing Melilite as the Grain Boundary Phase

N. CAMUSCU, H. MANDAL and D. P. THOMPSON

Materials Division, Department of Mechanical, Materials and Manufacturing Engineering, University of Newcastle upon Tyne, NE1 7RU, UK.

ABSTRACT

The area of composition defining the range of homogeneity for the α-sialon phase in metal sialon systems is in equilibrium with the nitrogen melilite phase (M), which extends into the volume of the Jänecke prism along a line of composition $Ln_2Si_{3-x}Al_xO_{3+x}N_{4-x}$ (designated M'). $\alpha + \beta$ sialon compositions containing grain-boundary melilite are therefore easy to prepare and the M' phase offers important advantages compared with other grain-boundary phases, namely:

. . . it does not oxidize catastrophically at 1000°C when present as a second phase in dense nitrogen ceramics, unlike its Al-free analogue.

. . . in rare earth systems (Ln = Nd, Sm), the M' composition with maximum Al substitution (x = 1) is very close to the high nitrogen limit of the liquid phase region in these systems,

. . . M' crystallises from grain-boundary liquid phase in the temperature range 1450–1650°C, rather than devitrifying from glass at lower temperatures,

. . . the eutectic temperature in the α–β–M' region is in excess of 1600°C.

   With correct compositional design, it is therefore possible to produce mixed $\alpha + \beta$ sialon ceramics containing M' as the grain boundary phase, with very little residual glass. These materials display good creep and oxidation resistance up to at least 1550°C and represent the most refractory family of sialon ceramics reported so far.

1. INTRODUCTION

An important strand of nitrogen ceramic research has been the development of materials which can be sintered to full density by pressureless sintering and which at the same time retain good mechanical properties up to high temperatures. These aims are to some extent mutually exclusive, and whereas sialon ceramics can be pressureless-sintered to full density at ≈1750°C, the eutectic temperature of ≈1350°C in the final ceramic represents a maximum temperature for strength retention. On the other hand, silicon nitride ceramics (i.e. materials which contain no aluminium) can retain their properties up to higher temperatures, but require pressure-assisted techniques for densification. Given the choice between a silicon nitride and a sialon ceramic, it is generally preferable to opt for the sialon, because of the ease of densification; the challenge therefore remains to find ways of achieving improved refractoriness of the resulting sintered ceramics. A considerable amount of know-how exists on the conversion of residual glass in sialon ceramics into crystalline grain-boundary phases. The system itself selects which phases are stable, even though depending on the devitrification temperature, there is some choice of final product. As a general rule, the resulting phases are either pure oxides (e.g. $Y_2Si_2O_7$, YAG, mullite in the Y–Si–Al–O–N system), or oxynitrides of high O:N ratio.[1] In combination with a sialon matrix, there is almost invariably a eutectic in between the matrix and grain-boundary phases, which exhibits melting at temperatures of $\approx 1350°C$. In previous work, this has represented a maximum operating temperature for heat-treated sialon ceramics.

A route for increasing the refractoriness of these materials is to replace the oxide/oxynitride phases produced during devitrification with alternative phases of higher N:O ratio *which have compositions on the nitrogen-rich side of the eutectic* in these systems. This principle is well-known in the $CaO–SiO_2$ system,[2] where compositions on the SiO_2-rich side of Ca_2SiO_4 exhibit liquid phase at temperatures above $\approx 1450°C$, whereas compositions on the CaO-rich side of Ca_2SiO_4 exhibit liquid at $\approx 2000°C$. In composite ceramics containing CaO and SiO_2, for good refractoriness it is important to adjust the $CaO:SiO_2$ ratio to lie in the correct phase region. In nitrogen ceramic systems, very little work has been carried out on oxynitrides with high N:O ratio. The only phase in M–Si–Al–O–N systems (M = Y, Ln), which offers promise is the nitrogen melilite phase, of general formula $M_2Si_{3-x}Al_xO_{3+x}N_{4-x}$. Recent work[3–5] has established that the melilite composition which exhibits maximum Al+O substitution (M') in low atomic number rare earth nitrogen melilites (x = 1 in the above formula) corresponds very closely to the maximum solubility limit of nitrogen in M–Si–Al–O–N liquids in these systems. By careful control of starting composition, it is therefore possible to sinter a sialon composition which retains a grain-boundary glass of this composition, which after subsequent crystallisation is converted into M'. The resulting sialon-M' microstructure would be expected to show good refractory properties, because the eutectic in this system would be close to the M' composition, which from previous experiments has been shown to melt at >1750°C. In fact, previous work[3–5] has shown that the M' phase crystallises out of grain-boundary liquid at temperatures above the eutectic temperature and in the range 1450–1650°C. Since the nitrogen content of the liquid at these temperatures is high, it is possible to convert almost all of it into M'. The principle of crystallising liquid into crystalline grain-boundary phases (as compared with devitrifying glass) has not been explored fully, but has the advantage that the process is much faster, and with careful selection of starting compositions, can be arranged to give a neglibible amount of residual liquid phase.

The above process cannot be applied as easily to β-sialon ceramics, because β-sialon is not in equilibrium with the M' phase at sintering temperatures. Previous work has therefore focused on α-sialon ceramics and (preferably) mixed α + β sialon ceramics as the relevant matrix phase, in which M' is then produced as the grain-boundary phase. The additional advantage of the latter system, is that two-phase α-β materials can be further tailored by selection of α:β ratio to give a pre-selected combination of mechanical properties.

An objection which might be raised, and is perhaps the reason why so little work has been carried out on M (M') phase as a grain boundary phase for silicon nitride ceramics is that extensive previous studies showed that dense Si_3N_4 ceramics containing ≈ 15 vol% of yttrium nitrogen melilite show catastrophic cracking on oxidation at $\approx 1000°C$.[6,7] Quantitative studies showed that this phenomenon arose because of the large amount of this phase in these materials, and also because of the 30% specific volume increase which takes place on oxidation. Further work has shown that M' phases exhibit much less specific volume change on oxidation, because of the increased amounts of mullite in the product compared with low density silica. Moreover, in the present materials, the volume of M' phase can be controlled to very low levels, because most of the densification additive can be incorporated into the α-sialon phase, leaving a relatively small volume of grain-boundary liquid to be crystallised into the M' phase. Recent studies have shown no evidence of catastrophic cracking in oxidation of $\alpha + \beta$ sialon – M' ceramics at 1000°C.

The present paper therefore describes experiments aimed at producing optimised $\alpha + \beta$ sialon – M' ceramics, and describes the success with which residual grain-boundary glass can be removed, and also describes their high-temperature properties. The difficulties of achieving the desired objectives are also discussed.

2. EXPERIMENTAL

Powders of Si_3N_4 (H.C. Starck-Berlin, Grade LC10), Al_2O_3 (Alcoa, Grade A17), AlN (H.C. Starck-Berlin, Grade B), and selected rare earths (Rare Earth Products, 99.9%), were mixed together by ball milling in appropriate amounts in isopropanol. The mixed powders were dried, sieved, and compacted into pellets by uniaxial and isostatic cold pressing. The resulting compacts, of mass 1–5g, were pressureless sintered at 1700–1800°C in nitrogen in a carbon resistance furnace in graphite crucibles lined with boron nitride powder. Subsequent heat-treatments were carried out at up to 1650°C for up to 168 hours in nitrogen, to study M' formation. Bulk and apparent solid densities of the resulting pellets were measured by flotation in mercury. Phase characterisation was carried out using a Hägg-Guinier focusing camera and $CuK\alpha_1$ radiation. Microstructures were characterised by SEM using a Camscan S4–80DV instrument. Hardness and fracture toughness were measured by indentation using 10 kg load.

3. RESULTS AND DISCUSSION

3.1 Densification and Phase Characterisation

Starting compositions were chosen to correspond to ones used previously,[5] and were based on a mix of 81.67% Si_3N_4, 16.08% AlN and 2.25% Al_2O_3 to which had been added equivalent amounts of the rare earth oxides Nd_2O_3 (N), Sm_2O_3 (S), Dy_2O_3 (D) and Yb_2O_3 (Yb) (see Table 1). Compositions were designed to give predominantly α-sialon materials with little or no β-sialon. All samples were sintered for two hours in a carbon resistance furnace in nitrogen at 1800°C and rapidly cooled (200°C

Table 1.   Compositions, sintered densities and phase compositions of samples used.

Sample	Compositions (%)				Density gcm^{-3}	Phase Content (XRD)			
	Si$_3$N$_4$	AlN	Al$_2$O$_3$	Ln$_2$O$_3$		α'	β'	21R	M'
N1000	72.78	14.32	2.01	10.89	3.21	75	25	mw	w
S1000	72.50	14.27	2.00	11.23	3.36	91	9	–	–
D1000	71.92	14.16	1.98	11.94	3.31	96	4	–	–
Yb1000	71.44	14.06	1.97	12.53	3.32	99	1	–	–

min^{-1} to below 1000°C). The X-ray diffraction results are shown in terms of percentage values of α- and β-sialon expressed relative to one another plus qualitative indications (s = strong, m = medium, w = weak) of the other phases present.

As expected, α-sialon was the predominant phase in all the samples, but despite the similarity in compositions, there was an increase in α content with increasing atomic number of rare earth element. This trend is consistent with work reported in previous studies.[8] Only in the Nd-containing sample were any other crystalline phases observed. In other samples the rare earth species was present in a glassy phase; in the neodymium sample, the M' phase is stable at the sintering temperature, and 21R remains to balance the overall composition.

The samarium sample sintered to the highest density (99.8% of theoretical). The lower density of the neodymium sample is linked to the retention of melilite in the product assemblage, because this phase prevents all the sintering additive going into the liquid phase and aiding densification. In the other samples, the higher α content shows that most of the densifying additive has entered the α structure, leaving very little residual grain-boundary liquid phase to aid the final stages of densification. Again, this is consistent with previous work.

3.2 Heat-treatment

The lowest eutectic temperature in these systems is in the range 1350–1450°C. For most sialon ceramics, heat treatment to devitrify the grain-boundary glass is carried out at temperatures between 1200 and 1400°C, i.e. below the eutectic temperature. In this case, there is no liquid phase involved in the process, and pores may appear at interfaces as a result of volume changes during the devitrification. Moreover, in most samples, residual glass is still present after devitrification. Previous work has shown that an alternative heat-treatment process which can be applied to samples with compositions of the type studied here, uses temperatures above the eutectic. In this process, residual liquid phase reacts with matrix grains to produce nitrogen-rich crystalline grain-boundary phase (M' in this case), and if the process is optimised, a negligible amount of glass remains in the final product, and this is generally present as isolated pockets. As a result, interfaces are glass-free (benefitting creep resistance), and the presence of liquid during heat-treatment allows further shrinkage, and prevents the development of pores in the microstructure.

Table 2. X-ray results of heat-treating samples for 24 hours at 1550°C.

Sample			Phase Content (XRD)				Unit Cell Dimensions (Å)			
	α'	β'	21R	A	G	M'	M' a	M' c	M a	M c
N1000	52	48	–	m	–	m	7.762	5.058	7.718	5.031
S1000	78	22	w	–	–	mw	7.726	5.018	7.683	4.988
D1000	81	19	–	–	–	w	7.653	4.959	7.609	4.932
Yb1000	91	9	–	–	m	–	–	–	–	–

$$A: LnAlO_3 \qquad G: Ln_3Al_5O_{12}$$

In the present work, heat-treatment was carried out above the eutectic temperature at 1550°C for 24 hours, and X-ray results of the final products are shown in Table 2.

M' was the major phase observed during heat-treatment. The unit cell dimensions given alongside confirm that it is certainly an M' phase which is being produced, and from previous work, the composition is close to that of maximum Al substitution ($Ln_2Si_2AlO_4N_3$ in Nd- and Sm- systems; slightly less Al content in the Dy system). M' is not stable in the Yb system. The $NdAlO_3$ phase observed in the heat-treatment of the neodymium sample is the stable phase during devitrification at lower temperatures on cooling, and faster cooling rates may be needed here to completely remove it. Another noticeable feature of the heat-treatment results is the increase in β content at the expense of α. This effect has been reported recently, and arises because of the reduced stability of the α phase at temperatures below ≈ 1600°C. As observed elsewhere, more transformation occurs for low atomic number elements, and less for higher atomic number elements. This effect is discussed further in the next section.

3.3 α ↔ β Sialon Transformation

A second set of samples, with the same compositions as those listed in Table 1, were sintered as before, and then heat-treated for increasing times at 1450°C. As noticed in the heat-treatment results reported in Table 2, α↔ β sialon transformation occurred, with increasing amounts with increasing time. Figure 1 shows the reduction in α content for the 4 samples studied as a function of time. Clearly the extent of α→β transformation is more marked in the case of the lower atomic number rare earths than with the higher ones. This has also been observed previously, and can be attributed to the increased amount of liquid remaining in the low atomic number samples due to the higher β content of these samples after sintering.[9] Just as in the transformation of α→β silicon nitride and α-Si_3N_4 → β-sialon, where the presence of a metal sialon liquid phase is essential to improve the kinetics of the process, the same principle is also true in the case of α→β sialon transformation. With ytterbium, the small amount of liquid remaining after sintering results in only a small amount of transformation, whereas in the case of neodymium, the large volume of liquid allows all the α-sialon in the sample to be transformed to β-sialon. The microstructures of as-sintered and heat-treated Nd and Yb samples are shown in Figs 2 and 3; the

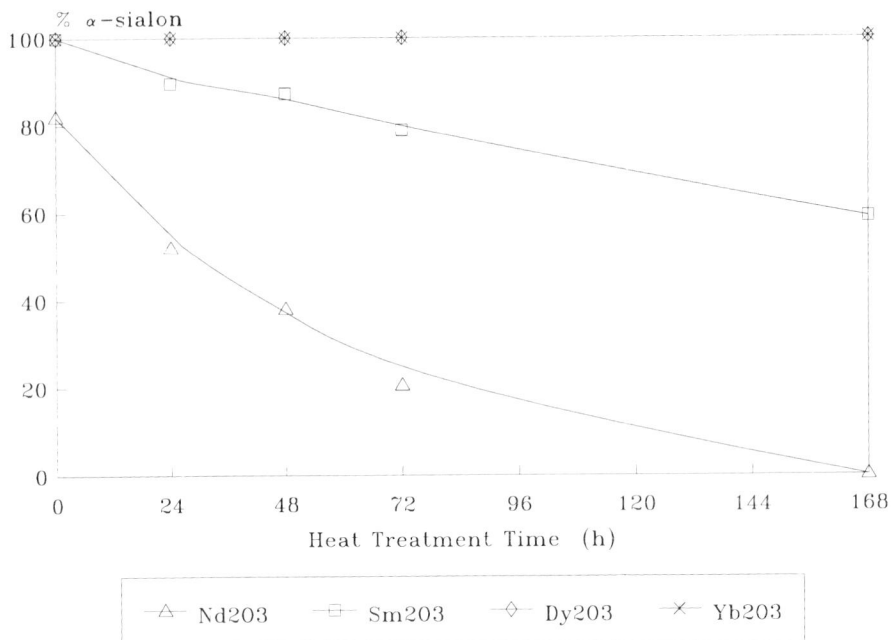

Figure 1.   Plot of % α-sialon against time for samples heat-treated for up to one week at 1450°C.

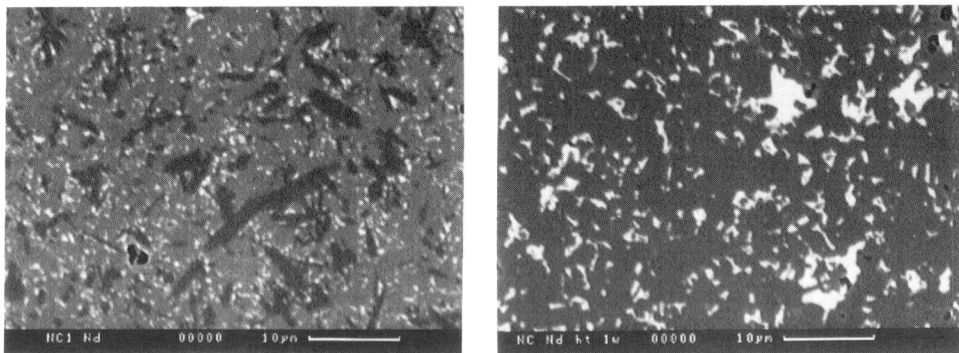

Figure 2.   Back-scattered SEM microstructures of Sample N1000 (a) after sintering at 1800°C and (b) after post-sintering heat treatment at 1450°C for 168 h.

contrast between the β matrix for neodymium and the almost pure α-matrix for ytterbium is clearly apparent. Another effect which comes out clearly from the micrographs is that the small residual glassy pockets in the ytterbium sample take on a distinctly spherical shape and are all of very similar size; moreover, they are not obviously located at the grain boundaries between α grains. Clearly the residual liquid phase in this sample is not wetting the grains, and as a result, the boundaries between adjacent α grains are almost invisible, giving the matrix a very uniform appearance. In contrast, in the neodymium sample, even though there is much more

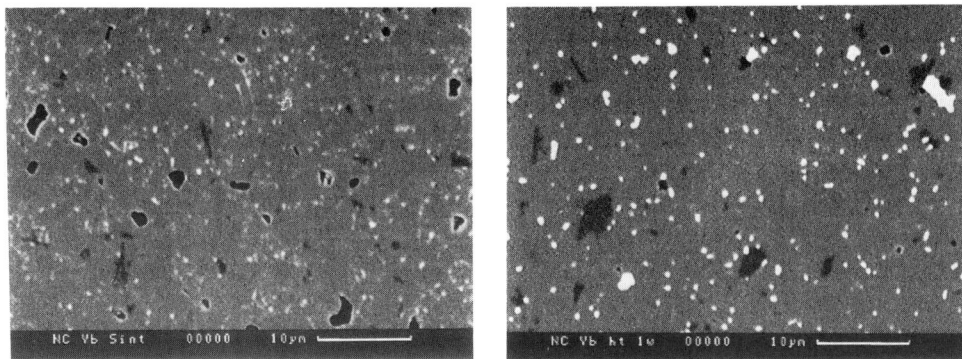

Figure 3.   Back-scattered SEM microstructures of Sample Yb1000 (a) after sintering at
1800°C and (b) after post-sintering heat treatment at 1450°C for 168 h.

second phase (because of $\alpha \rightarrow \beta$ sialon transformation), the shape is more angular, and the location of glassy pockets follows boundaries between β grains. The more wetting nature of the liquid phase in low atomic number rare earth sialon samples may be an additional factor influencing the greater extent of $\alpha \rightarrow \beta$ transformation occurring in this sample.

X-ray diffraction confirmed that the second phase in the heat-treated neodymium-containing samples was M'. Unfortunately it is impossible to distinguish on the micrographs between M' and residual glass, but it is believed from the intensity of M' lines on X-ray diffraction photographs that most of the liquid in the sample has been converted into M'. Since this has occurred simultaneously with the $\alpha \rightarrow \beta$ transformation, the overall reaction occurring can be represented by the equation:

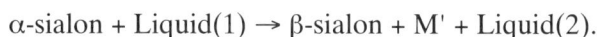

$$\alpha\text{-sialon} + \text{Liquid}(1) \rightarrow \beta\text{-sialon} + M' + \text{Liquid}(2).$$

It is not clear from the work carried out so far whether the main driving force for this reaction is $\alpha \rightarrow \beta$ sialon transformation or crystallisation of M'. In the low atomic number rare earth systems, both reactions occur very readily and to high levels of completion. However, for the purposes of the present study, $\alpha \rightarrow \beta$ transformation is only a complicating factor interfering with the main aim which was to transform the residual liquid phase into M' (plus additional α- or β-sialon to balance the equation). The fact that during heat-treatment, most or all of the α-sialon matrix phase converts to β, releases more liquid forming constituents, and not only does this give a much larger volume of M' in the microstructure, but it is much harder to avoid residual glass in the final heat-treated product. Further work is needed to clear up the uncertainties associated with the combination of chemical reactions taking place during heat-treatment of these samples.

3.4 Mechanical Properties

Table 3 shows the room temperature hardness and indentation fracture toughness of as-sintered and heat-treated materials determined using a 10 kg load, and carried out

Table 3.   Hardness and fractures toughness data for sintered and heat-treated samples

Sample	Type	HV10 (kgmm^{-2})	K_{Ic} (MPam$^{1/2}$)
N1000	Sintered	1487	6.43
	Heat-treated	1525	5.80
S1000	Sintered	1538	5.62
	Heat-treated	1680	4.74
D1000	Sintered	1608	5.43
	Heat-treated	1748	4.65

on samples heat-treated for 24 h at 1550°C. The higher hardness of the as-sintered materials with increasing atomic number reflects the higher α-content of the material. The increase in hardness with heat-treatment reflects the increased hardness of a crystalline grain-boundary phase compared with glass. It is interesting that the α→β transformation simultaneously occurring in the N1000 and S1000 sample does not prevent the increase in hardness due to crystallisation. Nevertheless it is clear that there is less difference in hardness between sintered and heat-treated samples for neodymium (where the sample has transformed from 75:25 α:β to 50:50 α:β) than for dysprosium, where the α:β ratio has changed from only 86:4 to 81:19. It is therefore not surprising that the heat-treated dysprosium sample shows the best hardness.

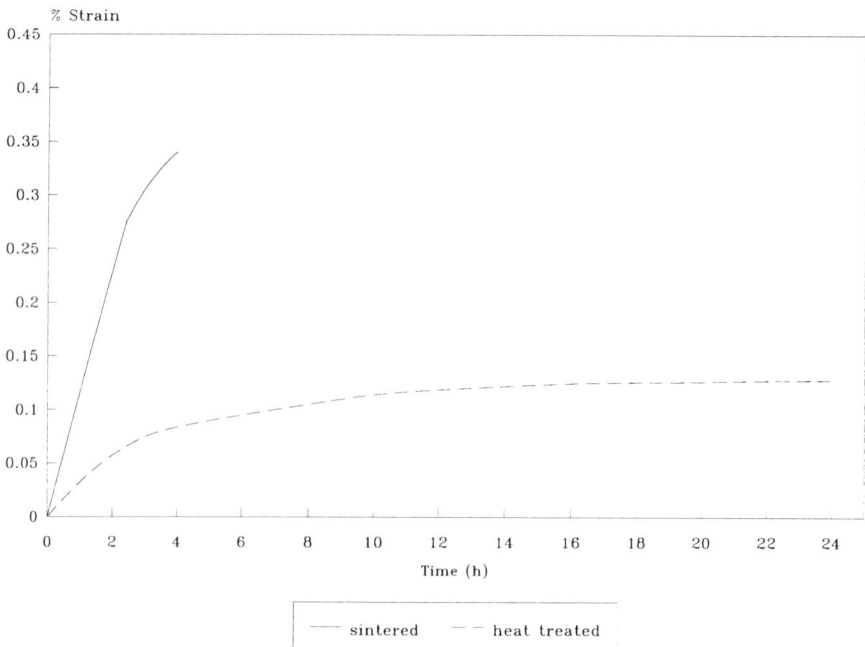

Figure 4.   Creep strain versus time for Sample S1000 after post-sintering heat treatment at 1650°C for 4 h.

Since the main aim of the present work was to prepare low-glass α-β sialon samples by high-temperature ($\approx$ 1550°C) post-preparative crystallisation, creep experiments were conducted on the heat-treated materials to assess residual glass levels. An advantage which has already been demonstrated for the M' phase is that it passivates on oxidation, and therefore shows good oxidation resistance at temperatures up to 1500°C.[3,4] This is an important advantage of these materials, because during creep testing in air, oxidation resistance is also being assessed. Creep testing of nitrogen ceramics is normally carried out at temperatures up to $\approx$ 1350°C, because residual glass in the sample is fairly soft at these temperatures, and eutectic liquid formation results in rapid fall-off in mechanical integrity. However, in the present case, heat-treatment of the sample above the eutectic temperature resulted in very little residual glass, and it was necessary to test the sample at higher temperatures. Figure 4 shows the creep behaviour of the heat-treated S1000 sample at 1600°C using an applied stress of 77 MPa. In this case, prior to testing, the sample had been heat-treated at 1650°C for 4 hours to convert grain boundary liquid into M'. The as-sintered sample crept rapidly at 1600°C as expected. But the heat-treated sample showed an amazingly good performance at this temperature. Clearly, there is very little residual liquid in the sample, and the initial increase in creep strain rate, which is probably due to initial oxidation of the sample, flattens off to show a very small almost constant strain rate after 12 hours.

4. CONCLUSIONS

Compositions have been identified in rare earth densified α-β sialon systems, which after initial sintering followed by heat-treatment above the eutectic temperature, yield mixed α-β sialon materials with the substituted melilite phase, M', as the only crystalline grain-boundary phase. Whereas in principle it might be possible to convert grain-boundary liquid into M' (plus additional α or β sialon), the situation during heat treatment at 1450–1600°C is complicated by the simultaneous additional process of α→β sialon transformation. This generates more M' in the final product and also more liquid phase, as a result of which it is more difficult to avoid final traces of liquid phase in the final product assemblage. The successful tailoring of a totally crystalline product assemblage therefore requires considerable care. However, creep measurements show that successfully-tailored materials can be produced which exhibit total creep strains of less than 0.12 after testing for 24 hours under 77 MPa at 1600°C. This is a formidable result, and confirms that these materials have definite promise for use at higher temperatures compared with more conventional sialon ceramics which contain alternative crystalline phases, and exhibit eutectic melting below 1400°C.

REFERENCES

1. D. P. Thompson: *Brit. Ceram. Proc.* **44**, eds. R. W. Davidge and D. P. Thompson, publ. Institute of Ceramics, 1990, 1.

2. Phase diagrams for Ceramists, **1**, eds. E. M. Levin, C. R. Robbins and H. F. McMurdie, publ. The *American Ceramic Society*, Columbus, USA, 1964, No. 237.

3. Y.-B. Cheng and D. P. Thompson: *J. Amer. Ceram. Soc.,* 1994, **77**, 143.

4. Y.-B. Cheng and D.P. Thompson: *J. Eur. Ceram. Soc.,* 1994, **14**, 13.

5. H. Mandal, D. P. Thompson and Y.-B. Cheng: *Proc. 5th Int. Symp.'Ceramic Materials and Components for Engines'*, Shanghai, 1994, eds. D. S. Yan, X. R. Fu and S. X. Shi, publ. World Scientific, Singapore, 1995, 202.

6. F. F. Lange, *Int. Met. Rev.,* 1980, **25**, 1.

7. J. K. Patel and D. P. Thompson in *Proc. 3rd Int. Symp. 'Ceramic Materials & Components for Engines'*, Las Vegas, 1988, ed. V. J. Tennery, publ. *The American Ceramic Society*, 1989, 987.

8. H. Mandal, D. P. Thompson and T. Ekström, *Special Ceramics,* **9**, ed. R. Stevens, publ. Institute of Ceramics, 1992, 149.

9. H. Mandal, N. Camuscu and D. P. Thompson, 'Comparison of the Effectiveness of Rare Earth Sintering Additives on the High Temperature Stability of α-Sialon Ceramics', *J. Mater. Sci.,* accepted for publication, 1995.

Microstructural Control of Properties in Transformed α-β Sialon Ceramics

H. MANDAL and D. P. THOMPSON

Materials Division, Department of Mechanical, Materials and Manufacturing Engineering, University of Newcastle upon Tyne, NE1 7RU, UK

ABSTRACT

Mixed phase α-β sialon ceramics have been prepared by pressureless sintering at 1775°C using oxides of Sm, Dy, Y and Yb as sintering additives. The resulting materials were then heat treated at 1000–1600°DC to devitrify the grain boundary glass into crystalline oxynitride phases, and at the same time α→β sialon transformation was monitored.

Results have shown that certain sialon compositions can undergo α↔β sialon transformation without further additions of oxides or nitrides merely by heat treatment at appropriately chosen temperatures. This transformation provides an excellent mechanism for optimizing phase content and microstructure, and in this way predetermined values of hardness, strength and toughness can be achieved from a single starting composition.

1. INTRODUCTION

Although the main motivation for the development of silicon nitride based ceramics during the last 30 years has been the prospect of the ceramic gas turbine,[1] the commercial realisation of this objective is still far from fulfilment and alternative applications have increased in importance. The unique combination of properties of sialon ceramics has led to them being applied to a wide range of lower-temperature applications; the most successful of which so far has been in the area of cutting tools.[2]

Pressureless sintered β-sialons were the first nitrogen ceramics to achieve large-scale commercial viability. This was because the presence of alumina in the starting mix lowers the eutectic temperature of the densifying liquid phase by some 200–300°C, making it possible to achieve full density by pressureless sintering and allowing complex shaped components to be prepared easily in large numbers and at lower cost. The resulting β-sialon materials have a Young's Modulus of ~300 GPa, strengths of up to 1000 MPa and K_{IC} values of up to 8 MPam$^{1/2}$. Refractoriness, which is poor in as-sintered materials (because of residual glassy phase) can be improved by heat-treatment, and hardness can be increased by modifying the starting composition to include a certain proportion of α-sialon in the final product. This is done by selecting as oxide sintering additives, those metals which are also α-sialon stabilizers (i.e. Li, Mg, Ca, Y or most Ln). Mixed two phase sialon ceramics offer possibilities of tailoring the microstructure and consequently the properties of the final product compared with monolithic α and β sialon ceramics for engineering applications.[3,4]

One of the main concerns of industry in the 21st Century will be the production cost and also versatility for different applications. Current technology for achieving different combinations of properties in sialon cutting tools requires the use of different starting compositions or replacement of the commonly used yttria sintering aid by other rare earth element oxides;[3,5] in future it will be important to develop technologies to achieve a wide range of properties from the same starting sialon compositions.

The present work describes the results of carrying out heat treatments on mixed phase α-β sialon ceramics, which result in phase changes in both the matrix and grain boundary phases as a result of which the microstructure reorganises itself and mechanical properties are affected. By careful control of the heat-treatment schedule it is possible to produce a precisely tailored combination of strength, hardness and toughness in the final product.

2. EXPERIMENTAL

Sialon compositions were prepared using powder mixtures of Si_3N_4 (HC Starck-Berlin, Grade LC1), Al_2O_3 (Alcoa, Grade A16SG), AlN (HC Starck-Berlin, Grade A), plus a constant molar amount of densifying additive e.g. 6 w/o Y_2O_3, 8.9 w/o Nd_2O_3, 9.3 w/o Sm_2O_3, 9.9 w/o Dy_2O_3, 10.5 w/o Yb_2O_3 to give overall compositions of the type $(Y,Ln)_{0.053}Si_{1.77}Al_{0.23}O_{0.33}N_{2.43}$ and $(Y,Ln)_{0.053}Si_{1.78}Al_{0.24}O_{0.26}N_{2.49}$, hereafter in this paper referred to as Compositions 1 and 2 respectively. Better representation of starting compositions are given in Fig. 1 in terms of equivalents of oxygen and aluminium added to silicon nitride.

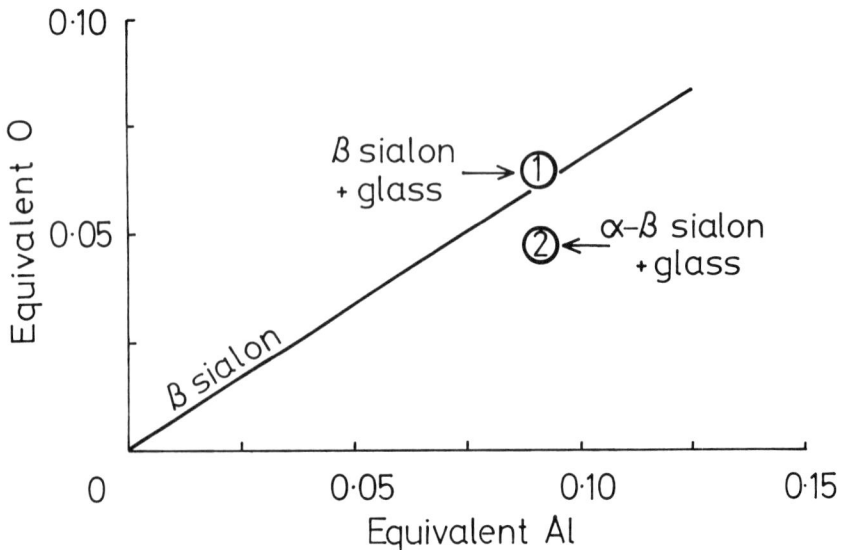

Figure 1.   Overall starting compositions represented in terms of equivalents of oxygen and aluminium added to silicon nitride.

The starting powders were mixed in water-free isopropanol and milled in rubber lined mills for 24 hours using sialon milling media. The slurry was then dried and compacts of size 16 × 16 × 6 mm made by die-pressing at 125 MPa. The specimens were subsequently embedded in micron-sized boron nitride powder in graphite crucibles and fired by pressureless sintering. Initial pressureless sintering was performed in a production furnace of 240 dm^3 volume at 1775°C under a slight nitrogen gas overpressure for 2 hours. Because the furnace was large, cooling rates were slow, and insufficient to prevent crystallization of the grain boundary glass. Samples were therefore subsequently re-sintered at the same temperatures in a small graphite-element furnace, and quenched by switching off the power at the end of the run. In this way, the crystalline grain-boundary phases were re-melted and then converted into a glass on cooling. The fully-dense, sintered and re-sintered materials were then subsequently heat-treated for 24 hours in nitrogen at 1000, 1150, 1300 and 1450°C in a molybdenum wire-wound alumina tube furnace.

Product phases were characterised by X-ray diffraction, using a Hägg-Guinier camera and CuKα$_1$ radiation. A computer-linked line scanner system (Type SCANPI LS-20) developed by Professor Werner (Arrhenius Laboratory, Stockholm University) was used for direct measurement of X-ray films and refinement of lattice parameters. The amounts of α and β sialon phases were found by quantitative estimation from the X-ray diffraction pattern using in-house calibration curves based on the integrated intensities of the (102) and (210) reflections of α-sialon and the (101) and (210) reflections of β-sialon.[6] The curves used were originally developed for α-silicon nitride rather than α-sialon, and therefore the amounts of α-sialon reported in this paper are overestimated by up to 10% compared with their true values, because of increased intensities of α-sialon lines for a given amount of sample compared with α-Si$_3$N$_4$ values.

Hardness (HV10) and indentation fracture toughness (K$_{IC}$) measurements were made at room temperature using a Vickers diamond indenter with a 98N (10 kg) load, and fracture toughness was evaluated according to the method of Evans and Charles.[7]

Microstructures were observed using a Camscan S4-80DV scanning electron microscope.

3. RESULTS AND DISCUSSION

3.1 Sintering and Heat-Treatment

Since pressureless sintering of these compositions was originally carried out in a large-scale furnace at 1775°C, X-ray analyses of the samples after sintering showed additional crystalline grain boundary phase(s). These were attributed to the slow rate of cooling in the furnace used. To avoid crystallisation, the sintered samples were re-heated to the sintering temperature (1775°C) for 15 minutes and then rapidly cooled to 900°C (2 minutes total cooling) followed by further cooling to room temperature at ~100°C per minute to avoid any crystallisation of oxynitride phases. X-ray diffraction showed that this indeed took place, but it was noticed that in all samples the

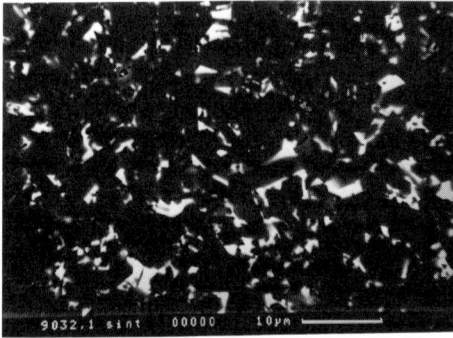
(a)

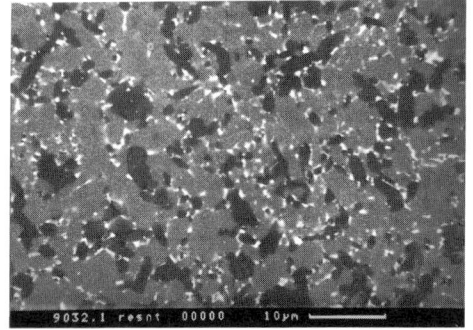
(b)

Figure 2.   Back scattered SEM micrographs of Yb_2O_3-densified pressureless sintered samples: (a) sintered in production furnace, and (b) re-sintered in laboratory and rapidly cooled to room temperature to prevent crystallisation of the grain boundary glass.

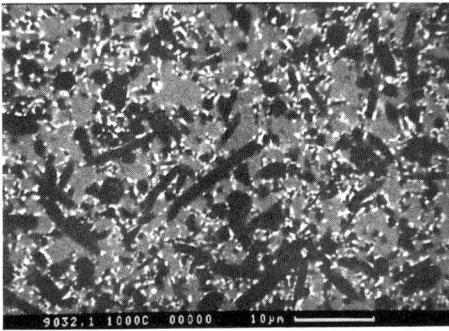
(a)

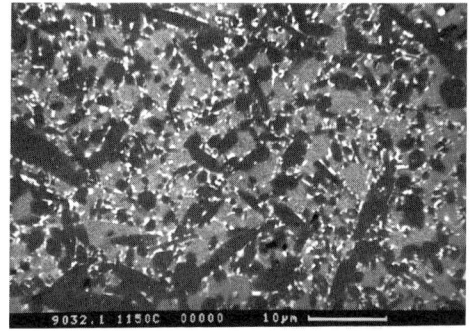
(b)

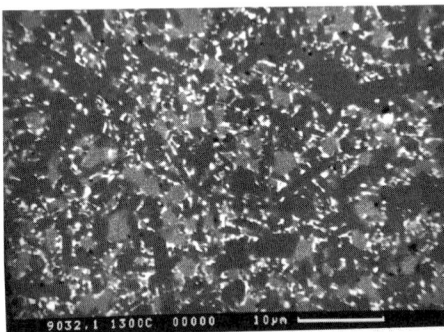
(c)

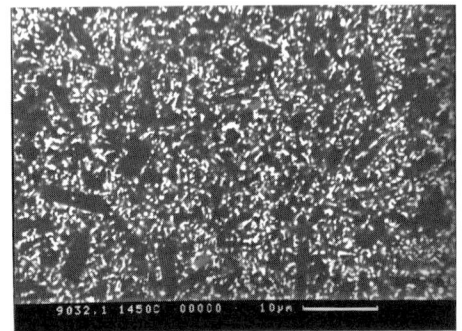
(d)

Figure 3.   SEM micrographs of the Yb_2O_3 densified sample shown in Figure 2 after further heat treatment at (a) 1000°C, (b) 1150°C, (c) 1300°C and (d) 1450°C.

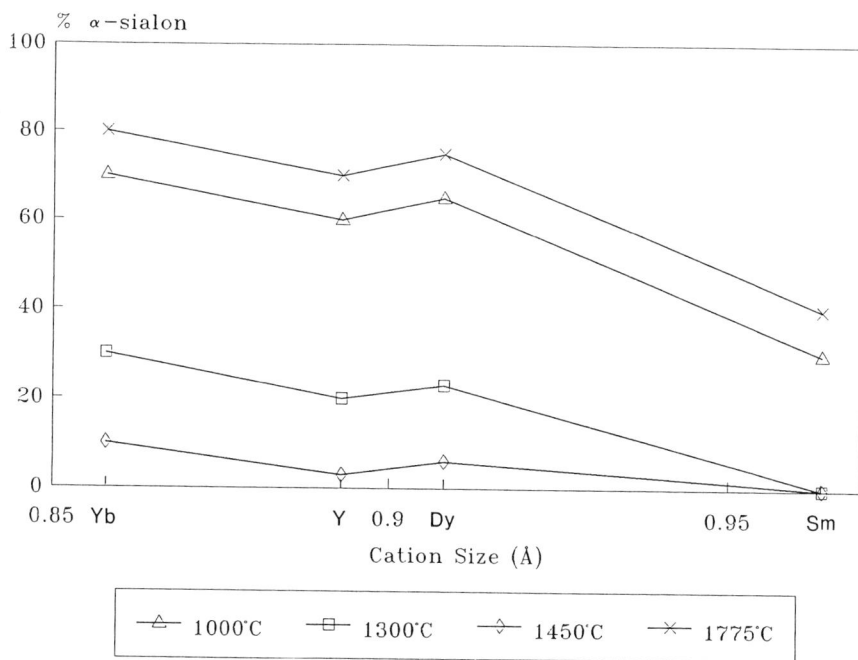

Figure 4.   α:(α+β) sialon ratio in sialon samples (composition 1) densified with different sintering additives and different heat treatment temperatures.

α:(α+β) ratio had increased. The variation in the amount of α-sialon in terms of sintering additive has been discussed elsewhere.[5,8] Thus, the original samples designated β-sialon, now contained a significant amount of α-sialon. The effect was more marked in the case of samples densified with higher atomic number rare earth oxides (e.g. Yb_2O_3, Dy_2O_3) since they contain the highest amount of α-sialon. This case is clearly apparent also from microstructures shown in Fig. 2, which shows the same sample sintered in the large production furnace and re-sintered in the laboratory type, high cooling rate furnace. Back scattered electron micrographs very clearly distinguish between the various phases; the β-sialon grains (which contain no rare earth element) are black and more needlelike, whereas the α-sialon grains (which contain a small amount of rare earth element) are grey and more equiaxed whilst rare earth rich crystalline or glassy phases appear fine grained and white, because of the high rare earth content.

Subsequent heat treatments of the re-sintered and fast cooled samples performed at 1000–1450°C and details of the crystalline phases produced and their high temperature performance are given elsewhere.[9] The noticeable difference in heat treated samples was the change in α:(α+β) sialon ratio as the heat treatment temperature increased. In this case the reverse reaction occurred (compared with that observed during re-sintering), i.e. the α:(α+β) sialon ratio in all samples dramatically decreased. This was again most marked for the higher atomic number rare earth sintering additives. Thus for example, the Yb_2O_3 densified β-sialon sample, which, after the

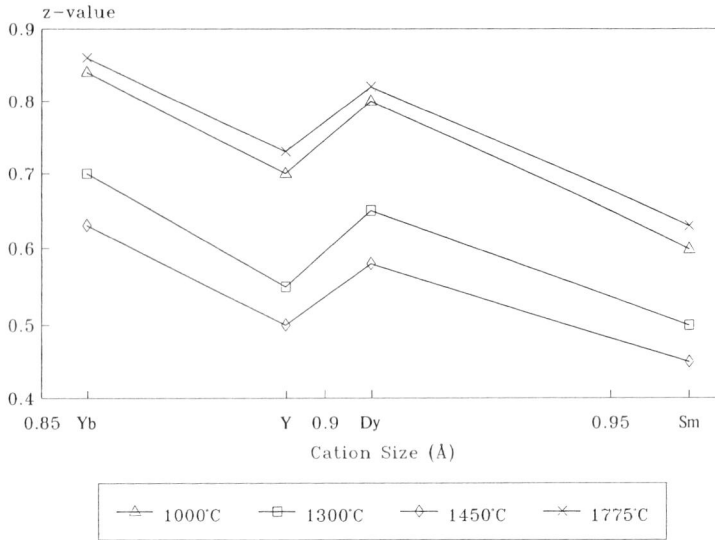

Figure 5.   Z-values of the β-sialon phase in mixed α-β sialon samples after further heat treatment as a function of sintering additive and heat treatment temperature.

initial production furnace pressureless sintering run at 1775°C was identified as containing 10% α-sialon and 90% β-sialon, and after re-sintering for 15 minutes at 1800°C followed by rapid cooling changed to 80% α-sialon and 20% β-sialon, dramatically changed to 10% α–90% β after 1450°C heat treatment, 40% α–60% β after 1300°C heat treatment, 50% α–50% β after 1150°C heat treatment and 70% α–30% β after 1000°C heat treatment. Back scattered electron images of these samples also clearly confirm the changes in α:(α+β) ratio that occur; Fig. 3 shows this kind of change for Yb_2O_3 densified material. This interesting transformation process was also observed for other sintering additives and the results of α:(α+β) transformation are given in Fig. 4.

   Another interesting feature of the microstructures is the change in α-sialon grain size. In the initial 1775°C sintered sample, the grain size is quite large (typically 2–3 μm) and this is substantially preserved at 1000°C, because this temperature is too low for much change to occur (even though the higher heat treatment temperature results show that α-sialon is probably unstable with respect to β-sialon, but the kinetics are too slow for any transformation to occur). At 1150°C, some α-sialon → β-sialon transformation occurred, and each α-sialon grain has split into smaller grains of β-sialon plus grain boundary phase. This process becomes more marked at the higher heat treatment temperatures, so that the microstructures of the 1450°C heat treated samples consist of large (up to 10 μm in length) β-sialon needles surrounded by a finer matrix of α-sialon plus other grain boundary phase(s).

   An additional effect which is clear from Fig. 3 is that as more α-sialon transforms to β-sialon, the β-sialon z-value falls from 0.84 to 0.63 (see Fig. 5) and there is a very small decrease in α-sialon unit cell dimensions from $\mathbf{a}$ = 7.805Å, $\mathbf{c}$ = 5.693Å after 1000°C heat treatment to $\mathbf{a}$ = 7.795Å and $\mathbf{c}$ = 5.682Å after heat treatment at 1450°C.

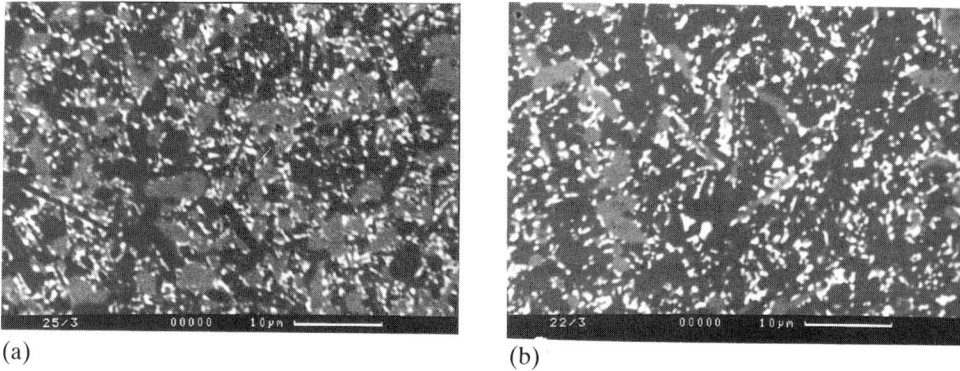

Figure 6. Back scattered SEM micrographs of Dy_2O_3 densified samples: (a) sintered and rapidly cooled, (b) heat treated at 1450°C for 60 hours.

These changes show that with increasing heat-treatment temperature, Si and N preferably come out of the α-sialon phase and go into β-sialon, whilst the residual Ln, Al and O are transferred to the grain-boundary phase assemblage.

The trend is also clearly apparent from microstructures of samples sintered with Dy_2O_3 (Fig. 6). After sintering followed by rapid cooling ≈63% α-sialon is observed, whereas after heat treatment, the α-sialon content dramatically changes to 18% α-82% β after 24 hours and 8% α-92% β after 60 hours.

Taking the now mainly β-sialon Dy-densified sample to 1800°C for a short time (≈ 15 min) once again causes the reverse (i.e. β → α sialon) transformation to occur (Fig. 7(a)), with reformation of the original α-sialon grains. Note that these are very similar in size to those present in the as-sintered samples. Again, taking the now mainly α-sialon Dy-densified sample to the heat treatment temperature of 1450°C for 60 hours, α-sialon transforms to β-sialon and very similar micrographs were obtained similar to Fig. 7(b) (see Fig. 6(b)). The same kind of behaviour was observed also with other sintering additives. These observations confirmed that the α ↔ β sialon transformation is reversible and similar microstructures can be obtained after several heat treatment cycles.

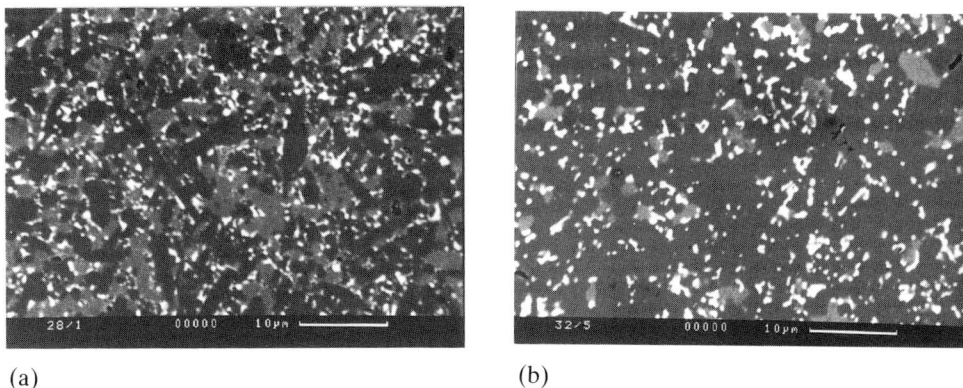

Figure 7. Back scattered SEM microgrpahs of Dy_2O_3 densified samples: (a) re-sintering of the sample in Figure 6(a) at 1800°C for 15 minutes, (b) heat treated at 1450°C for 60 hours.

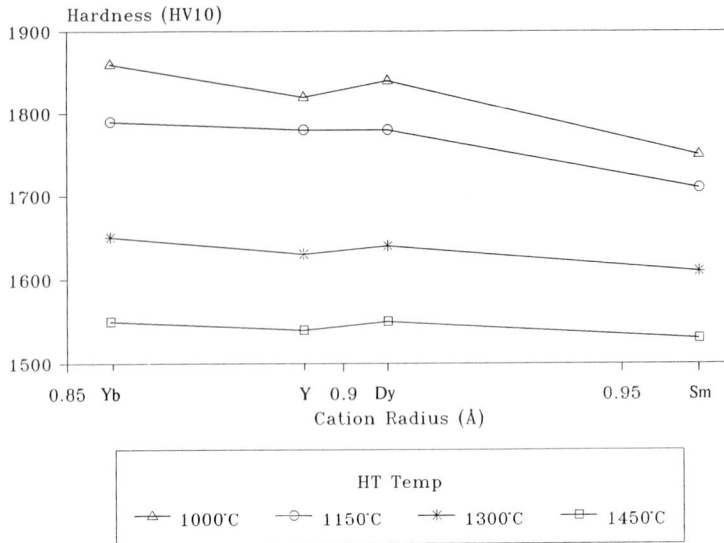

Figure 8.   Vickers hardness (HV10) of sialon ceramics (composition 1) as a function of
ionic size of rare earth cation and heat-treatment temperature.

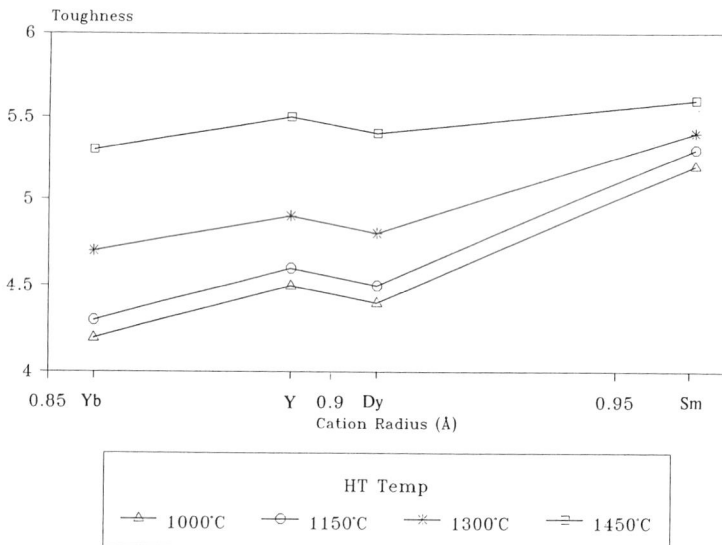

Figure 9.   Fracture toughness of sialon ceramics (composition 1) as a function of ionic size
of rare earth cation and heat-treatment temperature.

3.2 Mechanical Properties

Vickers hardness (HV10) and indentation fracture toughness (K_{IC}) measurements of the heat treated materials are shown in Figs 8 and 9 respectively. It is well known that an increase in α-sialon content leads to increased hardness and this trend is clearly visible in all samples prepared in the current work. A further reason for the higher hardness is the lower level of glass in α-sialon rich samples. Figure 8 shows an increase in fracture toughness after heat-treatment. Samples containing large β-sialon needles after sintering, retained these during subsequent heat-treatments, but these samples showed a higher K_{IC} at all temperatures, clearly attributable to the newly formed fine β-sialon grains. It is believed that the explanation for this behaviour lies in the increased possibilities for crack deflection in the fine grained microstructure of the heat treated samples. This effect cannot be separated from the increased proportion of tiny, acicular β-sialon grains, but is expected to provide a larger toughening contribution than that afforded by the relatively small aspect ratio of these transformed β-sialon grains.

The results shown in these figures confirm that hardness decreases and fracture toughness increases as the heat treatment temperature is increased up to 1600°C in direct proportion to the extent of α→β sialon transformation and therefore this transformation offers an excellent means of controlling the properties of the final ceramic.

3.3 Mechanism

It is not uncommon in mixed α-β sialon ceramics to find a different α:β sialon ratio in the final product compared with that expected from the starting composition, and also to find a different ratio in the centre of the sample compared with the outside. This is because the line of β-sialon solid solution and minimum rare earth content edge of the α-sialon region are very close to one another and the proportions of α- and β-sialon are therefore very sensitive to slight changes in composition which arise from milling additions, effects of furnace atmosphere, weight loss etc. However, the phenomenon described here is clearly not accountable on these grounds, because it is reversible.

For a better understanding of α↔β sialon transformation, samples were prepared in a more nitrogen rich area, Composition 2, in Fig. 1. Samples were prepared and heat treated in exactly the same way as Composition 1 and the results of XRD are given in Fig. 10 as a comparison with Composition 1 after 24 hours heat treatment at 1450°C. As can be seen from the figure, the amounts of α and β sialon are very nearly the same (73% α-sialon for Composition 1 and 82% α-sialon for Composition 2, in the Yb- sample) after sintering with Yb_2O_3 and Dy_2O_3 as the additive. However, the amounts of α and β sialon present in the two compositions were significantly different when Sm_2O_3 and especially Nd_2O_3 were used as the sintering additives ($\approx$ 51% and $\approx$ 20% α-sialon for Nd- Composition 1 and Composition 2, respectively). The reason for this difference between high and low atomic number rare earth sintering additives can be explained in terms of chemical stability of α-sialon in terms of ionic size.[5]

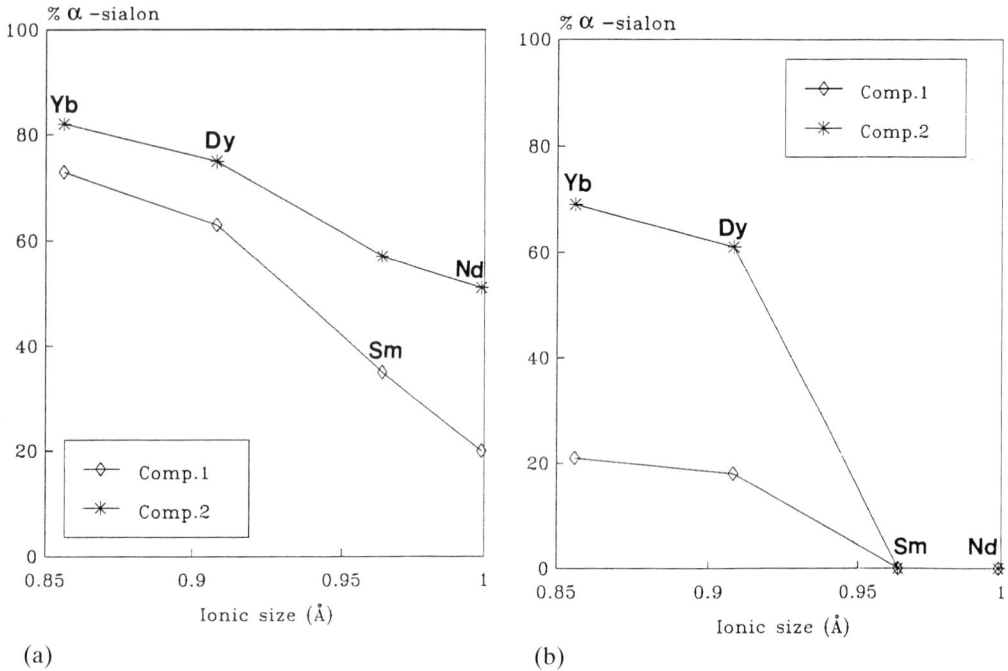

Figure 10.   α:(α+β) sialon ratio of (a) sintered and (b) 24 hours at 1450°C heat treated samples densified with different sintering additives.

Figs 11(a) and (b) show back scattered electron micrographs for Yb_2O_3 densified samples for which the amount of grain boundary phase in Composition 1 is nearly twice that in Composition 2. The difference between the two compositions can be more easily distinguished after heat treatment at 1450°C (see Figs 11(c) and (d)). It can be seen that there is a marked difference in the amount of transformation between Compositions 1 and 2 with Composition 1 showing almost complete transformation from α→β sialon whereas only a small amount of transformation is observed in Composition 2. Clearly, more oxygen rich α-sialons, such as present in Composition 1, transform to β-sialon more readily than nitrogen rich α-sialons, typified by the one present in Composition 2. The same kind of behaviour has been observed when Dy_2O_3 is used as the sintering additive.

The Nd- and Sm- systems show a greater tendency for α→β sialon transformation. Since Nd- and Sm- have a larger ionic radius compared to other rare earth elements such as Dy- and Yb-, formation of α-sialon becomes more difficult during sintering and therefore the amount of α-sialon obtained is less than that predicted.[10] The excess rare earth cations go into the liquid phase and increase the amount and reduce the viscosity of it. As a result, α→β sialon transformation becomes significantly faster and no α-sialon remained in Composition 2 after 24 hours heat treatment.

These results clearly show that a most important factor influencing transformation is the residual oxynitride liquid phase which is present in greater amount and with a higher oxygen content in Composition 1 compared with Composition 2. As a result, α→β sialon transformation proceeds more readily for Composition 1 than for Composition 2.

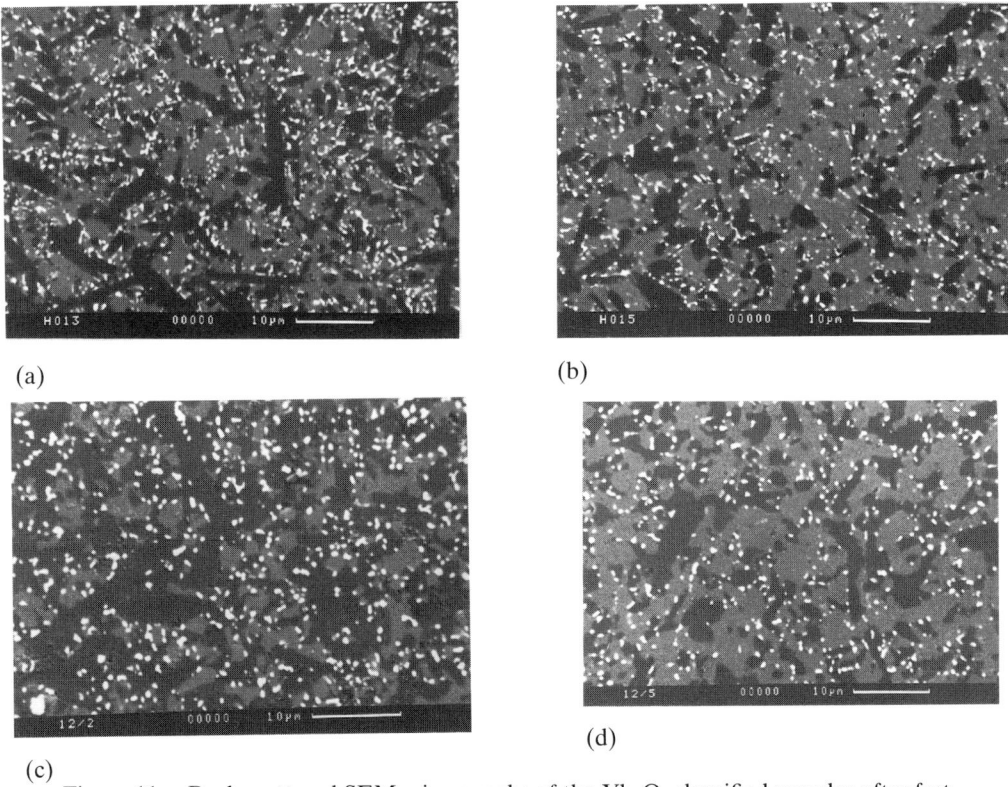

(a)

(b)

(c)

(d)

Figure 11.   Back scattered SEM micrographs of the Yb_2O_3 densified samples after fast cooling: (a) Composition 1, (b) Composition 2 and after heat treatment at 1450°C: (c) Composition 1, (d) Composition 2.

4. CONCLUSIONS

A combination of X-ray and microstructural observations on rare earth densified mixed α-β sialon ceramics has shown that the ease with which the α→β transformation proceeds depends mainly on the amount and viscosity of liquid phase present during heat treatment and therefore more oxygen rich α-sialons, such as the one present in Composition 1, are only stable at high temperatures (> 1600°C for Yb α-sialon) and transform to β-sialon more readily than nitrogen rich α-sialons, typified by the one present in Composition 2.

α→β and β→α sialon transformations are completely reversible.

The transformation offers an excellent mechanism for optimizing the properties of mixed-phase α-β sialon ceramics below the transformation temperature. This phenomenon has considerable similarities to the process of heat-treatment in metallurgy, where diffusion of the mobile species (in this case the rare earth cations) allows not only phase transformations in the matrix and the formation of different phases in the grain boundary, but also tailoring of the microstructure. This principle is seen as having important consequences for mechanical property optimization in other ceramic systems.

ACKNOWLEDGEMENT

One of us (H.M.) would like to acknowledge financial assistance from the Engineering and Physical Sciences Research Council (EPSRC) during the course of this work.

REFERENCES

1. K. H. Jack: 'Sialon Hardmetal Materials', in Proc. 2nd Int. Conf. Science Hard Mater., ed. E. A. Almond, C. A. Brookes and R. Warren, publ. Adam Hilger Ltd, Bristol, UK, 1986, 363–76.
2. N. E. Cother and P. Hodgson: 'The Development of Syalon Ceramics and their Engineering Applications', *Trans. J. Br. Ceram. Soc.*, 1982, **81**, 141–44.
3. T. Ekström and I. Ingelström: 'Characterization and Properties of Sialon Ceramics', In Non-oxide Technical and Engineering Ceramics, ed. S. Hampshire, Elsevier Applied Science Publishers, London, 1986, pp. 231–53.
4. G. Z. Cao, R. Metselaar and G. Ziegler: 'Microstructure and Properties of Mixed α-β Sialons', in Proceedings of the 4th International Symposium on Ceramic Materials and Components for Engines, ed. R. Carlsson, T. Johansson and L. Kahlman, Elsevier Applied Science Publishers, London, 1993, 188–95.
5. H. Mandal: 'Sialon Ceramics Sintered with Additions of Yb_2O_3, Dy_2O_3 and Sm_2O_3 or as Mixtures with Y_2O_3', in Proc. 8th International Metallurgy and Materials Congress, accepted for publication, Turkey, June '95.
6. K. Liddell: 'X-ray Analysis of Nitrogen Ceramic Phases', M.Sc. Thesis, University of Newcastle upon Tyne, 1979.
7. A. G. Evans and A. Charles: 'Fracture Toughness Determination by Indentation', *J. Amer. Ceram. Soc.*, 1976, **59**, 371–72.
8. H. Mandal, D. P. Thompson and T. Ekström: 'Heat Treatment of Ln-Si-Al-O-N Glasses', Proc. 7th Irish Forum Conference IMF7, Limerick, Ireland Sept. 1991, Editor: M. Buggy and S. Hampshire, Trans Tech Publication, Key Engineering Materials, Vol. 74–76, Zurich, 1992, 187–203.
9. H. Mandal, D. P. Thompson and T. Ekström: 'Heat Treatment of Sialon Ceramics Densified with Higher Atomic Number Rare Earth and Mixed Yttrium/Rare Earth Oxides', Special Ceramics 9, London, UK, Dec. 1990, Editor: R. Stevens, Institute of Ceramics, Stoke on Trent, UK, 1992, 97–104.
10. H. Mandal, N. Camuscu and D. P. Thompson: 'Comparison of the Effectiveness of Rare Earth Sintering Additives on the High Temperature Stability of α-Sialon Ceramics', accepted for publication in *J. Mater. Sci.* 1995.

Vacuum Heat Treatment of Sialon Ceramics

H. MANDAL and D. P. THOMPSON

Materials Division, Department of Mechanical, Materials and Manufacturing Engineering, University of Newcastle upon Tyne, NE1 7RU, UK

ABSTRACT

Recent work has shown that dense, glass-free, Si_3N_4 ceramics can be prepared by giving as-sintered samples, densified using the oxides of volatile metals, a vacuum heat-treatment (VHT) or a hydrogen heat-treatment (HHT) at carefully optimised temperatures below the original sintering temperature. In this way, residual M–Si–O–N glassy phase is removed as a gaseous mix of M, SiO and N_2, leaving a pure β-Si_3N_4 residue. Thermodynamic calculations show that magnesium oxide is the best sintering aid for subsequent removal in gaseous form, and other possible oxides are Li_2O, SiO_2, SrO, CaO and Yb_2O_3. Some shrinkage occurs during VHT, as a result of which densities in excess of 99% of theoretical can be retained. It is important to use a minimum amount of densifying additive (consistent with achieving full density after sintering) so as to maximize the final density after VHT. The resulting materials have excellent creep properties and show good oxidation resistance up to temperatures in excess of 1600°C.

The present paper describes the application of the same procedures to sialon ceramics. Sialons have the advantage over silicon nitride that they can be pressureless sintered to full density, and therefore it is easier to produce the initial fully-dense sample. However, it is impossible to fully remove Al-rich glassy residues by VHT (or HHT), because the only available volatile aluminium-containing species, Al_2O is appreciably volatile only at temperatures in excess of ≈ 1800°C. The high temperature properties of VHTed sialons are therefore not as good as Si_3N_4-based materials, but are nevertheless significantly better than the properties of as-sintered materials.

The techniques described here offer a completely different approach to the task of producing sialon ceramics which retain good mechanical properties up to higher temperatures. Instead of using refractory oxides to promote the formation of high-viscosity M–Si–Al–O–N liquids and high T_g residual grain-boundary glass, this approach uses oxides of volatile metals which form low-melting, low-viscosity liquids, and which are then subsequently removed. The method offers considerable promise for the production of a more refractory generation of nitrogen ceramics.

1. INTRODUCTION

Aluminium-containing additives are used extensively along with the other oxide additives, as sintering additives for silicon nitride because of the marked effect they have on densification. Indeed, in the absence of alumina or aluminium nitride, it is very difficult to achieve a fully dense silicon nitride product without the application of pressure (i.e. using GPS, HP or HIP). The most important role of aluminium is not that it substitutes for silicon in silicon nitride to form a β-sialon product, but rather that it forms a much lower melting Si–Al–O–N or M–Si–Al–O–N liquid (M = additional metal cation present as a sintering additive) than in M–Si–O–N systems, with reduced viscosity and improved wettability, which greatly assists the rearrangement and $\alpha \rightarrow \beta$ solution-reprecipitation stages of densification.[1] On cooling, aluminium remains in both the sialon matrix grains and also in the grain boundary phase, where it may be present either as a glass, a crystalline phase or more commonly a mixture of both. Glass-containing nitrogen ceramics lose strength at about 900–1000°C, due to the glass softening; in materials with crystalline grain boundaries, strength is retained up to the eutectic temperature in the appropriate M–Si–Al–O–N system. This is typically 1300–1350°C, which is some 300°C higher than softening temperatures for M–Si–Al–O–N glasses.[1] However, this is still a low temperature compared to the decomposition temperature of β-sialon (≈ 1900°C). Clearly the best way to produce a glass-free product would be to avoid the necessity of having glass-forming constituents present in the first place, but it has been established beyond dispute that sialon ceramics requires a liquid phase in order to achieve complete densification. The present paper explores an alternative avenue for producing glass-free materials, namely the removal of glass by post-preparative vacuum or hydrogen heat-treatment.

Recent research[2,3] has described the application of this process to silicon nitride ceramics, and shown that the glassy phase in aluminium-free silicon nitride ceramics densified by hot-pressing with $\approx 2\%$ MgO, can be totally removed by post-preparative vacuum heat-treatment at 1570°C. This process is aided by surrounding the sample with a carbon environment, so that the Mg–Si–O–N glass is effectively reduced to a mixture of $Mg(g)$, $SiO(g)$, $CO(g)$ and $N_2(g)$. The glass boils out of the sample in gaseous form, and during this process the sample shrinks slightly. Thus, instead of the (say) 7 w/o loss (calculated from 2 w/o MgO, 3–4 w/o SiO_2 and some nitrogen as the glass-forming constituents), resulting in a 7% reduction in density, densities up to 99% can be retained. This figure is larger if less MgO is used for the original densification.

Most of the previous work in this area was carried out using MgO as the densification additive, because this is the easiest, cheapest and most convenient oxide additive which can subsequently be removed by vacuum heat-treatment. Thermodynamic calculations have shown that the other metal oxides which can be used for densification and subsequently removed by VHT are Li_2O, SiO_2, CaO, SrO and Yb_2O_3, and successful vacuum heat-treatment has been demonstrated for the most of these additives[2,3].

This paper applies the same procedure to β-, α- and O-sialon ceramics, and describes the effectiveness of the process in removing grain-boundary glass.

2. EXPERIMENTAL

Starting powder mixes corresponding to β-sialon ($z = 1$), α-sialon and O-sialon final products were made up using silicon nitride (H.C. Starck Berlin, Grade LC10), alumina (Alcoa, Grade A17), aluminium nitride (H.C. Starck-Berlin, Grade B), Li_2CO_3 (BDH, AnalaR grade), CaO (BDH, AnalaR grade), MgO (BDH, AnalaR reagent), $SrCO_3$ (Laboratory Chemical Grade, Griffin & George) and Yb_2O_3 (99.9%, Rare Earth Products). Powders were mixed by ball milling for 1 day in isopropanol, dried and compacted into pellets by uniaxial and cold isostatic pressing.

They were then pressureless sintered by heating in BN-lined carbon crucibles, to temperatures in the range 1700–1800°C, in a resistance-heated carbon element furnace. Samples were then given post-preparative heat-treatments in the same furnace at ≈1650°C in the vacuum provided by a rotary pump (typically 10^{-1} torr). Samples were characterised both before and after heat-treatment by weight change, density measurement (immersion in both mercury and hot-water), X-ray diffraction (Hägg-Guinier focusing camera with $CuK\alpha_1$ radiation) and Scanning Electron Microscopy (Camscan S4–80DV with EDX facilities). In the latter case, EDX scans were taken over relatively large areas of the microstructure to observe the average reduction in sintering additive concentration after heat-treatment.

3. RESULTS AND DISCUSSION

3.1 β-Sialon

Table 1 shows the results of vacuum heat-treating β-sialon ceramics.

Consideration of the table shows that significantly different results are obtained when different powder media are used to surround the specimen. Boron nitride effectively protects the sample against the severity of the reducing atmosphere provided by the carbon furnace. In this case, weight losses during VHT are low, even when quite high temperatures (1750°C) are used. Using no powder bed results in some weight loss during VHT, but again, higher temperatures need to be used if all the additive is to be removed. In contrast, using carbon as the powder bed results in efficient removal of volatile constituents. Clearly the carbon establishes a highly reducing atmosphere in the immediate vicinity of the sample.

Table 1. Weight loss and X-ray diffraction data for z = 1 β-sialon ceramics after VHT with the sample (a) open to the furnace atmosphere and (b) packed in carbon powder.

				(a)				
Sample	Wt. % additive	T (°C)	t (h)	Density (gcm⁻³) Starting	After VHT	ΔW (%)	% T.D.	X-ray results
1	1 MgO	1625	3	3.18	3.15	3.85	99.0	β-sialon
2	0.5 MgO	1625	3	3.18	3.16	2.65	99.4	β-sialon
3	0.7 Li₂O (*)	1550	4	3.09	3.03	5.25	97.4	β-sialon
4	2.6 SrO (*)	1700	4	3.26	3.23	2.50	99.1	β-sialon

				(b)				
Sample	Wt. % additive	T (°C)	t (h)	Density (gcm⁻³) Starting	After VHT	ΔW (%)	% T.D.	X-ray results
1	1 MgO	1625	2	3.18	3.13	4.3	98.4	β-sialon
3	0.7 Li₂O (*)	1550	1	3.09	3.00	5.6	97.0	β-sialon
4	2.6 SrO (*)	1650	2	3.26	3.18	7.2	97.5	β-sialon

* Oxides added as carbonate.

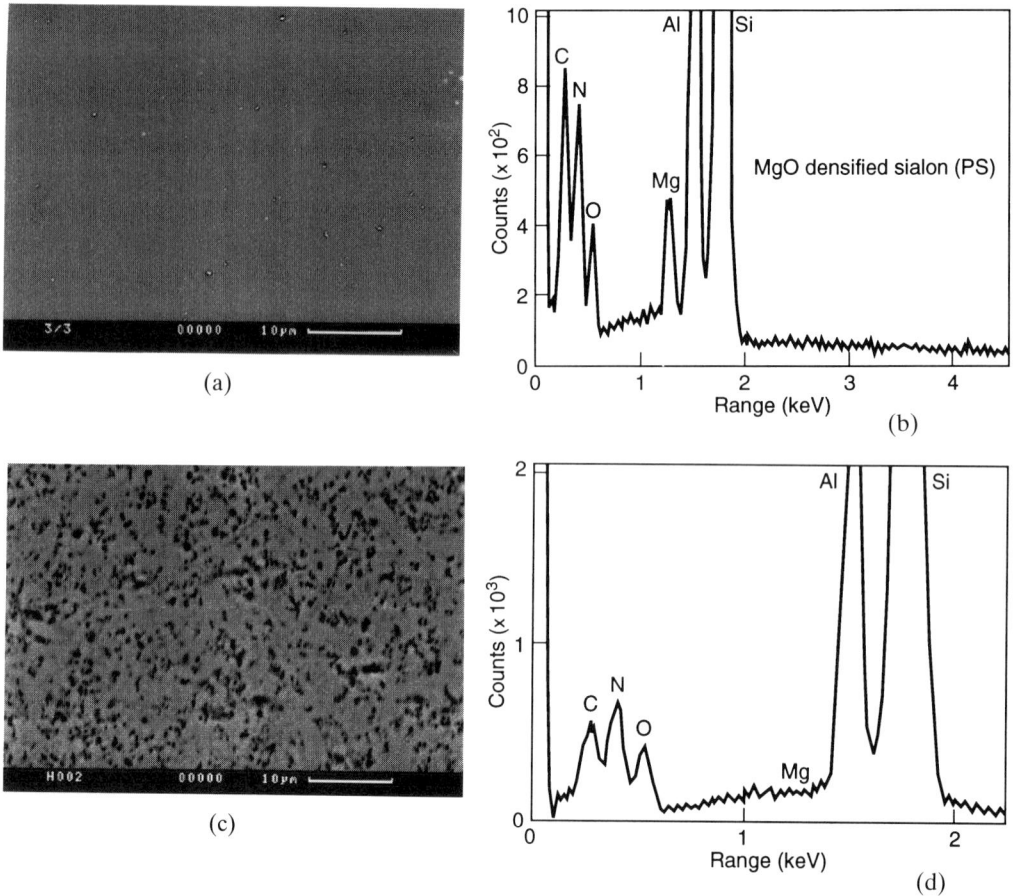

Figure 1.   (a) SEM micrograph and (b) EDX spectrum of z = 1 β-sialon pressureless-sintered with 1% MgO at 1750°C for 1 h; (c) SEM micrograph and (d) EDX spectrum of the same material after VHT at 1625°C for 3 h; (e–h) Similar micrographs and EDX spectra for samples densified using 0.5% MgO.

Compared with results reported previously for silicon nitride ceramics,[2,3] weight losses for samples densified with MgO are considerably smaller. Attempts to achieve higher losses by using higher temperatures resulted in decomposition of the β-sialon phase. Figure 1 shows back-scattered scanning electron micrographs of the 1% MgO and 0.5% MgO materials, and the associated EDX traces. It is clear that Mg has been totally removed from the samples after vacuum heat treatment, and that reducing the level of MgO in the starting mix results in less porosity in the final vacuum heat-treated material (consistent also with the density data shown in Table 1). The results are consistent with loss of MgO from the glass, plus loss of some of the silica, but an aluminium-rich glassy phase must remain. Oxidation studies carried out on the vacuum heat-treated 1% MgO sample showed that oxidation commenced at ≈ 1400°C, and Fig. 2 shows a comparison of the oxidation behaviour of the VHTed 1% MgO z=1 β-sialon material with a Syalon 201 material oxidised at this temperature. The latter

(e)

(f)

(g)

(h)

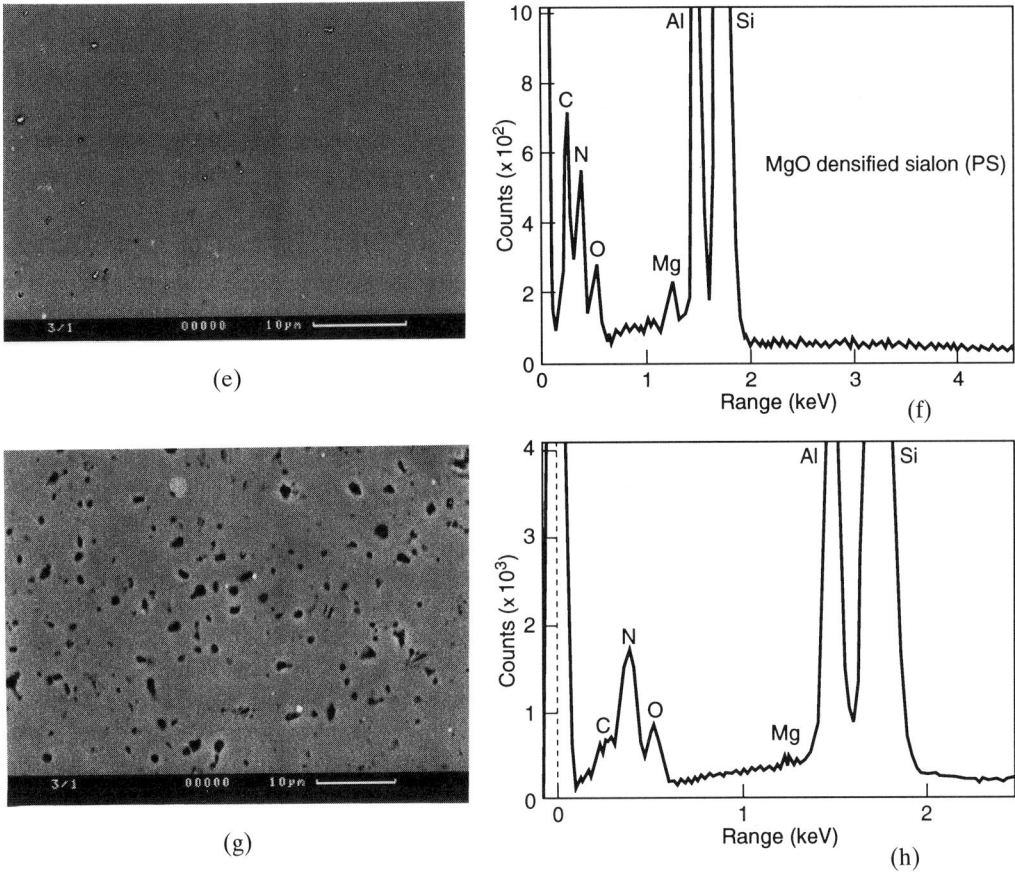

Figure 1(e), (f), (g) and (h).

consists of a phase assemblage of β-sialon + YAG, but the eutectic in the Y_2O–Al_2O_3—SiO_2 system at ≈ 1350°C results in rapid oxidation of this material above that temperature. The difference in oxidation behaviour at 1400°C is clearly apparent, with the Syalon 201 material showing a non-protective white crust, the residue of a surface layer of liquid through which nitrogen gas has bubbled out, and which has subsequently crystallised. In contrast, the VHTed material shows a surface glaze, which is essentially passivating under these conditions. This behaviour is not as good as the excellent oxidation resistance (negligible weight loss and good surface appearance after 24 h of oxidation at 1650°C) reported for vacuum heat-treated silicon nitride ceramics. In the case of the β-sialon material, there is still an Al-rich glassy residue, but the absence of the densifying cation (in this case magnesium), results in a higher eutectic for the residual Si–Al–O–N liquid phase, and the resulting material retains good oxidation behaviour up to ≈ 1500°C.

Table 1 also lists data for β-sialon samples densified with Li_2O (originally added as Li_2CO_3), and SrO (originally added as $SrCO_3$). The same story is broadly true in that

Figure 2.   Samples of (a) VHTed 1% MgO densified β-sialon and (b) Syalon 201, after oxidation for 24 h in air at 1400°C.

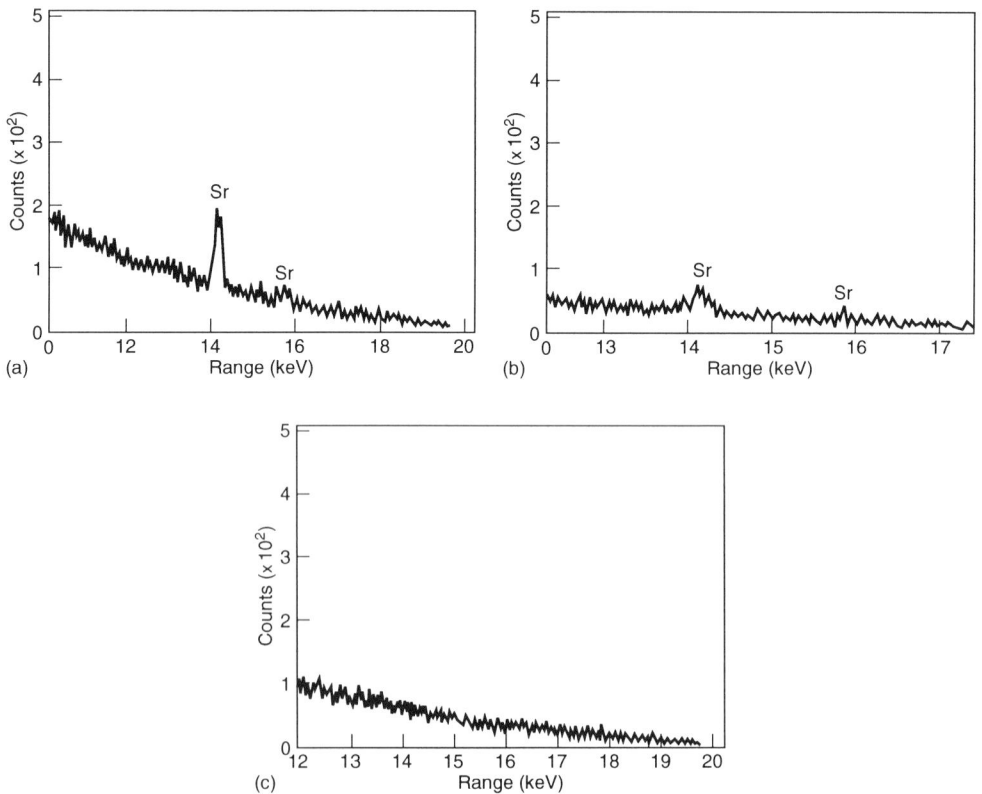

Figure 3.   EDX spectra of z = 1 β-sialon materials densified with 2.59% of SrO, (a) after hot-pressing for 1 h at 1800°C, (b) after VHTing for 4 h at 1700°C with no surrounding powder bed, and (c) after VHTing at 1650°C for 2 h in a carbon powder bed.

Figure 4.   β-sialon samples hot-pressed with 2.59% SrO and oxidised in air for 12h at 1350°C, (a) with an intermediate VHT stage, and (b) without VHT.

some glass is retained in both cases after VHT, but the densifying cation is totally removed. EDX spectra for the Sr-densified samples (Fig. 3) showed that the Sr was harder to remove than Mg (as predicted from thermodynamic calculations). Even after 4 h at 1700°C without packing powder in the crucible, some Sr remained. By using carbon packing powder (Figure 3(c), under the same conditions, the strontium was totally removed. Figure 4 compares the appearance of the VHTed Sr-densified sample shown in Fig. 3(c) with the as-sintered material after oxidation. Whereas the as-sintered material shows extensive development of oxide scale on the surface at 1350°C, the VHTed sample is almost unmarked after 12 hours of oxidation at this temperature.

3.2 α-Sialon

α-sialons are stabilized by metal cations (Li, Mg, Ca, Y and most Ln), which occupy the large interstices in the three-dimensional network of SiN_4 tetrahedra. Vacuum heat-treatment offers several possibilities in this case: by using volatile cations which can also enter the α-structure (Li, Mg, Ca, Yb), the cation can be removed from residual grain-boundary glass, or removed from the α-sialon; removal of all the glass is unlikely because of the difficulty of removing aluminium in volatile form.

Initial attempts to produce dense magnesium α-sialon by pressureless sintering were unsuccessful. The products were highly porous and even though experiments

Table 2.   VHT results for α- and O-sialon samples packed in carbon powder.

Sample	Wt. % additive	T (°C)	t (h)	Density (gcm⁻³) Starting	Density (gcm⁻³) After VHT	ΔW (%)	% T.D.	X-ray results
1	CaO	1650	3	3.19	3.18	7.7	99.2	α-sialon
2	Yb_2O_3	1650	3	3.60	3.54	8.9	98.5	α-sialon
3	$3 Y_2O_3 +$ $5.2 Yb_2O_3$	1650	2	3.36	3.27	6.7	97.3	α+β-sialon
4	1 MgO	1625	3	2.98	2.95	2.6	99.1	O-sialon

showed that Mg could be removed, this work was discontinued because of the high porosity of the final material.

However, calcium α-sialons can be easily densified to full density by pressureless sintering.[4] Figure 5(a) shows an EDX spectrum of a fully dense m = 1.5, n = 1.5 Ca α-sialon material. After vacuum heat-treatment, the intensity of calcium peaks in the EDX spectrum decreases sharply corresponding to the removal of calcium from grain-boundary glass. Under the VHT conditions used (1650°C for 3 h) it was not possible to remove the calcium from the α-sialon phase itself, as indicated by the small residual Ca peak on EDX spectra of the VHTed material (Figure 5(b)).

Previous calculations showed that Yb_2O_3 was a candidate oxide for vacuum heat-treatment because of the low boiling point of ytterbium metal (960°C). Figure 6 shows the microstructure of a fully dense sample of m = 1.5, n = 1.5 Yb α-sialon, in which residual grain-boundary glass shows up as a uniform dispersion of white grains, and trace amounts of 21R appear as clumps of dark grains. After vacuum heat treatment, it is noticeable that the majority of white grains in the sample lose their contrast, consistent with removal of Yb from the glass. The two shades of grey in the residual α-sialon matrix correspond to slightly different levels of Yb in the α-sialon phase as has also been reported by other researchers.[5] A sample prepared in a previous research programme[6] and containing equal amounts of Y and Yb was given

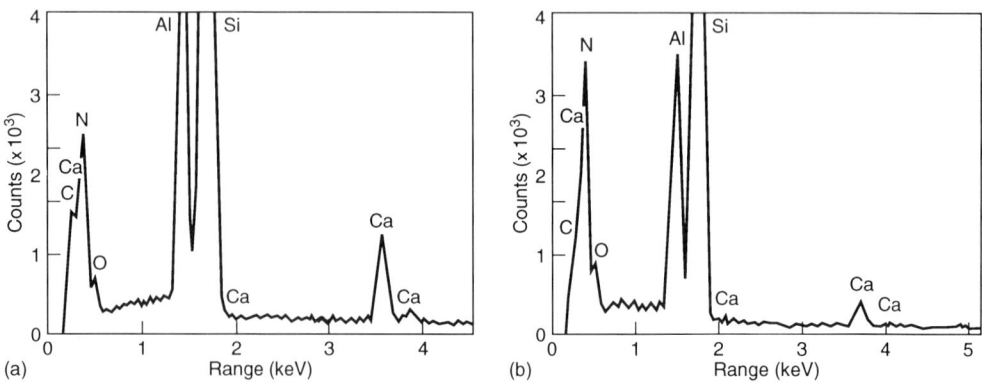

Figure 5.   EDX spectra for CaO-densified α-sialon (a) as-sintered, and (b) after VHT for 3 h at 1650°C.

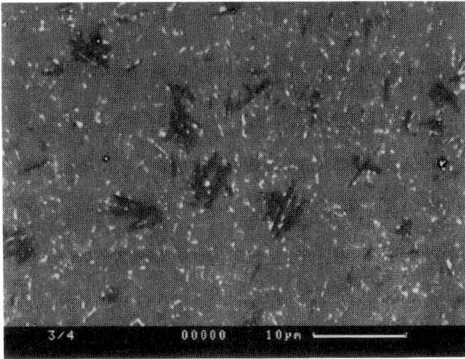

(a)

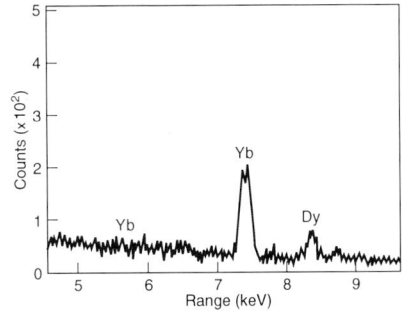

(b)

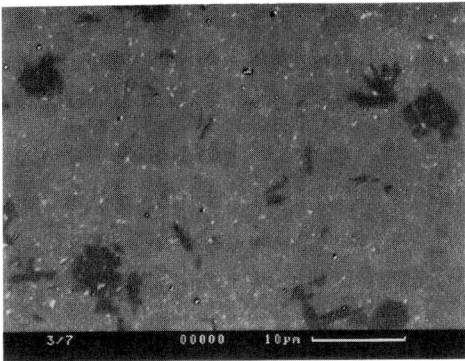

(c)

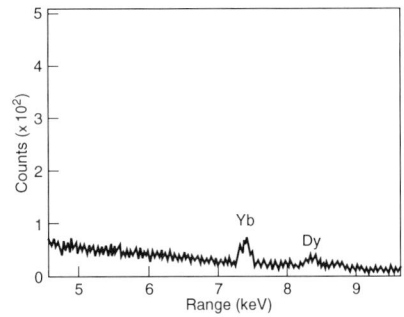

(d)

Figure 6.   Hot-pressed m = 1.5, n = 1.5 α-sialon densified with Yb_2O_3. (a) SEM micrograph, and (b) EDX spectrum of the as-sintered sample; (c) shows the same material and (d) the EDX spectrum, after VHT for 3 h at 1650°C.

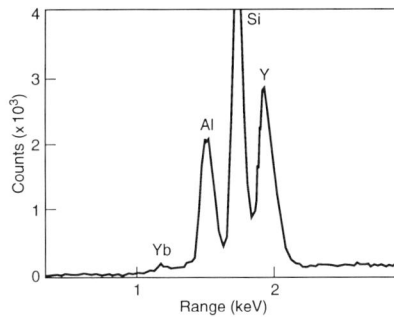

Figure 7.   EDX spectrum for a mixed pressureless-sintered Y_2O_3–Yb_2O_3 densified α-sialon after VHT at 1650 for 2 h.

(a)

(b)

(c)

(d)

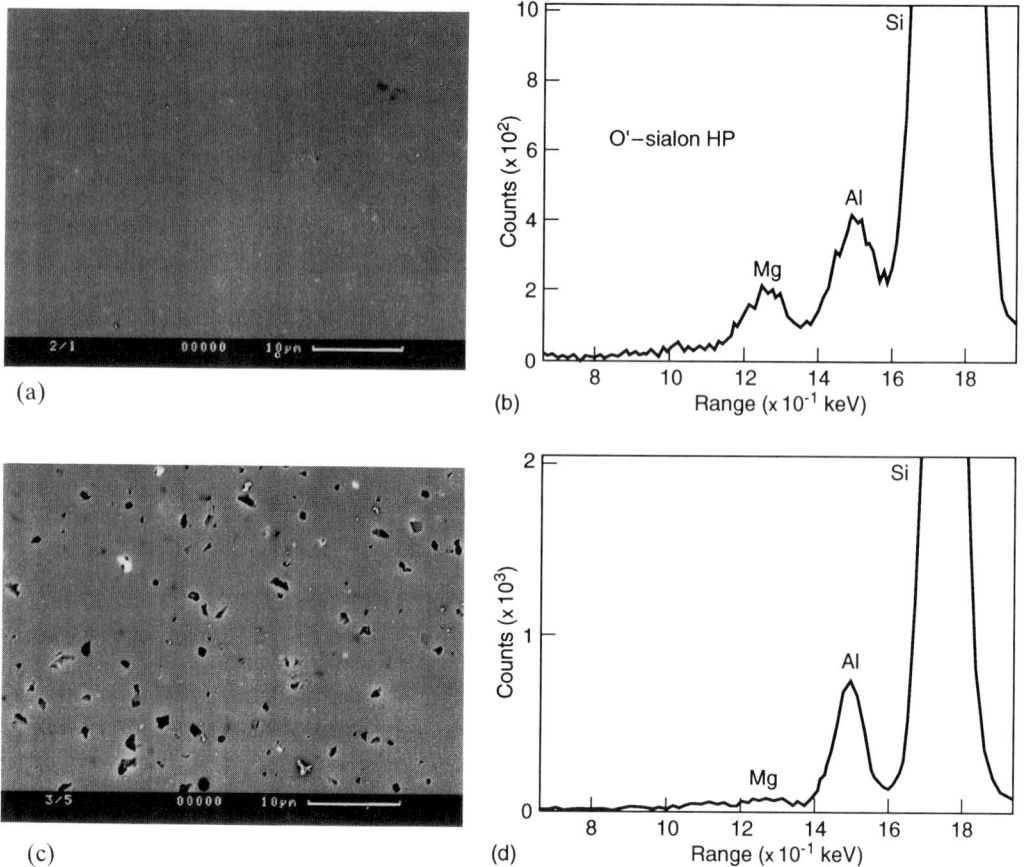

Figure 8.   x = 0.15 O'-sialon densified with 1% MgO. (a) SEM micrograph and
(b) EDX spectrum of the as-sintered sample; (c) and (d) show micrographs
and EDX spectra of the same material after VHTing for 3 h at 1625°C.

a vacuum heat-treatment at 1650°C, surrounded by carbon powder. Table 2 shows
the results. The starting density (3.36 gcm^{-3}) was consistent with full density for this
starting composition, and the weight loss of 6.7% is consistent with most of the Yb
being lost. EDX analysis of the sample after VHT (Fig. 7) showed a strong yttrium
peak, but an almost negligible Yb peak consistent with weight loss observations.
X-ray diffraction showed no change in the phases present, either in intensity of lines
or peak positions compared with the sample before VHT. It is concluded that during
VHT, Yb was removed from the glass, but not from the α-sialon phase. During the
initial densification, it is believed that the yttrium and some of the ytterbium entered
the α-structure, leaving the remaining Yb in the glass. Vacuum heat treatment is able
to remove Yb from the glass but not from the α-sialon, consistent with the stronger
bonding and greater stability of this phase.

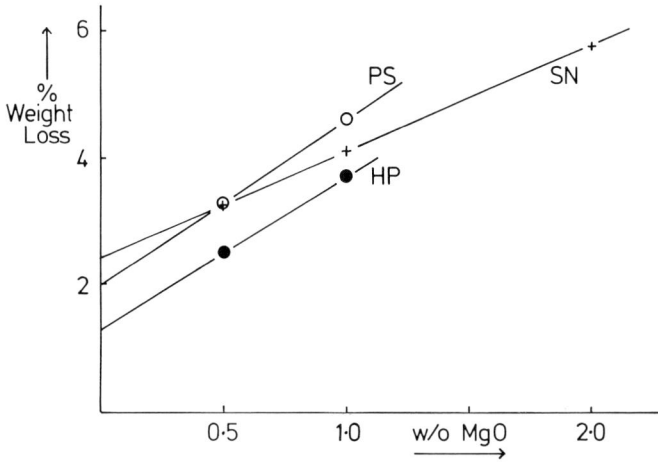

Figure 9(a). Weight loss versus % MgO, and (b) weight loss versus density for vacuum heat-treated silicon nitride (SN), pressureless-sintered β-sialon (PS) and hot-pressed β-sialon (HP) materials.

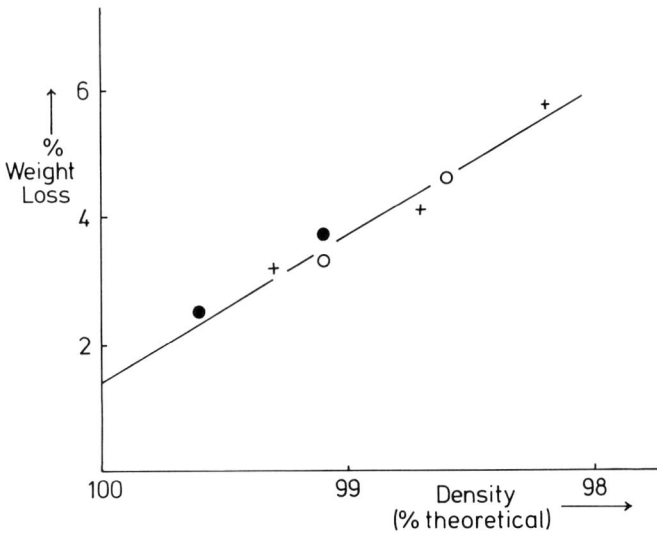

Figure 9(b).

3.3 O-Sialon

Previous results have already been reported showing that the magnesium present in grain-boundary glass in x = 0.15 O-sialon compositions ($Si_{2-x}Al_xO_{1+x}N_{2-x}$) can be removed by vacuum heat treatment (see Table 2). Figure 8 shows micrographs and EDX spectra of this sample before and after heat-treatment. As with the β-sialons reported above, it is possible to totally remove magnesium from the sample, but it is suspected from the weight loss data than an Al-rich glassy residue remains.

3.4 Comparison of Weight Loss Data

For MgO-densified materials, weight losses during VHT increase with increasing level of additive in the starting mix. Figure 9 compares this for a range of different materials, and the results are similar. The curves extrapolate to just below ≈ 2% loss for zero additive, which is less than the 3–4% which might be expected on the basis of the surface silica content of the starting silicon nitride. However, some MgO and SiO_2 is lost during initial densification and this must then be excluded from the weight loss calculations during VHT.

It is interesting that the same points on Fig. 9(a) give a very good linear relationship when plotted in the form of weight loss versus density (Fig. 9(b)). Again the graph extrapolates to just less than 2% loss for fully dense materials. These graphs show that the good densities of β-sialon ceramics after initial pressureless sintering are substantially preserved after vacuum heat treatment.

5. CONCLUSIONS

Vacuum heat-treatment is successful method of removing volatile cations from residual glass in sialon ceramics, but in contrast to the application of the same process to dense silicon nitride ceramics, it is impossible to remove all the glass. In particular, Al-rich residues remain, which give relatively poor creep and oxidation resistance to the final material but with a significantly improved performance compared with the as-sintered materials.

Present results have focused on sialon ceramics containing relatively high aluminium contents. Further work should concentrate on sialons containing very small Al levels, at which content, it may still be possible to retain the benefits of easier densification with improvements in properties resulting from vacuum heat-treatment if only a small amount of Al-rich glass remains in the final product.

REFERENCES

1. M. H. Lewis: in *New Materials and their Applications*, Ed. S. G. Burnay, publ. Institute of Physics Con. Series No. 111, IOP Publishing Ltd., 1990 427–34.
2. H. Mandal and D. P. Thompson: in *Proc. 5th Int. Symp. 'Ceramic Materials & Components for Engines'*, Shanghai '94, Ed. D. S. Yan, X. R. Fu and S. X. Shi, pub. World Scientific, Singapore, 1995, 45–52.

3. H. Mandal and D. P. Thompson: *Preparation and Characterisation of Glass-Free Silicon Nitride Ceramics*, in *Proc. ECerS IV*, C. Galassi, ed., publ. gruppo editoriale Faenza editrice, Italy, 1995, 217–24.
4. S. Hampshire, H. K. Park, D. P. Thompson and K. H. Jack: *Nature,* **274** 1978, 880–1.
5. Z. J. Shen and T. Ekström: 'Temperature Stability of Rare Earth Doped α-Sialon Ceramics' in *Proc. 5th Int. Symp. 'Ceramic Materials & Components for Engines',* Shanghai '94, Ed. D. S. Yan, X. R. Fu & S. X. Shi, publ. World Scientific, Singapore, 1995, 206–10.
6. H. Mandal, D. P. Thompson and T. Ekström: *Special Ceramics* **9**, Ed. R. Stevens, publ. Institute of Ceramics 1992, 149–162.

Prospects for Structural Ceramic Composites

A. HENDRY

Metallurgy and Engineering Materials Group, University of Strathclyde, Glasgow, UK

ABSTRACT

The benefits and disadvantages of ceramics in structural applications and the difficulties for the design engineer in utilising monolithic ceramics are well known. Several materials science approaches designed to improve the mechanical behaviour of engineering ceramics have been proposed and of these the use of relatively simple composite methods has had the greatest exposure. However, compared to polymer or metal composites the fabrication of ceramic composites presents much greater problems and these in turn lead to consequent uncertainty in predicting behaviour under load.

The methods of producing ceramic matrix composites are briefly reviewed and discussed in the context of the difficulties which are specific to this family of materials and have implications for possible applications. These are illustrated by an example showing the benefits of a simple approach to the manufacturing technology and the confidence which this can convey to the design engineer.

In general, the expectations for structural ceramic composites have been overstated and insufficient regard paid to the difficulties inherent in these materials for both the fabricator and the designer. A case is made for the future application of ceramic matrix composites which balances the technical potential of the materials against the financial cost and the aspirations of the users.

1. INTRODUCTION

Surveys of the economic prospects for structural ceramics, as opposed to electronic applications – see for example reference 1, Fig. 1 – predict a relatively slow rate of growth to the end of the century and the reason most frequently given is the lack of confidence in the mechanical behaviour of such materials among design engineers. At the same time in technical articles on the future for ceramic artifacts, the ability to develop high strength, hardness, oxidation resistance or other attractive properties is emphasised and the unfortunate tendency of structural ceramics to fail in a brittle and catastrophic manner is at best minimised and at worst, totally ignored.

Nevertheless, in papers which seek to link applications to the best properties of engineering ceramics, the word 'toughness' frequently appears as a desirable property – see for example Richerson et al.[2] on silicon nitride. The resistance to crack propagation is therefore of paramount importance if designers are to become convinced of the value of ceramics as engineering materials. A second property which must also be addressed is the intrinsic variability of ceramic mechanical properties as measured by parameters such as the Weibull modulus. This lack of certainty, as it is seen by engineers, is again a major impediment to development of the potential for application of structural engineering ceramics. Putting the above two factors together

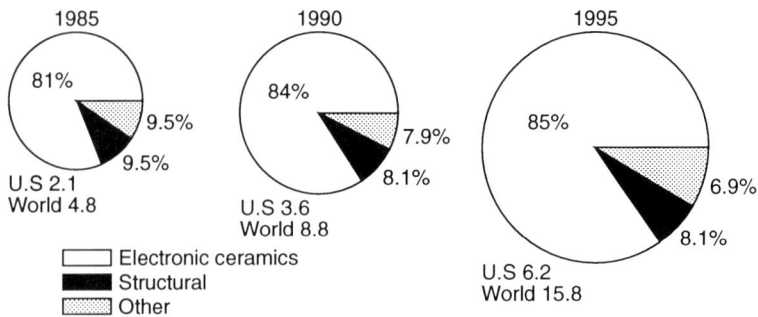

Figure 1.   US and World markets for advanced ceramics 1985–1995.[1]

it can be seen that the ability to inhibit crack nucleation and retard (or prevent) crack propagation is the prime objective of the ceramic scientist and engineer.

From these considerations and the success in preventing brittle failure of polymers by fibre reinforcement, has come the search for methods of applying classical composite action in a ceramic matrix. It could be argued of course that many of the traditional ceramic compositions are composites and that this recent development is not new, but that is to miss the point of the reasons outlined above for the modern move toward ceramic matrix composites. It is now proposed that ceramics be used in applications which are truly load-bearing and are therefore much more mechanically demanding than uses of traditional ceramics where the main stresses are thermally induced or entirely compressive. If modern engineering ceramics are to be used in replacement of metals, for example, in combustion engineering or in engines, then the stress systems will be much more complex and the reliability factors much more demanding.

A great deal has been written in the recent past on the subject of ceramic matrix composites, and it is not the purpose of the present paper to give a review of the state of the art; for the interested reader this can be found in conference proceedings such as those at Riso Laboratories in 1988 – Metallic & Ceramic Composites[3] – and 1990 – Structural Ceramics[4] – or in the 1993 meeting on silicon nitride.[5] In this presentation the factors which govern the close correlation of processing, microstructure and properties in the science and engineering of ceramic composites will be discussed with a forward look at what the future holds.

2. PROCESSING OF COMPOSITES

The theoretical benefits of composite action on the toughening of materials can be shown as in Fig. 2 where the influence of shape of reinforcement is clear. Thus long fibre or rod-like shapes give the greatest relative effect, but significant toughening can also be achieved simply by increasing the internal surface area for crack deflection. This is of importance in ceramic composite manufacture because there are considerable problems in manufacturing, for example, long-fibre composites,

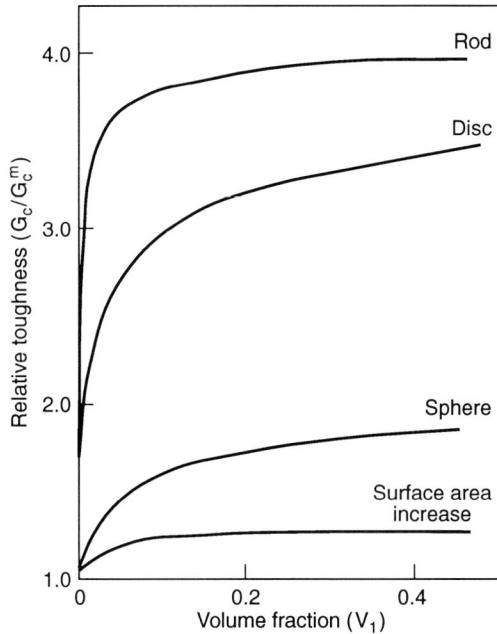

Figure 2.   Relative toughening by reinforcement of differing physical forms. (After Pompe Ref. 4).

whereas the ability to incorporate a particulate reinforcement is much more amenable to traditional ceramic processing and is therefore more likely to be economic. The important point is that there is a balance between the property requirement and the expense to which the manufacturer can go in processing the component.

The manufacture of particulate, whisker or chopped fibre composites can be achieved by many of the traditional ceramic methods. Slip casting, die pressing and cold isostatic pressing have all been applied to these materials and the methods of greenforming and binder removal are similar to those for monolithic bodies. Injection moulding has also been used for these materials but the binder loadings are of course much greater and therefore greater care is required during removal and burn-out. Sintering of ceramic matrix composites is however more complex than for monolithic or traditional compositions as there is, in the majority of cases, a problem of chemical incompatibility at the interface between matrix and reinforcement. At the firing temperature most ceramic composite systems will react and the firing gas atmosphere may also play a part. It follows therefore that the rate of pore closure and shrinkage will have a pronounced effect on the chemical equilibria at the interface, particularly in cases where gas is generated by interfacial reaction. Examples of this behaviour are in chopped-carbon fibre reinforced sialon[6] and silicon carbide whisker reinforced alumina.[7] With control of both the interface and the matrix however it is possible to obtain good high-temperature properties in the composite as is shown in Fig. 3 for the latter system.

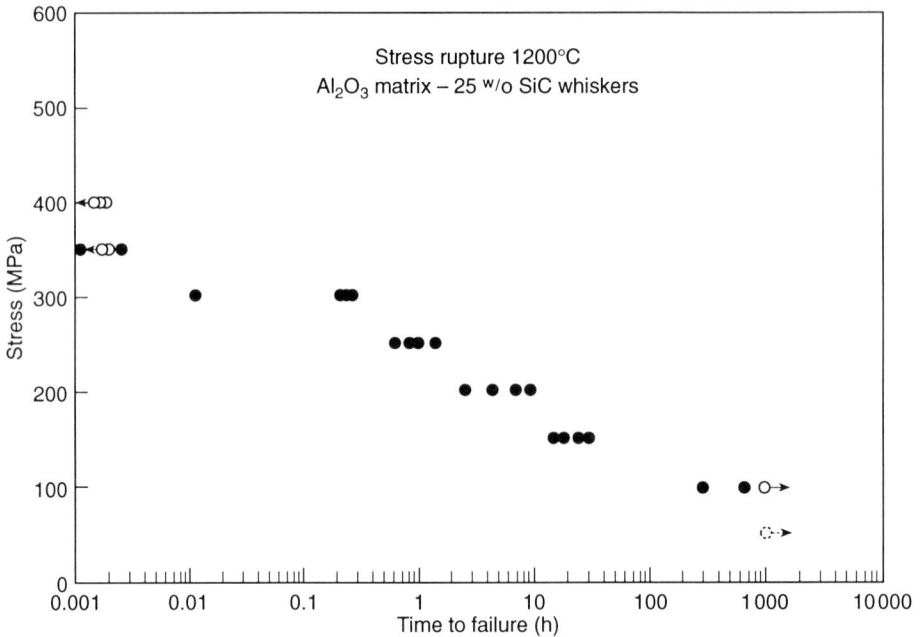

Figure 3.   Stress-rupture data for alumina-silicon carbide CMC at 1200°C.[7]

A different type of interfacial reaction involving the transfer of elements by diffusion occurs in steel particle or short fibre reinforcement of silicon nitride where the pick-up of silicon into the metal, driven by the thermodynamic potential gradient, results in melting of the steel at firing temperatures below that of the unreacted metal.[8] Here again, however, control of the reactions during processing gives a useful composite which can be used as an interlayer in ceramic to metal joining.[9]

Long-fibre reinforcement is much more difficult to achieve due to the problems of obtaining good infiltration of fibre bundles or tows by the matrix precursor, and the consequent pore concentrations which result in poor mechanical reproducibility. Several processing methods have been proposed of which the most promising is sol-gel technology where the equivalent particle size of the ceramic matrix is very small and infiltration can be achieved in the sol state before gelation.[10] Again there are problems of drying and binder removal as well as of interfacial incompatibility on firing which are common to almost all systems but successful products have been obtained as in Fig. 4 which shows the stress-strain curve for a long-fibre unidirectional composite of silicon carbide (Nicalon) fibre in CAS glass.[11] This type of mechanical behaviour is often held to exemplify the 'graceful failure' behaviour of CMC as there is a high proportion of the total failure strain after matrix cracking. However note should be taken of the strain units (microstrain) when using such examples to convince design engineers of the 'increased' fracture strain of CMC!

The problem of residual porosity after firing of long-fibre composites can be addressed by either nanotechnology powders or hot isostatic pressing but both

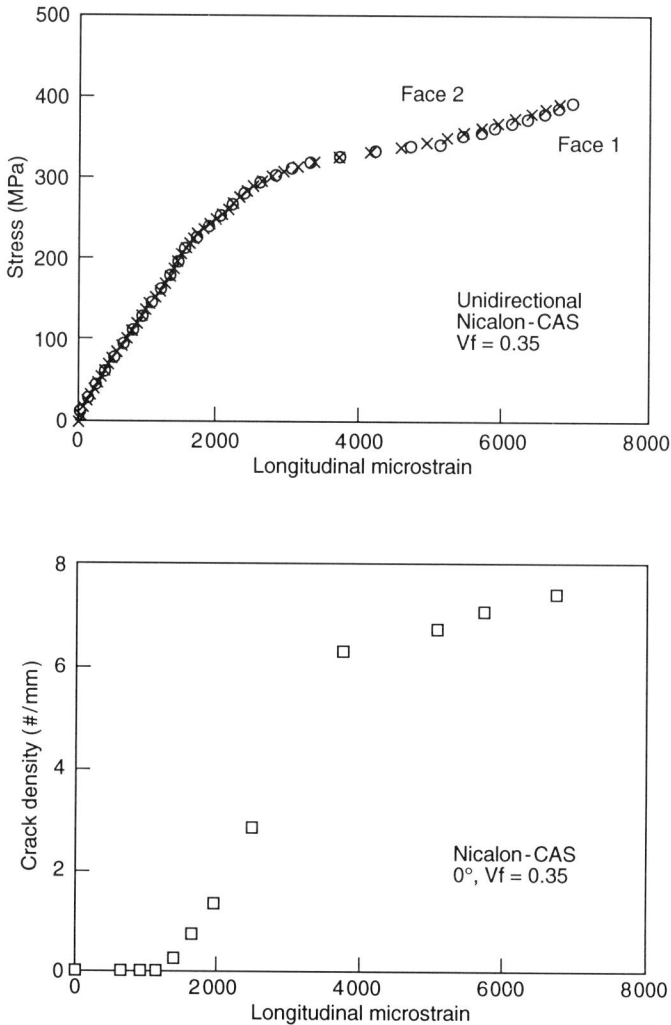

Figure 4.   Stress-strain curve and crack detection in a unidirectional fibre reinforced CMC.[11]

approaches are extremely expensive and at present can only be justified for high added value components.

The aim of ceramic matrix composite processing development must therefore be to understand the aspects which are of specific concern, such as interfacial reaction during firing, residual porosity or cost, and to seek means of achieving the desired property specification reproducibly. An additional aspect of the long term feasibility of these materials is to develop confidence in the family of CMC by exploring applications where (i) they offer particular benefits to the designer, (ii) the processing can be carried out by well-established routes and (iii) they can be shown to be reliable. The following section gives an example of this approach in combustion engineering where a CMC tube is designed to replace an existing steel component. The processing route is a conventional cold isostatic press and sinter and there is an element of microstructural development which offers further promise for the future.

3. SILICON CARBIDE REINFORCED SIALON-MATRIX TUBES

The following example serves to illustrate the manner in which conventional and relatively inexpensive ceramic processing technology can be utilised to produce a CMC material which has a specific set of property requirements for use in a gas-fired burner application. The specification had three elements as follows; (i) maximum possible thermal conductivity through the wall thickness (this also implies minimum wall thickness in a ceramic design), (ii) zero linear shrinkage on firing and zero gas permeability after densification, and (iii) minimum creep deflection in bending during high-temperature application. The configuration of the ceramic components in the burner is shown in Fig. 5. There are additional design constraints such as ceramic to metal joining but these are not considered here.

The development programme began with the selection of the material. To give the required thermal conductivity dictated the use of either silicon carbide or carbon, with the former being chosen because of its higher resistance to oxidation. This is then used to form the reinforcing phase and the concentration was chosen to give sufficient material to obtain the required conductivity. The reinforcement is chosen in particulate form as shown below for ease of processing. The matrix material was then selected on the basis of several criteria including the shrinkage, porosity and creep requirements but also because of the desire to develop a low-cost processing route. It was therefore decided to use a sialon matrix and that this should be formed

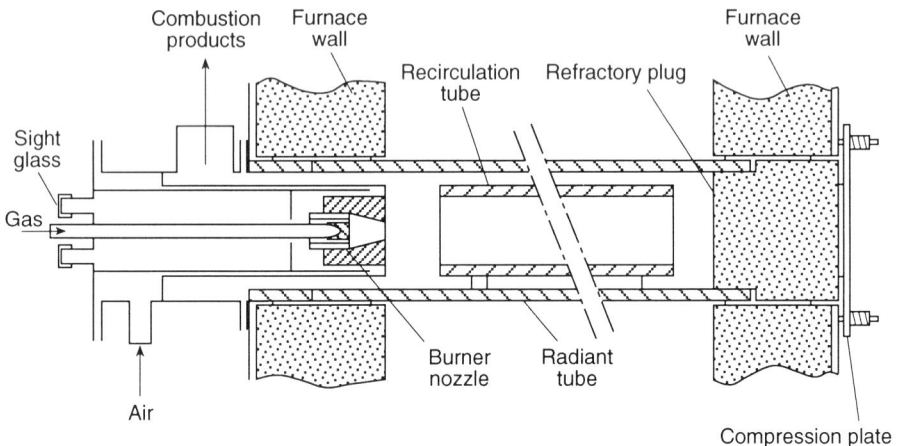

Figure 5.   Ceramic composite radiant tubes in a gas-fired burner assembly (British Gas plc).

by in-situ reaction of a sinterable mix of silicon nitride with high-alumina cement and clay,[12] the latter components also confer on the mix a degree of plasticity which is imperative in pressing thin-wall tubes. The requirement for zero linear shrinkage on firing is obtained by carefully controlling the size fractions of the carbide reinforcement to give a rigid skeleton of reinforcement particles around which the matrix material may react and densify but without necessarily achieving full density. Indeed, provided that the sintered tube has fully closed porosity (zero permeability) there is no absolute requirement for full density.

The forming process is conventional cold isostatic processing and the detailed formulation of the binders and lubricants, and the schedule for their removal were developed by Thor Ceramics Ltd. who collaborated in the project. The final stage was a controlled atmosphere firing cycle which developed the sialon matrix and gave the final tube form with zero shrinkage and permeability.

This example shows the materials science and engineering development of a ceramic matrix composite designed to satisfy a particular set of engineering criteria but which has two additional features which highlight a potential way forward for ceramic matrix composite technology. First, the material has satisfied the design criteria in a demanding application but in a manner which maximises the benefits to be derived from the ceramic construction. The component was a direct replacement for a metallic material but offers additional value in terms of temperature capability and corrosion resistance in a configuration which minimises the disadvantages of ceramics in terms of brittle fracture. In so doing an increased degree of confidence is achieved in the viability of CMC materials from the designers point of view. Secondly, the materials development part of the programme has led to further options in both the matrix and reinforcement technology. There is therefore an opportunity to further build confidence in the capability of CMC to give even higher temperature limits to the operating specification using the same manufacturing route. A gradual increase in property specification – and therefore of burner efficiency – is possible.

4. CONCLUSIONS

The only conclusion possible from a review of the materials science literature on ceramic matrix composites is that although a large amount of extremely valuable, not to say elegant, work has been done, there is still a considerable gap between that and the materials engineering of components. It is undoubtedly the case that this arises mainly because of the over-optimistic claims made for CMC and the excessively demanding applications to which the materials have been directed. It is the high-technology applications fixation which has held up the development of confidence in the materials more than any intrinsic difficulty in materials processing. The ability to manufacture acceptable components which are fit for purpose by a conventional ceramic processing route is more likely to lead to increased interest in these materials than the quest for the holy grail of full density by exotic materials processed at enormous cost.

Ceramic components are brittle and have tiny fracture strains and no amount of ceramic composite engineering will alter that fact, but if the intrinsic benefits of

ceramics in terms of hardness, high-temperature capability and resistance to corrosion are allied to the advantageous aspects of composite behaviour and developed in
components manufactured by tried and tested economic routes, then the future for
structural engineering ceramics will improve considerably. Within that family of
materials, ceramic matrix composites will have an increasing role to play.

ACKNOWLEDGEMENTS

The author gratefully acknowledges the help of his colleagues in the Ceramics Group
in preparing this paper as well as that of Thor Ceramics in their support of ceramic
composite development. The award of a SERC/ACME grant to the work on thin-
wall ceramic composite tubes is also acknowledged.

REFERENCES

1. Anon: 'An analyses of the market for advanced ceramics', Arthur D Little Inc., New York,
   1990.
2. D. W. Richerson and P. M. Stephan: 'Evolution of applications for silicon nitride based materials', *Mat. Sci. Forum*, 1989, **47**, 1.
3. S. I. Anderson, H. Lilholt and O. B. Pedersen (eds): *Mechanical behaviour of metallic and
   ceramic composites*, Proc. of 9th Riso Symposium on Metallurgy & Materials Science, Riso Nat.
   Lab. Denmark, 1988.
4. J. J. Bentzen, J. B. Bilde-Sorensen, N. Christiansen, A. Horsewell and B. Ralph: *Structural
   Ceramics – Processing, microstructure and properties*, Proc. of 11th Riso Symposium on Metallurgy and Materials Science, Riso Nat. Lab., Denmark, 1990.
5. M. J Hoffman, P. F. Becher and G. Petzow: *Silicon nitride 93*, Proc. of Conference on Silicon
   Nitride-based Ceramics, Trans Tech Publications, Switzerland, 1994.
6. P. Dupel and A. Hendry: 'Interfacial reaction in carbon fibre-sialon ceramic composites', *J.
   Eur. Cer. Soc.*, 1995. In the press.
7. C. A. Tracey and M. J. Slavin: *Characterisation of SiC whisker reinforced Al$_2$O$_3$ for ambient
   and high temperature applications*, Annual Meeting of Amer. Cer. Soc., Am. Cer. Soc., Westerville, USA, 1989.
8. A. Smith, A. Abed, H. J. Edrees and A. Hendry: 'Ceramic matrix composites of silicon nitride
   with conducting particles'. *Silcon Nitride 93*, Proc. of Conference on Silicon Nitride-based
   Ceramics, Trans Tech Publications, Switzerland, 1994, 423.
9. A. Abed and A. Hendry: 'Application of ceramic composites in joniing', *Silic. Ind.*, 1995. In the
   Press.
10. R.S. Russell-Floyd, B. Harris, R. G. Cooke, J. Laurie, F. W. Hammet, R. W. Jones and T.
    Wang: 'Application of solgel processing for the manufacture of fibre-reinforced composites', *J.
    Am. Cer. Soc.*, 1993, **76**, 2635.
11. R. Talreja: 'Fatigue of fibre-reinforced ceramics', *Structural Ceramics – Processing, microstruc-
    ture and properties*, Proc of 11th Riso Symposium on Metallurgy and Materials Science, Riso
    Nat. Lab., Denmark, 1990, 145.
12. H. J. Edrees and A. Hendry: 'Influence of atmosphere and sintering additives on clay-silicon
    nitride densification', *Br. Cer. Trans.*, 1995, **94**, 15.

The Effects of Sintering Temperature on the Mechanical and Electrical Properties of a Novel 5 Mole % Yttria-Zirconia Material

G. P. DRANSFIELD, T. A. JENNETT and E. J. A. WILLIAMS*

Tioxide Specialties Ltd., Haverton Hill Road, Billingham, Cleveland, TS23 1PS, UK.
*Brunel University of Materials Technology, Uxbridge, Middlesex, UB8 3PH, UK.

ABSTRACT

The composition studied lies in the middle of the tetragonal + cubic mixed region of the ZrO_2—Y_2O_3 phase diagram. The effects of sintering temperature and alumina addition on the mechanical and electrical properties of a novel 5 mole% yttria additioned zirconia were studied. It was found that superior strengths and toughness values were obtained at lower sintering temperatures and increasing alumina level. The causes of the improved mechanical properties at lower sintering temperatures are discussed. Electrical properties of the material, at temperatures up to 1000°C, using a novel DC conductivity method, are also reported.

1. INTRODUCTION

Yttria tetragonal polycrystals (Y–TZP's) containing 2–3 mole% yttria, are noted for their high strength and toughness, owing to transformation toughening of the tetragonal phase. 'Fully stabilised zirconias' (FSZ's), stabilised in the cubic form by >8 mole% yttria addition, have high ionic conductivity, but low strength and toughness. Intermediate compositions, when fired, consist of a mixture of tetragonal and cubic phases and therefore have intermediate mechanical and electrical properties.

Typical properties for the materials are summarised in Table 1.

In order to achieve high ionic conductivity it is necessary to have high oxygen vacancy concentration, i.e. yttria levels of ~8 mole%. The resultant material is predominantly cubic phase[1] and has lost any degree of transformation toughening.

Table 1. Changes in mechanical and electrical properties of yttria stabilised zirconia with increasing yttria level.

Yttria Level/ Mole%	Phase	K_{IC}/MPa m$^{1/2}$	MOR/MPa	Ionic conductivity/ S m^{-1}
3	Tetragonal	6–10	1000–1200	4–5
5	Tetragonal + Cubic	4–6	600–800	8
8	Cubic	2–4	200–400	12–15

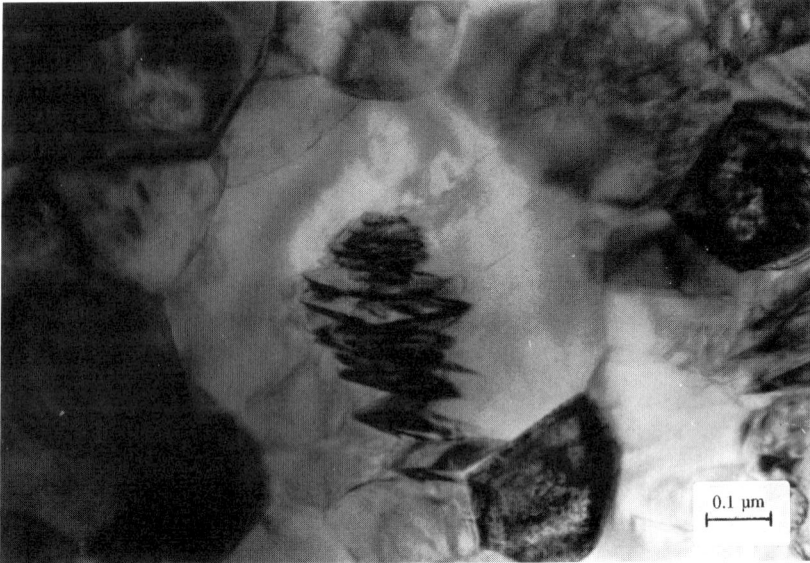

Figure 1. 'Core-shell' effect in Y-TEP material (Courtesy M. Rainforth)

It has therefore not yet been possible to make an yttria stabilised zirconia material with high fracture toughness (>6 MPa m$^{1/2}$) and high conductivity (>10 S/m). It has, however, been shown that compositional zoning of yttria stabilizer can lead to a 'core-shell' effect whereby different zirconia phases are observed within the same grain of a Y-TZP material. An example is shown in Fig. 1.

This effect has only been observed, using coated starting powders. It would be difficult to achieve in co-precipitated materials, where main phase and stabiliser are intimately mixed on a nano-scale.

This paper will describe a novel coated yttria-zirconia material, made by DC plasma synthesis and subsequent powder coating with 5 mole% yttria. The variation of the mechanical and electrical properties with sintering temperature and alumina addition will be reported.

2. EXPERIMENTAL

Zirconia powder with a crystal size of ca 70 nm was prepared by plasma oxidation of zirconium tetrachloride and then coated with 5 mole% yttria stabiliser, according to a method, which has been described previously.[2] Additions of 0.3 wt% hafnia and 0–0.5 wt% alumina (BACO RA207LS) were made by attrition milling, using 0.6 mm zirconia beads. The powder was then filtered and oven dried at 110°C.

The resulting powders were then pressed by uniaxial pressing at 25 MPa, then isostatic pressing at 80 MPa and sintered at temperatures in the range 1400–1600°C, with a soak time of 2 hours. Sintered densities were measured by a water immersion method and strengths by the biaxial disc flexure method of Sivill.[3] Ionic conductivity at 1000°C was measured by a '4 Terminal' DC technique, which had been developed in-house, with assistance from Aberdeen University. Puck shaped sintered samples of 8–10 mm diameter were coated at each end with platinum paste and platinum foil electrodes attached, in the form of tags. Platinum wires were then wrapped around the tags and clamped to ensure a good electrical contact. The samples were then placed in the hot zone of a tube furnace and heated to 1000°C and resistance measurements made, using a Solartron Schlumberger 7150 digital multi-meter. Resistance was measured by a four terminal technique to eliminate the resistance of the wires connecting the meter to the sample. Errors may have been generated by the contact resistance between the wires and the platinum foil strip, but results were in good agreement with those obtained when using two terminal a.c. impedance measurements, using a Hewlett Packard 4192A impedance analyzer.[4]

3. RESULTS AND DISCUSSION

Sintered densities of >99% theoretical were obtained for all compositions between 1400 and 1500°C, but decreased markedly thereafter, as shown in Fig. 2.

Strengths increased with alumina addition and decreased with increasing sintering temperature, as shown in Fig. 3.

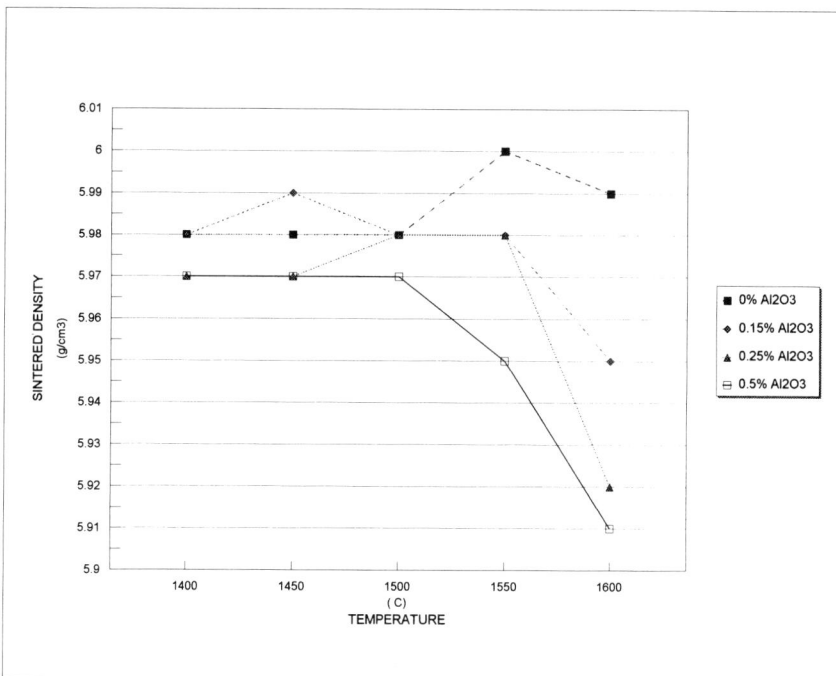

Figure 2.   Variation in sintered density with sintering temperature and alumina content.

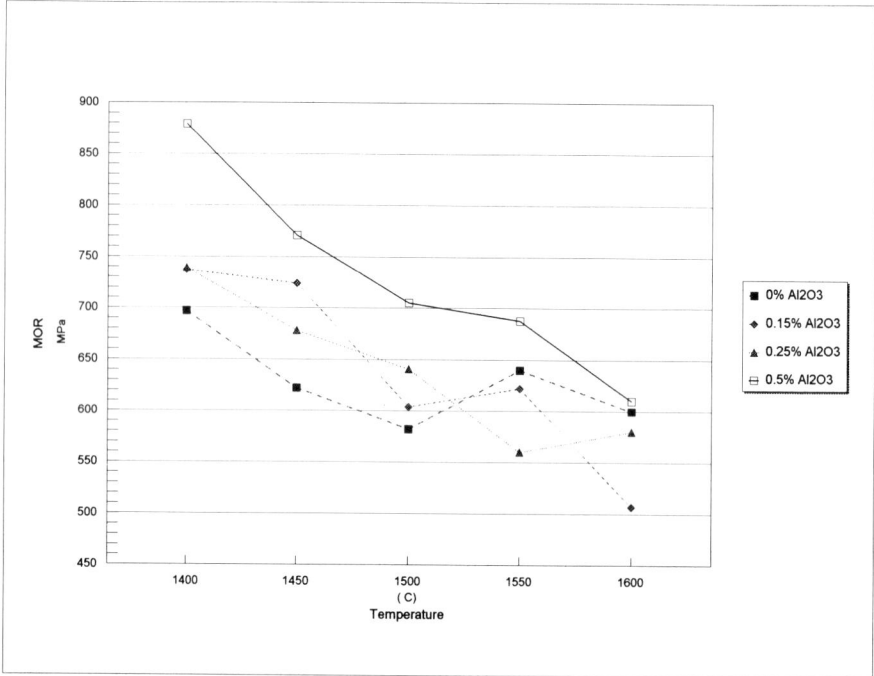

Figure 3.   Variation in MOR with sintering temperature and alumina content.

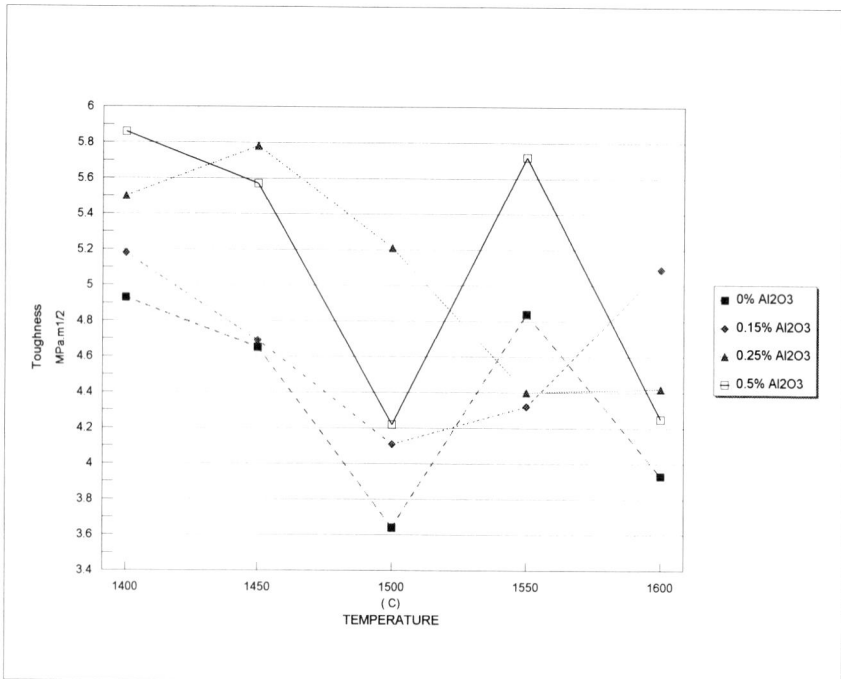

Figure 4.   Variation in fracture toughness with sintering temperature and alumina content.

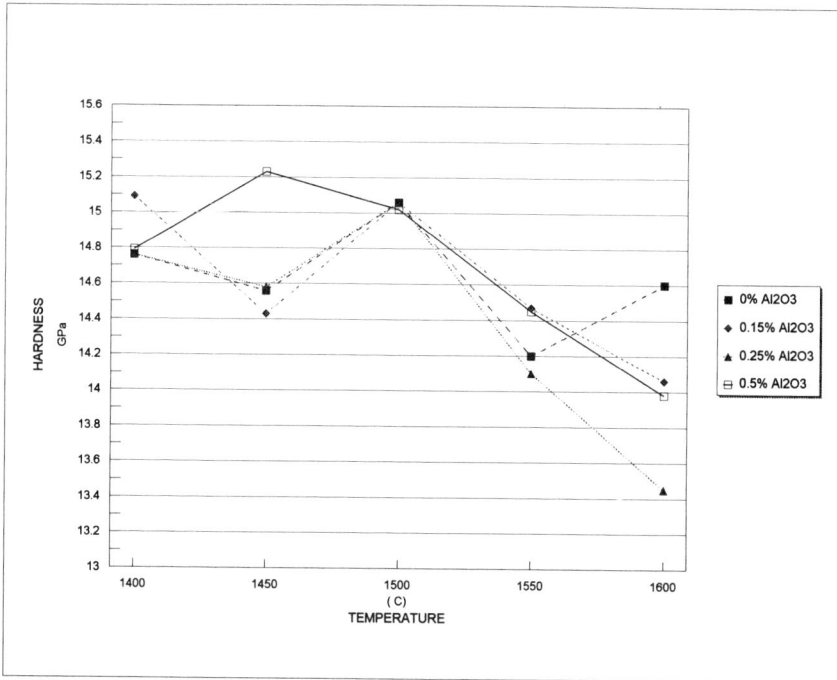

Figure 5.   Variation in hardness with sintering temperature and alumina content.

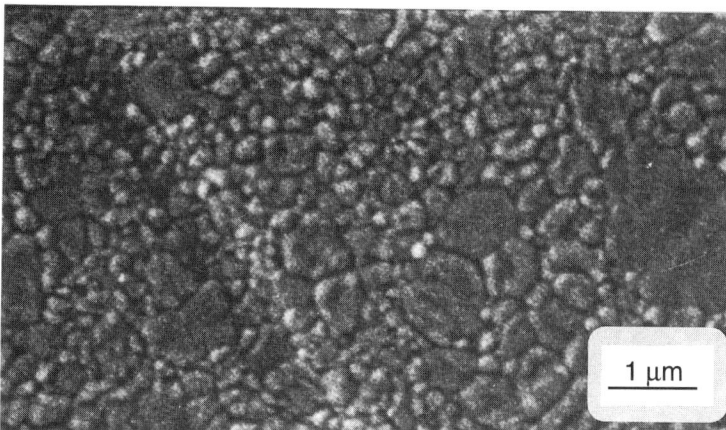

Figure 6.   5 mole % Yttria – Zirconia, no alumina addition, sintered @ 1400°C.

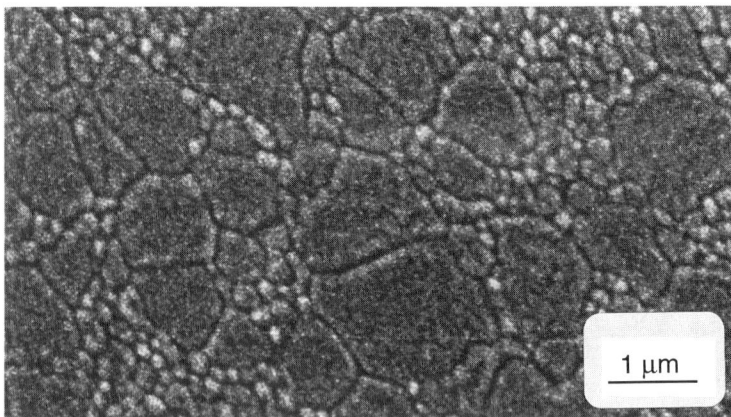

Figure 7.   5 mole % Yttria – Zirconia, 0.5% alumina addition, sintered @ 1400°C.

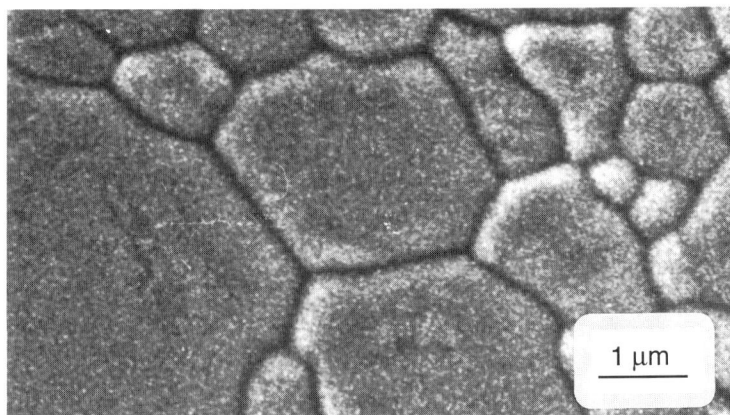

Figure 8.   5 mole % Yttria – Zirconia, no alumina addition, sintered @ 1600°C.

The relationship between fracture toughness and sintering temperature shown in Fig. 4 was more complex.

Hardness went through a maximum between 1450 and 1500°C, but fell away sharply thereafter, particularly when alumina was added, as shown in Fig. 5.

SEM micrographs (Figs 6 and 7) show that addition of alumina leads to a more clearly defined, bimodal microstructure at 1400°C, while it is clear that significant grain growth has taken place when the material is sintered at 1600°C. Figure 8 shows the microstructure for the unadditioned material.

Ionic conductivity was higher than expected from literature values,[5] with a maximum value at 1550°C, as shown in Table 2.

The high densities obtained were to be expected given the known high sinterability of coated, plasma derived Y–TZP powders.[2] The high fracture toughness at low sintering temperature is believed to be due to high transformability of grain cores depleted in yttria. With increasing equilibration of yttria with increasing temperature,

Table 2. Effect of sintering temperature on ionic conductivity at 1000°C.

Sintering temperature/°C	Ionic conductivity/S m^{-1}
1,450	8.6
1,500	8.7
1,550	9.7
1,600	7.8

fracture toughness is expected to drop, as has been observed for 3 mole% coated yttria–zirconia powder.[6] The subsequent rise in toughness above 1500°C is more difficult to explain, but may be due to increasing cubic content. The formation of more cubic grains will deplete the matrix of yttria, thus rendering it more transformable. The overall effect on strength, however, is liable to be detrimental, due to the large size of cubic grains and their tendency to have porosity associated with them.[7] The decrease in strength with increasing sintering temperature can thus be explained. The role of alumina in enhancing strength and fracture toughness is not fully understood, but it is known that fine alumina particles act as grain boundary pinning points preventing the development of large grains of cubic phase.[8] Decreasing hardness at higher sintering temperatures can be attributed to increasing grain size.

The relatively high conductivity of the material is the subject of investigation, but it may be that high yttria levels around the outer region of the grain may provide a high vacancy pathway for ionic conductivity.

4. CONCLUSIONS

The following conclusions can be drawn from this work for 5 mole% yttria-coated zirconia:
1. Superior mechanical properties could be obtained at lower sintering temperatures, than would be the case for co-precipitated materials.
2. Mechanical properties were improved through alumina addition.
3. The electrical properties were intermediate between 3 and 8 mole% yttria–zirconia materials, but were overall higher than expected.

The improvements in mechanical and electrical properties may be due to compositional zoning of the stabiliser. An overall conclusion was that this represented a promising material, worthy of further investigation. Specific properties to be investigated in the future include the ageing properties of the material under autoclave conditions, the effects of sintering termperatures <1400°C, the effects of alumina on electrical properties and the effects of higher alumina levels.

ACKNOWLEDGEMENTS

The authors wish to thank Tioxide Group Ltd for their kind permission to publish this article. Thanks must also go to Iain Gibson of Aberdeen University for setting up the DC ionic conductivity test and to Sue Lucas of Tioxide Specialties for preparing the SEM micrographs.

REFERENCES

1. R. A. Miller, R. G. Smialek and Garlick: 'Phase Stability in Plasma sprayed Partially Stabilized Zirconia-Yttria', *Advances in Ceramics* Vol. 3, Science and Technology of Zirconia, 1981, 241.
2. G. P. Dransfield, K. A. Fothergill and T. A. Egerton: 'The use of plasma synthesis and pigment coating technology to produce an yttria stabilized zirconia having superior properties', *Euro Ceramics*, Vol. 1 de With, G., Terpstra, R. and Metselaar, R. eds, pub. Elsevier (London) 1989, 275–279.
3. A. D. Sivill: Ph.D. Thesis, Department of Mechanical Engineering, Nottingham University, 1974.
4. R. D. Duncan: M.Sc. Thesis, Department of Chemistry, Aberdeen University, 1993.
5. R. Maenner, E. Ivers-Tiffee, et al.: 'Characterisation of YSZ Electrolyte Materials with Various Yttria Contents', *Euro-Ceramics II*, Vol. 3, G. Ziegler and H. Hausner eds, Deutsche Keramische Gesellschaft, Cologne, 1992, 2085–2089.
6. S. Lawson, G. P. Dransfield and W. M. Rainforth: 'Enhanced performance of Y–TZP materials through compositional zoning on a nano-scale, presented at 8th CIMTEC, Florence, Italy, 1994.
7. S. Lawson: unpublished work.
8. R. Stevens: *Introduction to Zirconia*, Magnesium Elektron, 1986, 24.

Sliding Wear of Y–TZP's

I. BIRKBY, P. HARRISON*, M. RAINFORTH#
and R. STEVENS@

Dynamic Ceramic Limited, Bournes Bank, Burslem, Stoke-on-Trent, UK.
* *Caradent, White Cross Industrial Estate, Tideswell, Derbyshire, UK.*
# *Department of Engineering Materials, University of Sheffield, Mappin Street, Sheffield, S1 3JD, UK*
@ *School of Materials, University of Leeds, Leeds, LS2 9JT, UK.*

ABSTRACT

For the last ten to fifteen years tetragonal zirconia polycrsytals (TZP's) have been hailed as one of the great hopes for the advanced ceramic community. Although these materials have received great reviews due to their combination of high strength and toughness, they have been subject to much more mixed reviews in the area of sliding wear behaviour.

The aim of this paper is to consider some of the fundamental, mechanical and physical properties of these materials in relation to sliding wear behaviour. The affect of transformation toughening is discussed in relation to tribological performance together with the significant impact of thermal conductivity. Pin-on-disc tests are related to micro-structure property relationships and thermal characteristics.

1. INTRODUCTION

In 1975 the late Ron Garvie heralded the arrival of transformation toughened zirconia ceramics with his seminal work entitled 'Ceramic Steel?'.[1] Using the scientific principles established by Garvie, many zirconia based ceramics were investigated to ascertain the natural boundaries of the transformation toughening phenomenon. In the late 1970's, a zirconia–yttria combination was found to display extremely high values of toughness and strength. Due to the homogeneous tetragonal phase and fine grained microstructure, the material was given the title, Yttria Tetragonal Zirconia Polycrystals, or as it is now known throughout the world, Y–TZP.

Following on from the initial discovery of Garvie, the use of transformation toughened ceramics was evaluated in a number of industrial applications such as automotive engine cylinder liners, valve guides, pump liners and extrusion dies.[2] The material stimulated great interest within the engineering and scientific communities. The interest of engineers was aroused by the potential provided by a ceramic material with high strength and toughness. Scientists were fascinated by the mechanics and microstructure/property relationships of the transformation toughening phenomena.

A common assumption throughout the early investigative work was that due to the relatively high hardness value of the material, good wear properties would ensue. This belief was based on the principle that wear resistance was proportional to hardness; a belief supported by several hundred years of experience with metals.

Unfortunately, but perhaps not surprisingly, wear studies showed that the performance of ceramics did not display the same correlation with hardness as metal components. Consequently, it was realised that the relatively young science of 'tribology', derived from the Greek 'tribos' (rubbing or attrition), was fundamental to the successful implementation of an engineering ceramic solution, particularly one involving Y–TZP, with its unique and variable mechanical properties.

As Y–TZP's are well suited to microstructure property tailoring, the primary aim of this work was formulated to evaluate the effect of transformability on the sliding wear performance of Y–TZP's. A tri-pin-on-disc tester was used under conditions of unlubricated sliding. Worn surfaces and wear debris was characterised by SEM, TEM and XRD analysis.

1.1 Reported Sliding Wear Data for Y–TZP's

A considerable number of wear tests have been performed under a wide range of test conditions with many different TZP compositions. Unfortunately, the interpretation of this body of data is not straightforward due to the considerable variability in test techniques, environmental conditions, material compositions and processing parameters. Consequently, there is a high variability in recorded wear data which is superimposed on the inherent statistical variability of any single wear test. Although an individual wear test may cast light on the effect of a particular variable, it is incorrect to draw widespread inferences from such a result as the tribological performance of any wear couple is influenced by many factors.

Throughout this paper the quoted wear coefficients unless indicated otherwise are calculated from eqn (1).[3]

$$V = kPX \tag{1}$$

where V .. worn volume
 P .. load
 X .. sliding distance
 k .. wear coefficient

The relative physical significance of this wear factor can be illustrated by reference to a well known 'low wear' application i.e. metallic piston rings running lubricated against a piston bore display a wear factor of approximately 10^{-9} mm^3 Nm^{-1}. One of the lowest wear factors for ceramic on ceramic unlubricated contact, 1.4×10^{-10} (mm^3 Nm^{-1}), was recorded for 99.7% Alumina pairs at 7.5N load and a sliding speed of 0.24 m s^{-1}.[4]

Table 1 is a collection of significant and representative test results for Y–TZP/Ceramic sliding wear tests performed under nominally dry conditions (R.H. 50% ± 10%). Only the results for tests conducted at room temperature are reported to provide a useful comparison with the data in this study. Where the wear coefficient was not reported in the 'standard' format i.e. k (mm^3 Nm^{-1}), but sufficient information was contained in the reference to allow its computation, the results are indicated with (#). The coefficient of friction value is the steady state coefficient of friction.

Table 1.   Wear Factors for Y–TZP Sliding Pairs.

Test type + Ref.	Load (N)	Speed (m^{-1})	Sliding distance	Interface materials	Coeff. of friction	Wear coefficient k (mm^3/Nm^{-1})
Hemisphcl. Pin-on-plate[5]	4.9	0.001	–	Y–TZP on Y–TZP	–	8.8×10–9
Reciprocating[6] Pin-on-plate	3.7	0.08	0.864	3Y–TZP on 3Y–TXP	0.62–0.69	1.6×10–5
,,	,,	,,	,,	3Y–TZP on Mg–PSZ	0.66–0.75	$1.4 \times D\,10$–5
Unidirectional Pin-on-disc[7]	9.8	0.001	0.120	3Y–TZP on 3Y–TZP	0.35	4×10–7
Unidirectional Pin-on-disc[8#]	10	0.4	1.256	3Y–TZP on 3Y–TZP	0.57	8×10–4
Pin-on-disc[9#]	9.8	0.18	–	3Y–TZP on 3Y–TZP	0.65	2.2×10–6
,,	,,	,,	–	3Y–TZP on SiC	0.3	1.0×10–8
,,	,,	,,	–	3Y–TZP on Si$_3$N$_4$	0.4	1.2×10–7
Pin-on-disc[10]	10	0.1	0.150	3Y–TZP on 3Y–TZP	0.46	1.2×10–4
,,	,,	,,	,,	3Y–TZP on RBSN	0.87	1.9×10–5
,,	,,	,,	,,	99.9% Al$_2$O$_3$ on 3Y–TZP	0.69	4.7×10–5
Ball-on-three flats[11#]	10	0.001 9	0.0014	3Y–TZP on 3Y–TZP	0.37	7×10–6
,,	2	,,	,,	,,	–	2.5×10–7
,,	200	0.38	2.166	,,	0.64	2.5×10–4
Pin-on-disc[12]	33.47	0.3	3.0	99.7% Al$_2$O$_3$ on 3Y–TZP	0.15	1.6×10–3
,,	,,	0.7	,,		0.13	1.5×10–3
,,	,,	1.0	,,		0.135	3.0×10–4
Pin-on-disc[13]	19	0.24	11.1	3Y–TZP on ZTA	0.12–0.44	8.9×10–4

It is apparent from Table 1 that wear tests[5] conducted at relatively low load (<5N) and speed (~ 0.001ms^{-1}) have shown low wear factors for like on like TZP combinations in the region of 10^{-9} to 10^{-8} mm^3 Nm^{-1}.

Stachowiak and Stachowiak[6] conducted dry sliding wear tests on a high frequency wear test rig with pin on plate geometry and reciprocating motion. The applied normal load was 3.7N at a velocity of 0.08 ms^{-1}, reciprocating at 7.5Hz. The wear factor for the Y–TZP combination was 1.6×10^{-5} mm^3 Nm^{-1}. The friction coefficient

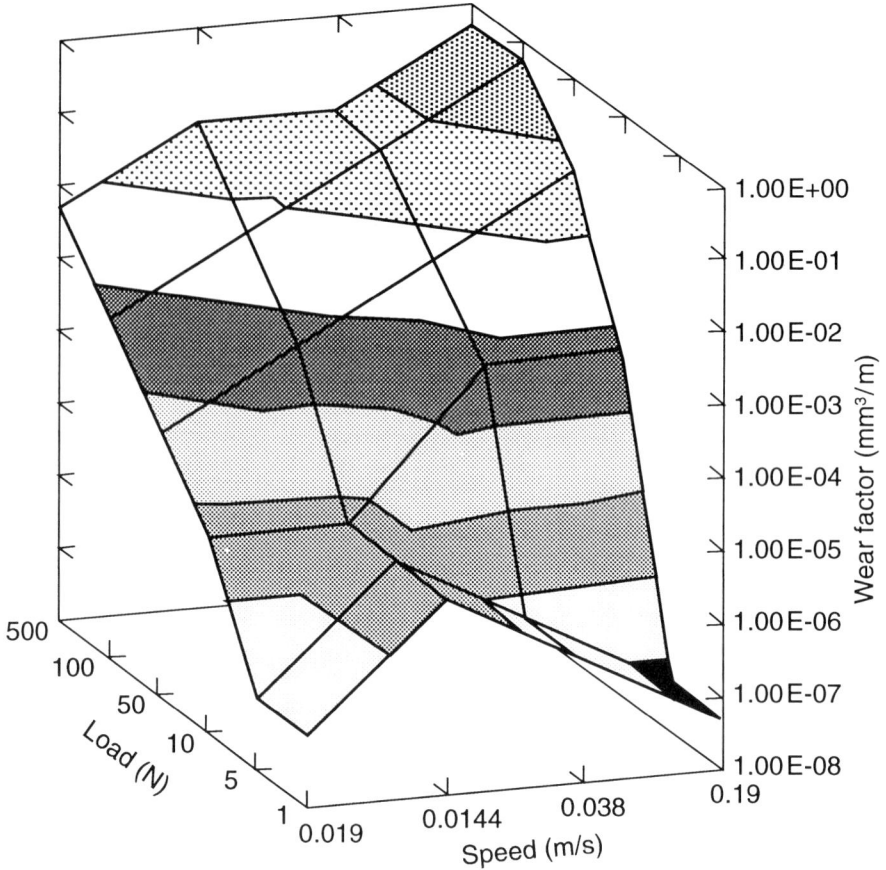

Figure 1.   Wear Map for Y–TZP sliding in dry air (after Lee et al.[11])[1]

was in the range 0.62–0.69. During the initial wear stage, visible wear scars were formed on the plate surface and a substantial amount of material was removed. The ridges of the as-ground surface were levelled by plastic deformation due to the high pressure in the loaded zone. Some of the ridges were removed in a sheet form and pressed into neighbouring grooves, creating a smooth surface layer. The presence of this layer was confirmed by a reduction in the surface roughness, from the as-ground value of 1 μm Ra to 0.2 μm Ra following the test. Some of the wear particles remained in the wear zone contributing to the formation of a heavily deformed wear surface. The surface layer was subsequently deformed by delamination, generating small flat wear particles. SEM analysis of the wear particle morphology indicated that the debris was generated by plastic deformation.

Lee et al.[11] have conducted a comprehensive study of the sliding wear of Y–TZP under dry and lubricated conditions. They also confirm the low wear factors for low load low speed conditions and investigated the effect of load and speed on sliding

---

[1] Wear factor was measured in mm³ m⁻¹

wear, leading to the production of a series of 'wear maps'. An example of the wear map produced by Lee *et al.* is shown in Fig. 1. For pin loads between 2 and 380N, and speeds from 0.0019 to 0.57ms^{-1}, the recorded wear factors for dry sliding varied from 10^{-8} to 10^{-3} mm^3 Nm^{-1} (these values were calculated from the results of Lee *et al.* who measured wear rate in mm^{3}Nm^{-1}). The coefficient of friction ranged between 0.45 to 0.7. They found that the low wear region was limited to loads below 10N. The wear rate increasing linearly with load. The wear mechanism in this region was plastic deformation and microcutting. The wear rate displayed anomalous behaviour in relation to the sliding speed at low loads as shown in Fig. 1.

The anomalous behaviour in the low load region of the map, was attributed to the attainment of a critical level of surface stress with increasing velocity through thermo/mechanical effects, the tetragonal to monoclinic transformation follows and the surface is placed in compression. The wear rate subsequently reduces during higher velocity sliding due to the crack stopping abilities of the compressive surface layer. It should be noted that the large amount of data required to compile the wear maps necessitated the use of a step loading wear test procedure. For a constant test velocity the load was increased in 20N stages, the test proceeded for 5 minutes and the wear scar measured. Consequently the total sliding distance during the low speed test is only 0.57 m, increasing to 171 m for the 0.57 ms^{-1} test. As both of these sliding distances are relatively short (sliding wear tests often running for several hundred km), the anomalous low load, low speed behaviour could also be attributed to an initial 'running in' period and mating of the two surfaces.

As the sliding speed increased and the interface temperature rose, the wear resistance of the TZP/TZP couple decreased. This behaviour has been attributed to the decrease in material properties such as hardness, yield stress and fracture toughness with increasing temperature, aided by the low thermal conductivity of zirconia. At high sliding speeds the asperity flash temperatures become significant and the wear mechanism becomes dominated by fracture mechanisms aided by the inoperative, (at best, reduced) transformation toughening process. Lee *et al.*[11] reported surface material which was removed by chipping, together with large debris particles squeezed into layers and exhibiting a 'fish scale' texture.

A tri-pin-on disc tester was used by Rainforth[13,14] to measure the relative performance of 3Y–TZP pins in dry sliding against a zirconia toughened alumina (ZTA) disc. The initial test was conducted at 0.24 ms^{-1} sliding speed, 19N per pin load for a total sliding distance of 11.1 km. The friction coefficient varied from 0.12–0.44. The wear factor was relatively high at 8.9×10^{-4} mm^3 Nm^{-1}. This is approximately 10^5 times higher than would be expected for an automotive cam follower in service. During the test, faint orange/white sparks were visible, emanating from the interface, indicating flash temperatures of approximately 1400°C. The wear mechanism was predominantly fracture dominated with a severe break-up of the ZTA disc surface. The wear debris was composed of an amorphous phase containing variable amounts of alumina and zirconia, and a second category which was microcrystalline. This was found to be fully tetragonal with a crystallite size in the range 5–50 nm. No evidence was found for the generation of monoclinic phase in the debris. TEM examination of the 3Y–TZP pin provided significant information of

Worn Surface

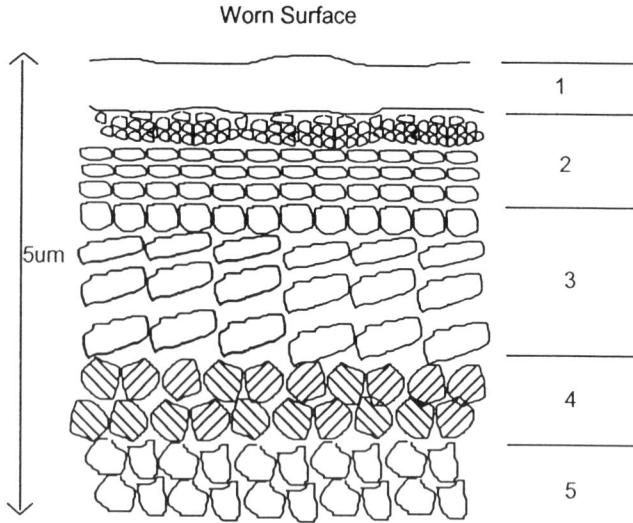

Figure 2.   Schematic of TEM cross section from worn 3Y–TZP pin following sliding against a ZTA disc (after Rainforth[13])

the nature of the wear process. The wear surface was divided into a series of zones as illustrated in Fig. 2.

The outer layer (1) was predominantly amorphous with numerous small fractured particles of alumina. This layer was generally dense and adherent to the surface. Region 2 consisted of fine (50 nm) tetragonal particles with heavily microcracked grain boundaries. Within region 3 the tetragonal grains were noticeably elongated and aligned parallel to the sliding surface. Monoclinic phase was not detected until region 4. The monoclinic band extended in some areas to 10 μm below the surface, before the parent structure was detected in region 5. The following mechanism was proposed to account for the structural modification of this near surface area.

At the surface the flash temperatures were high enough to allow complete mixing of the zirconia and alumina on an atomic scale leading to an amorphous mixture. The order of magnitude reduction in grain size in region 2 was believed to have been a result of intragranular microcrack coalescence during the grain elongation process of region 3. Tetragonal grains were found in region 3 with an aspect ratio of 30:1 corresponding to a shear strain of 8. Such a shape change could have only occurred by dislocation flow. Grain boundary sliding was also identified which will have further increased the surface strain. The behaviour is similar to the superplastic forming of Y–TZP's described by Wakai et al.[15] The similarity in the structure of region 3 to the structure reported by Wakai et al., would appear to indicate a temperature of approximately 1500°C. The monoclinic phase in region 4 was generated either during cooling of the pin or during TEM preparation.

The 3Y–TZP/ZTA test was repeated at a lower sliding speed of 0.02 ms^{-1} and a reduced load of 10 N per pin.[16] The recorded wear factor for this test reduced to 2.4 × 10^{-7}; the friction coefficient was in the range 0.18–0.39. In this instance the wear mechanism was a plastic abrasion process, with minor amounts of

grooving on the surface attributed to abrasion by alumina grains or grain fragments. The wear debris was found to be amorphous with no crystalline material
present in any areas. The reduced wear at the lower load of 10N per pin agrees well
with the threshold load for severe wear in the dry sliding of TZP couples proposed
by Lee *et al.*[11]

Tucci and Esposito[12] assessed the dry sliding behaviour of a 99.7% Alumina pin
against 3Y–TZP discs. Although the reported results (Table 1) are in general agreement with the results of Rainforth,[13,14] a drop in the wear factor is recorded at a
relatively high sliding velocity of 1.0 ms[-1]. The authors attribute this behaviour to a
reduction in the thermal shocking of the wear track, presumably as a consequence of
air cooling. This line of reasoning is difficult to comprehend as frictional heating
increases with sliding velocity. Perhaps it is the formation of a dense and adherent
alumina–zirconia amorphous layer which is providing protection with a more compliant load bearing surface. At the beginning of the test the higher pin hardness led to
the removal of material from the softer discs. Some of the debris was thrown away
from the wear surface while some remained in the wear area where it was crushed by
compressive forces and fractured during rolling to produce very fine sub-micron
particles. The high surface energy of these smaller wear particles leads to a desire to
reattach to the surface to lower the overall surface energy. The compacted layers
adhere to the contact surfaces of the sliding pairs changing the initial zirconia/
alumina contact to one of zirconia/zirconia contact. The debris transfer film increases
the contact area which reduces the contact pressure and hence the wear rate. No
evidence of the stress assisted tetragonal to monoclinic phase transformation was
found on the worn disc surfaces.

Breznak *et al.*[15] tested Y–TZP self mated couples in reciprocating sliding at relatively high loads (44N) and speeds (1.5ms[-1]) and recorded a wear factor of 2×10^{-4}
mm[3] Nm[-1]. XRD studies on the wear debris confirmed the presence of large amounts
of tetragonal phase and small amounts of monoclinic phase. The wear debris was of
the order of 5–50 nm and was found by selected area diffraction to be tetragonal. This
debris was attributed to a microfracture process of the tetragonal grains. It is likely
however, that under such a load and speed regime a degree of recrystallisation has
occurred in the wear debris and surface, in a similar manner to the mechanism
described above by Rainforth.

The dry sliding wear performance of both 3Y–TZP and a fully stabilised 8 mol%
yttria zirconia were assessed by Fischer *et al.*,[7] using a pin on plate test geometry. The
test conditions were 9.8N per pin load at a sliding speed of 0.001 ms[-1]. As the
combination of load and speed was relatively mild, it is a little surprising to note that
a relatively high average wear factor of 5×10^{-7} mm[3] Nm[-1] and a friction coefficient
of 0.35 were recorded for the 3Y–TZP. The fully stabilised 8Y–YSZ (Yttria Stabilised Zirconia) displayed poor tribological behaviour with an average wear factor
of 2.5×10^{-4} mm[3] Nm[-1] and a coefficient of friction of 0.7.

In common with many advanced ceramics, Y–TZP's appear to exhibit a transition
from mild to severe wear at a critical load value. Similarly, such transition behaviour
has been observed in relation to sliding speed. Dufrane[16] has applied the following
equation[17,18] for critical sliding velocity to TZP couples.

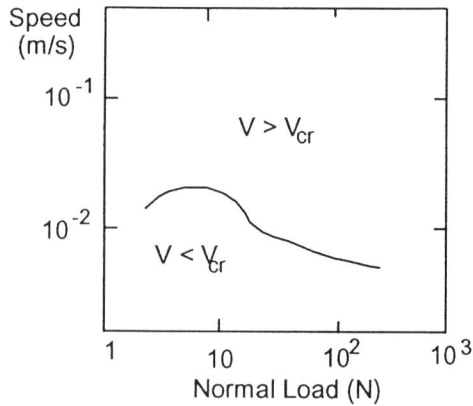

Figure 3.   Critical Sliding Velocity for TZP couples in dry air (after Lee et al.[11])

$$V_{cr} = \frac{4C^2}{(\mu\alpha E)^2\,\pi k_c z}$$

(2)

where V_{cr}.. critical sliding velocity
         C ..   thermal conductivity
         μ ..   coefficient of friction
         α ..   coefficient of thermal expansion
         E ..   Young's modulus
         K_c ..   thermal diffusivity (C/pw)
         z ..   width of slider
         ρ ..   density
         w ..   specific heat capacity

Applying the above equation to 3Y–TZP in a ring on flat test geometry which simulates the piston ring/cylinder liner contact reveals a critical sliding velocity of 0.08 ms[-1]. The calculation used a valuation of 0.12 for the coefficient of friction of TZP in mineral oil. Accordingly, a coefficient of friction of 0.04 increases the critrical velocity to 0.94 ms[-1]; a maximum critical velocity of 1.5 ms[-1] can be postulated for a friction coefficient of 0.01.

The above analysis has also been applied to the wear data produced by Lee *et al.*[11] The following diagram (Fig. 3) indicates their derivation of the critical velocity threshold which is in good agreement with the experimental evidence.

2. EXPERIMENTAL

2.1 Materials

In this series of tests, only one generic powder type was used, (Tosoh[1] – grades TZ3YB and TZ2Y), to minimise the secondary effects of grain boundary phases,

---

[1] Tosoh Europe B.V., Crown Building – South, Hullenbergweg 359, 1101 CP Amsterdam, Holland

Table 2. Processing conditions and mechanical properties of wear test specimens

Material → Property ↓	2Y–TZP	3Y–TZP	3Y–TZP (HIPed)
Sintering Temp. (deg. C)	1575	1450	1450
Sintering Time (hrs)	2	6	6
HIPing Temp. (deg. C)			1480
HIPing Times (hrs)			0.5
HIPing Pressure (MPa)			150
M.O.R. (3-point) (MPa)	950	110	1300
Fracture Toughness (MPam1/2)	12	6	6.5
Hardness (HV50)	1100	1450	1400
Grain Size (mm)	0.85	0.4	0.45

powder purity etc. To modify the transformability of the test specimens, two powder grades and three sintering schedules were employed.

3. WEAR TESTING

3.1 Pin-on-Disc Wear Tester

A series of wear tests was performed on a tri-pin-on-disk tribometer[1], Fig. 4 shows the general layout and the strain gauges attached to the lever arms for measuring the dynamic coefficient of friction. Each of the tests was performed under nominally dry conditions at a temperature of approximately 20°C ± 1°C. The apparatus was covered by a glass dome to prevent the ingress of foreign bodies and moisture to the contact faces. In this series of tests the load per pin was limited to a lightly loaded configuration of 4.8N at a sliding speed of 0.24 ms^{-1}. The test conditions were held constant

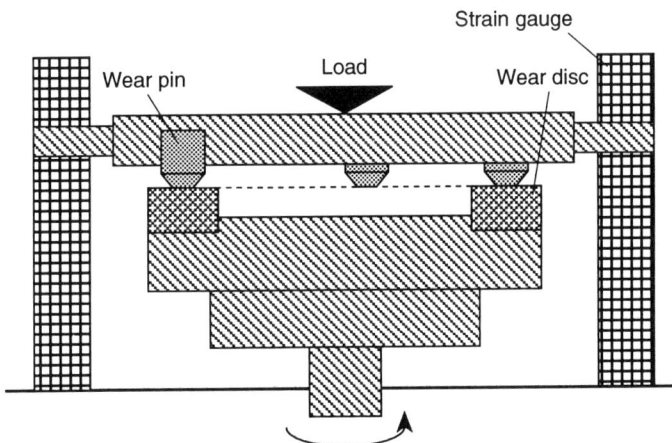

Figure 4.   Tri-Pin-on-Disc-wear tester

[1] Department of Mechanical Engineering, University of Leeds

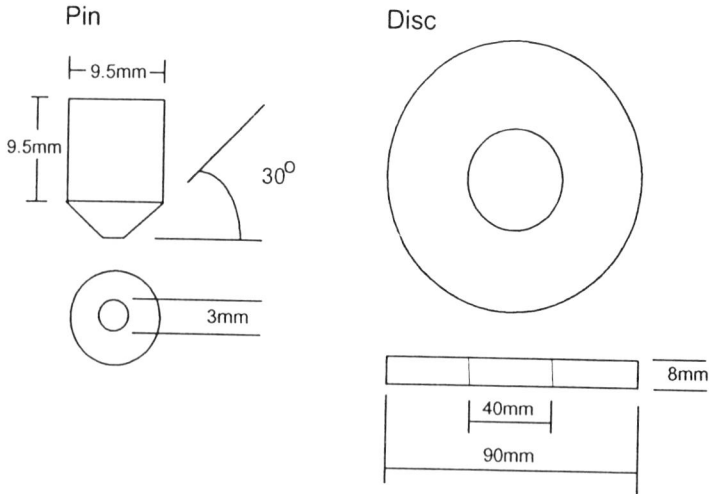

Figure 5.   Tri-Pin-on-Disc test pin geometrics

whilst the transformability of the test specimens was varied. The geometry of the individual test pins and discs is shown in Fig. 5. Both the pins and discs were manufactured by die pressing at 93 MPa followed by sintering under the conditions shown in Table 2. The exact geometries were generated by diamond grinding. The mating surfaces of the pins and discs were diamond lapped using 15μm diamond paste to a surface roughness of 0.5 μm Ra. The mechanical properties of the individual pin and disc materials were measured prior to testing, the relevant values are also shown in Table 2.

3.2 Test Conditions

Prior to testing, each pin was numbered, weighed on a Mettler balance and the surface roughness of the mating surface measured on a Rank Taylor Hobson Talysurf 5. The pins were cleaned ultrasonically and located within the pin holder. The pins were clamped into the head by means of collets which were tightened by grub screws. Only a few mm of each pin projected from the pin holder to maximise the stiffness of the assembly. The pin position was adjusted until all three pins were level to within 20μm of each other. The tests were conducted for a sliding distance up to a maximum of 500 km. Each individual test was stopped at regular intervals and the pins weighed to determine a wear coefficient. When the pins were replaced in their respective holders, care was taken to ensure correct alignment to the wear track. The weight losses were normalised by reference to an un-tested control pin. The wear coefficients were calculated from eqn (1).

The nominal contact area of the pin face on the disc was deduced by measurement, using a talycontour trace to determine the lengths of the major and minor axes of the elliptical wear face formed during the wear process. From these values the nominal contact area and hence the nominal contact stress was calculated. The coefficient of friction was also recorded at regular intervals from the strain bridge readings.

Table 3. Tri-Pin-on-Disc test, Sliding Wear Factor, Nominal Contact Stress and Coefficient of Friction

Wear data→ Material ↓	Wear coefficient (mm^3 Nm^{-1})	Nominal contact stress (MPa)	Friction coefficient
2Y–TZP			
30 km sliding	9.73×10^{-6}	0.54	~ 0.3
250 km sliding	7.38×10^{-6}	0.175	~ 0.3
3Y–TZP			
30 km sliding	1.54×10^{-5}	0.37	~ 0.3
250 km sliding	9.75×10^{-6}	0.16	~ 0.3
3Y–TZP (HIP)			
39 km sliding	3.22×10^{-5}	0.38	~ 0.3
250 km sliding	8.51×10^{-6}	0.14	~ 0.3

4. RESULTS

Wear performance is reported with reference to worn surface microstructures, debris analysis and phase content. The effect of transformability and thermo elastic instability are also discussed.

4.1 Wear Factors

The results for the three pin on disc tests are shown in Table 3.

The wear coefficient and contact stress are shown graphically against sliding distance in Figs 6 and 7.

It can be seen from these figures that the wear coefficient falls as the contact stress reduces. Although the wear factor varies considerably in the initial stages it appears to stabilise at a value of approximately 7×10^{-6}–1×10^{-5} mm^3 Nm^{-1} and a contact

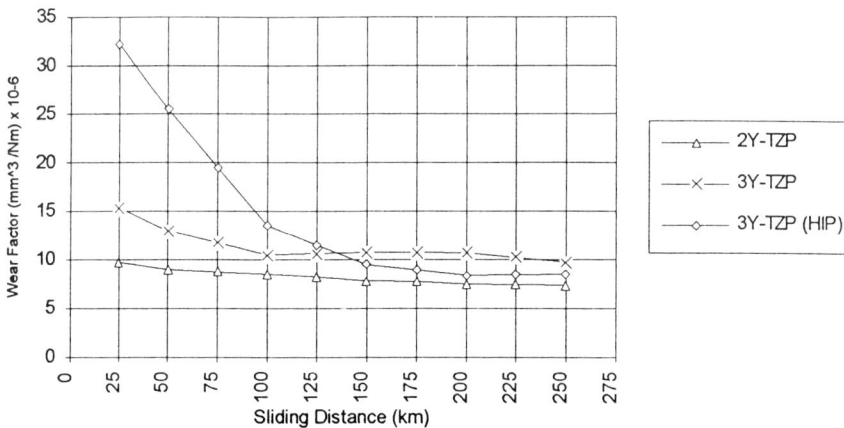

Figure 6. Sliding wear factor for TZP/TZP wear couples

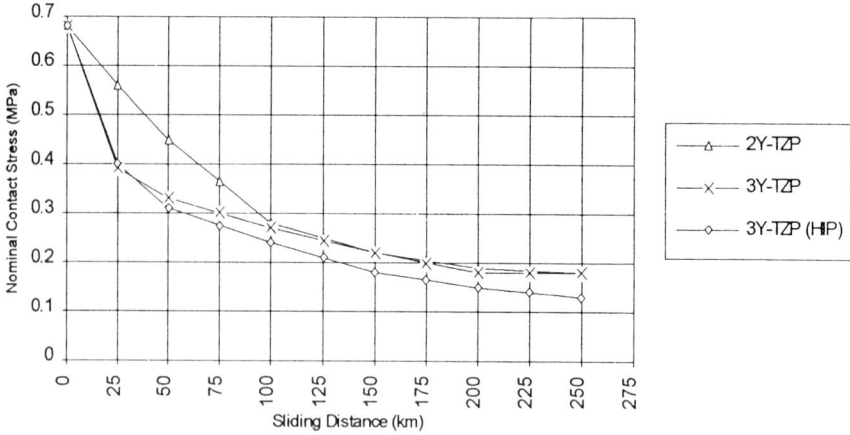

Figure 7.   Nominal contact stress for Y–TZP pins on Y–TZP disc

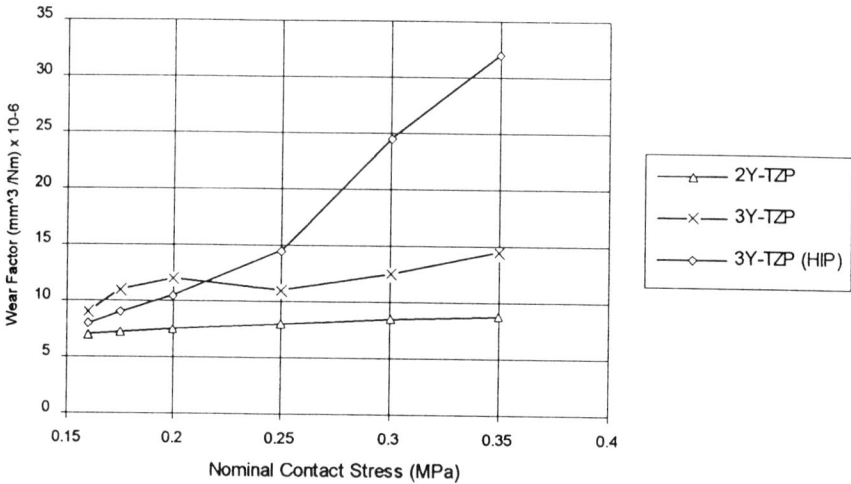

Figure 8.   Wear Factor vs. contact stress for Pin on Disc tests

stress between 0.14 MPa and 0.17 MPa after a sliding distance of 250 km. Figure 8 shows the relationship between contact stress and wear with an apparent transition from mild to severe wear at a contact stress between 0.2 to 0.3 MPa for the 3Y–TZP and 3Y–TZP hipped materials. No such behaviour is apparent for the 2Y–TZP ceramic.

The above result is significant in that it contradicts the commonly held view that it is the load which determines wear rate, not the apparent contact stress. This view is based on the concept that the real area of contact increases on initial contact to a value at which the load is supported. As the true contact area is lower than the apparent contact area, the nominal contact stress based on the apparent contact area is held to be irrelevant. However, the above results may suggest that at very small values of apparent contact area, the rate at which the true contact area can increase to support the load is slow. Consequently, during 'running-in' periods the apparent contact stress is important and will be determined by microstructural features of the appropriate scale, such as grain size, porosity etc. as well as by the final machining and polishing operations.

The coefficient of friction was measured dynamically for each of the tests and varied significantly in the range 0.1 to 0.3. These high values were not recorded at the beginning of each test when the contact surfaces were smooth, initially values of approximately 0.05 were recorded. Chatter occurred occasionally and was reflected in the spread of coefficient of friction values obtained during the life of a wear test.

The wear of the pins and discs at the end of each test was excessive and therefore accurate roughness values and talysurf traces proved difficult to obtain, although a surface roughness value of approximately 1.0 μm Ra was recorded. The talycontour trace shown in Fig. 9 shows the maximum deviation from the original shape to be approximately 0.5mm.

The recorded wear coefficients determined in the present series of tests are in agreement with the literature values for tests conducted under similar load and speed conditions.[6,9]

During the initial stages of the tests the performance of the 2Y–TZP was superior to both 3Y–TZP's. As the 2Y–TZP material is considerably tougher than the other two materials and increasing strength appears to have little effect, one could postulate that in addition to the contact stress effects, it is the increased resistance to thermal shock induced crack propagation which accounts for this behaviour, as indicated by the thermal shock resistance parameter[19] R''',

$$R''' = \frac{E\gamma}{\sigma^2 (1 - v)} \tag{3}$$

where γ .. fracture surface energy
σ .. tensile strength
v .. poissons ratio
E .. Elastic Modulus

A simple comparison of the 2Y–TZP and 3Y–TZP (HIP) materials reveals the R''' parameter for the 2Y material to be 2.8 times greater than for the 3Y–TZP (HIP) material. This is in close agreement with the recorded relationship of the wear rate at the beginning of the test. This argument is only valid where the region which is arresting a thermal shock crack is at a relativley low temperature (< 300°C) as otherwise transformation cannot occur. Such differences in temperature between 'flash' regions and relatively cool bulk regions which retain a transformation capability, are more likely at the beginning of the test where the number of asperity contacts is relatively low. As the test proceeds and the true area of contact increases, the bulk temperature will increase with a consequent reduction in potential for transformation toughening. This transformation dependent behaviour would account for the small differences in wear factors for all three materials at the end of the test.

Original Lapped
Surface

1 mm

Max Depth = 0.52 mm

0.2 mm

Ra = 0.98 µm

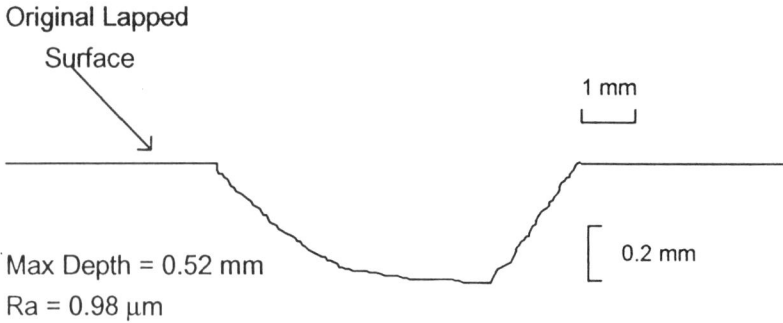

Figure 9.   Depth of wear scar on 3Y–TZP disc after 250km sliding

4.2 SEM Analysis of Worn Surfaces

A large amount of wear debris was produced during the tests which tended to be lightly
bonded to the leading edge of the pin thus suggesting that the wear process was aggra-
vated by the presence of a third body. The wear on both pins and discs was directional in
nature with numerous ridges running parallel to the sliding direction. Examination of the
worn surfaces showed a series of plastically deformed areas and 'gouged' ridges. Typical
SEM micrographs of the worn surfaces are shown in Figs 10 and 11.

In Figs 10 and 11 there is substantial evidence of cracking in a direction perpendicular to the sliding direction. The type of lateral cracking shown in Fig. 11 is similar
to the thermal shock cracks generated at the onset of thermo-elastic instability (TEI).
In this case the TEI would appear to relate to the final stages of wear plate compaction and smearing. A monolithic wear plate having formed during sliding (aided by

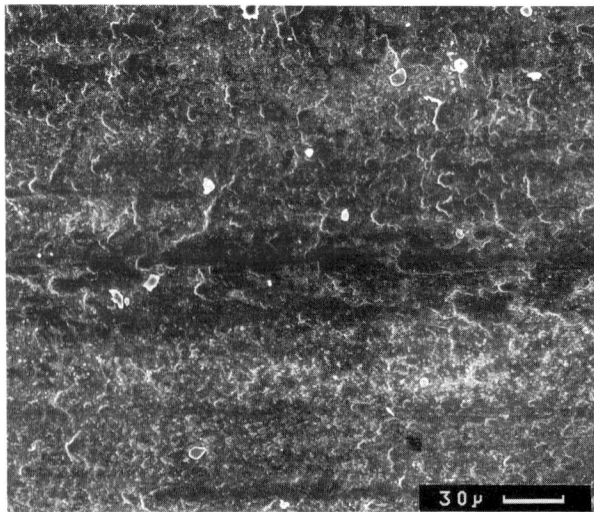

Figure 10.   SEM Micrograph of 2Y–TZP wear pin showing heavily compacted and
smeared wear debris plates.

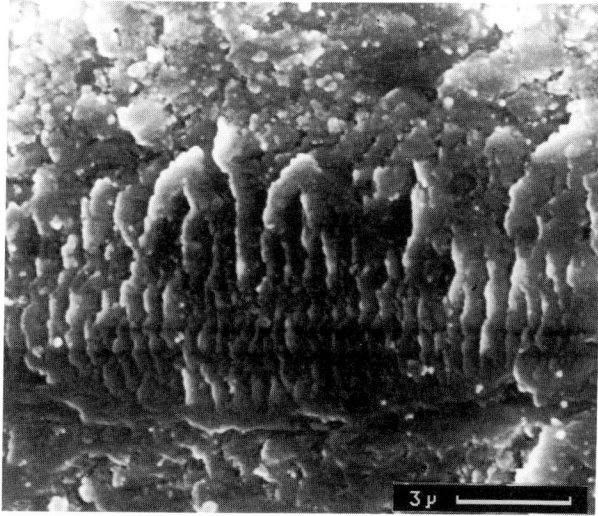

Figure 11.   SEM Micrograph of 2Y–TZP wear pin showing fractured wear debris plate and 'fish scale' texture.

Figure 12.   3Y–TZP Wear Disc Surface (base of wear track) showing elongated granular morphology indicative of high interface temperatures.

the very high interface temperature) and subsequently fractured as the counterface moved and rapid cooling ensued. In many areas intimate mixing of the counterface surfaces appears to have occurred, with compaction and smearing of wear debris. Based on the work of Rainforth[13] in assessing flash temperatures on TZP surfaces, the frictional heating would have been sufficient to cause flash temperatures greater than 2000°C. The wear disc surface (Fig. 12) shows evidence of high temperature sliding with elongated and rounded grain features. No discernible differences in the wear surfaces were visible between the 2Y and 3Y materials. With no firm evidence of grain pop-out or removal visible, it is unlikely that the t-m stress induced transformation played a significant role in the wear process of these samples.

4.3 Wear Debris Analysis

The X-Ray Diffraction traces of the collected wear debris are illustrated in Figs 13 and 14. It is apparent from the traces that substantial peak broadening has occurred, indicating a very fine crystallite size and a highly strained crystal lattice structure.

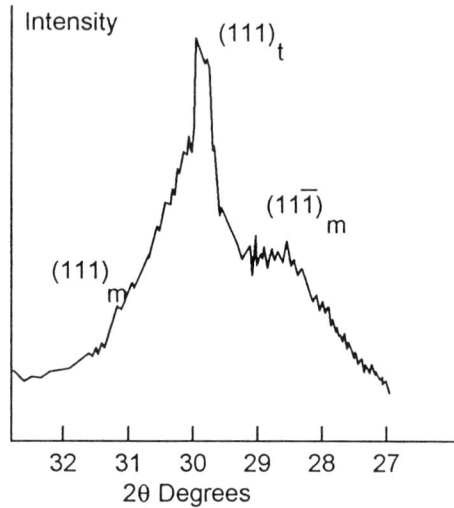

Figure 13.   X-Ray Diffraction trace for 2Y–TZP pin 3Y–TZP disc wear debris

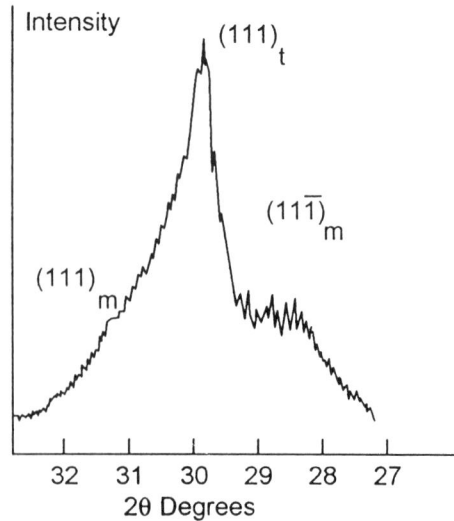

Figure 14.   X-Ray Diffraction trace for 3Y–TZP Pin 3Y–TZP disc wear debris

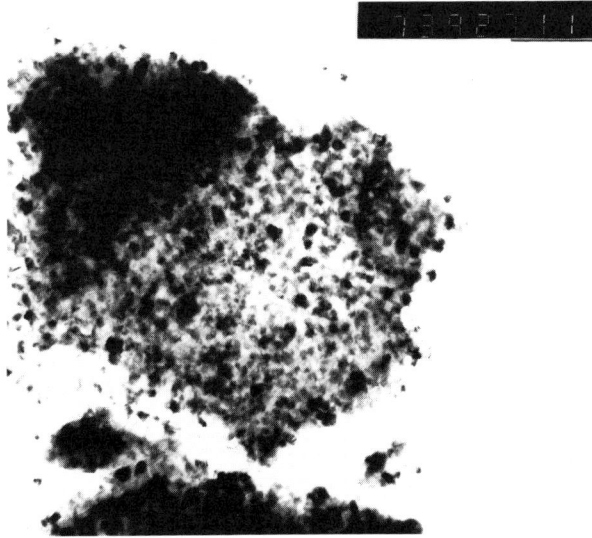

Figure 15. TEM micrograph of 3Y–TZP pin 3Y–TZP disc wear debris (x 73,000) showing recrystallised tetragonal crystallites

Figure 16. Selected Area Diffraction pattern of wear debris from 3Y–TZP Pin 3Y–TZP disc

This means the exact determination of the monoclinic peaks is difficult. Indeed the monoclinic (111) peak is barely discernible.

Figure 15 shows a bright field image of a worn particle in which the individual crystallites are easily visible. This crystalline nature is depicted in Fig. 16 in which the rather 'spotty' electron diffraction pattern is shown.

Further analysis of this diffraction pattern confirmed that it was derived from a monoclinic crystallite. An apparent contradiction exists in the conflicting evidence for the presence of the monoclinic phase i.e. the SAD results indicate monoclinic phase but conversely, the X-Ray Diffraction traces indicate the presence of a significant amount of tetragonal phase. It is possible that the tetragonal and monoclinic phases exist in different size distributions. X-Ray Diffraction will not generate discrete peaks if the particle size is reduced much below 50 nm. On the other hand electron diffraction enables characteristic patterns to be generated from crystals an order of magnitude smaller ($\approx$5 nm). Nevertheless, there is still firm evidence for the presence of monoclinic phase which indicates that some of the material has undergone transformation. The size of the particles could be of significance in that very small particles are stabilised in the tetragonal form due to a higher thermodynamic surface energy of the tetragonal form compared to the monoclinic. At a critical particle size spontaneous transformation to the monoclinic takes place. The effect was first noted for pure zirconia on calcining zirconium hydroxide and could well be the reason why the 3Y–TZP in the form of finely divided wear debris gives only a small amount of monoclinic phrase.

A further important feature to note from the X-ray diffraction traces is the development of a preferred orientation effect. This is discernible by the more pronounced monoclinic ($1\bar{1}1$) peak in comparison to the (111) peak. This type of preferred orientation effect has been noted in the fracture surfaces of bend strength specimens.[20] Any process which develops a shear stress as opposed to a tensile stress, would be expected to enhance the preferred orientation effect. This is certainly the case for sliding wear.

The lack of microcracking in the wear debris and the substantial line broadening in the XRD trace is indicative of a tetragonal structure; it is likely that some form of dynamic recrystallisation of the strained tetragonal phase has occurred during the wear process, presumably due to the high flash temperatures. Similar behaviour has been reported by Rainforth[13] in the dry sliding of Y–TZP on ZTA.

5. DISCUSSION

From the results presented above the following wear mechanism is proposed:
- Microfracture of Y–TZP asperity contacts leads to the generation of third body wear particles.
- High interface temperatures lead to thermo-elastic instability and thermal shock cracking.
- Transformation of 2Y–TZP material may occur in the vicinity of a propagating thermal shock crack.
- Compaction/smearing of wear debris and intimate mixing of the counterfaces leads to the delamination of plates from the pin surface.
- High Temperature dynamic recrystallisation of the wear debris occurs.

The sliding speed and load conditions for this test are within the region of thermo-elastic instability as described by Lee *et al.*[11] In this wear environment, a high propensity for t-m transformation may improve the running in behaviour and the resistance to thermal shock crack propagation, but is generally ineffective due to the high interface temperatures.

Although several wear mechanisms were recorded in this study, in the case of Y–TZP interfaces it is the tribo-chemical surface reactions which appear to be the predominant influence in both ceramic/ceramic and ceramic/metal contacts. In the *tri-pin-on-disc* tests, a small amount of t-m transformation occurred and possibly played a minor role in minimising the initial surface damage due to thermal shock cracking. Although the SAD analysis of the wear debris indicated the presence of monoclinic phase, the lack of any conclusive XRD information on the formation of monoclinic phase indicated that transformation related wear played a minor role in Y–TZP like on like wear combinations. From the SEM evidence it appears that the major events are the generation of high interfacial flash temperatures, solid solution bonding of the Y–TZP asperity surfaces, followed by the breaking of the asperity junction, followed by generation and smearing of the wear debris. A minor amount of third body abrasive also occurs. The predominance of each mechanism would depend on the operating conditions and the microstructures of the wear couple. The complexity of the possible combination of variables no doubt accounts for the many different results reported in the literature.

In both test types, fluctuations in the measured coefficients of friction reflect the evolution of wear debris in the sliding interface and the dynamic equilibrium between production and elimination from the wear track. The accumulation of wear debris between ceramic sliding interfaces has been shown[21] to lead to an increase in friction and a reduction in wear rate for the following reasons:

(i) Wear debris interacts with the sliding surfaces by ploughing and abrasion, increasing the friction coefficient.

(ii) Wear debris presents a load carrying capacity which decreases the wear rate by minimising the contact area.

In summarising one of the key elements of this study, namely, the effect of transformability, it would appear that although an optimised combination of strength and toughness is desirable for structural applications, it is of little if any benefit in sliding wear applications. In common with most of the literature reports (except those at very low sliding speeds), this study has found high transformability to be of limited significance in sliding wear.

The only area where high transformability may have played a beneficial role, was in the initial stages of the tri-pin-on-disc test for the TZ2Y/TZ3YB test, the high transformability possibly acting to minimise the propagation of thermal shock cracks. However, under higher load conditions the likelihood of this behaviour continuing would diminish. From previous studies.[22] The t-m transformation has been shown to have either a negative influence on wear behaviour or has been inoperative due to the high interfacial temperatures. Consequently, it is suggested that the closest one could get to an 'ideal' Y–TZP structure for self mated sliding wear would be a strong and tough bulk material, with a dense, fine grianed, untransformable external surface. A technique for producing such a composite structure was described by Whalen,[23] for the prevention of ageing in Y–TZP's. The technique involved grinding the surface of a Y–TZP test bar under controlled conditions and annealing the bar at approximately 1300°C. The net effect of this thermomechanical treatment is to recrystallise the surface grains, to produce a fine (<0.1 μm) grain structure, whilst retaining the bulk properties of the Y–TZP. Considerable work would be required to optimise the characteristics of the bulk microstructure, the grinding conditions and the heat treatment schedules, but the proposal is worthy of further consideration.

6. CONCLUSIONS

The unlubricated self mated sliding wear of Y–TZP's has been investigated by tri-pin-on-disc tests. Wear occurred by the microfracture of asperity contacts and the generation of third body wear particles. High interface temperatures led to thermo-elastic instability and thermal stress cracking. Compaction of wear debris and intimate mixing of the counterface materials caused delamination of smeared wear plates from the surface. High temperature dynamic recrystallisation of the wear debris occurred. A highly transformable 2Y–TZP displayed superior wear performance in the initial stages of the test. It was postulated that this behaviour was due to the increased resistance to thermal stress cracking. Once steady state sliding was attained, it appeared that the sliding interface temperatures had prevented the operation of the transformation toughening mechanism. No firm evidence was found for surface transformation. The self mated sliding wear of Y–TZP's is dominated by the low thermal conductivity of the material. Except for low load, low speed conditions or lubricated contact, self mated Y–TZP pairs do not represent a good tribological couple.

REFERENCES

1. R. C. Garvie, R. H. J. Hannink and R. T. Pascoe: 'Ceramic Steel?', *Nature*, 1975, **258** (5537), 703.
2. R. C. Garvie: 'Structural Applications of Zirconia Bearing Materials', *Advances in Ceramics 12,* Proc. 2nd Int. Conf. on the Science and Technology of Zirconia, Am. Ceram. Soc., 1983.
3. J. K. Lancaster: 'The influence of substrate hardness on the formation and endurance of Molybdenum Disulphide films', *Wear*, 1967, **10**, 103.
4. N. C. Wallbridge and D. Dowson: *Wear Behaviour of Ceramics*, Inst. of Tribology, University of Leeds, Internal Research Report, 1984.
5. H. G. Scott: 'Friction and Wear of Zirconia at very Low Sliding Speeds', *Proc. Int. Conf. On Wear of Materials*, Reston VA, ASME, 1983.
6. G. W. Stachowiak and G. W. Stachowiak: 'Unlubricated Friction and Wear Behaviour of Toughened Zirconia Ceramics', *Wear*, 1989, **132**, 151.
7. T. E. Fischer, M. P. Anderson, S. Jahanmir, R. Salher: 'Friction and Sear of Tough and Brittle Zirconia in Nitrogen, Air, Water, Hexadecane and Hexadecane Containing Stearic Acid', *Wear of Materials 1987,* ASME, 1987, 257.
8. S. Sasaki: 'The Effects of the Surrounding Atmosphere on the Friction and Wear of Alumina, Zirconia, Silicon Carbide and Silicon Nitride', *Wear*, 1989, **134**, 185.
9. M. Iwasa, Y. Toibana: 'Friction and Wear of Ceramics Measured by a Pin-on-Disc Tester', *Yogyo-Kyokai-Shi*, 1986, **94**(3), 336.
10. G. J. Wright: 'Aspects of the Tribological Properties of Engineering Ceramics', *19th Annual Symposim Proceedings – Durability of Ceramic Products 9/9/87,* S.A. Ceramic Society, 1987, 37.

11. S. W. Lee, S. M. Hsu, M. C. Shen: 'Ceramic Wear Maps: Zirconia', *J. Am. Ceram. Soc.*, 1993, **76**(8), 1937.
12. A. Tucci, L. Esposito: 'Microstructure and Tribological Properties of Zirconoia Ceramics', *Wear*, 1994, **172**, 111.
13. M. Rainforth: *Ceramic Metal Wear Mechanisms*, Ph.D. Thesis, University of Leeds, 1990.
14. F. Wakai, S. Sakaguchi, Y. Matsuno: 'Superplasticity of Yttria-Stabilised Tetragonal Zirconia Polycrystals', *Adv. Ceram. Mater.*, 1986, **1**, 259.
15. J. Breznak, E. Breval, N. H. Macmillan: 'Sliding Friction and wear of Structural Ceramics pt. 1, Room Temperature Behaviour', *J. Mat. Sci.*, 1985, **20**, 4657.
16. K. F. Dufrane: 'Sliding Performance of Ceramics for Advanced Heat Engines', *Ceram. Eng. Sci. Proc.*, 1986, **7**(7–8), 1052.
17. T. A. Dow, R. A. Burton: 'The Role of Wear in the Initiation of Thermoelastic Instabilities in Sliding Contact', *J. Lubr. Tech. Trans.*, ASME series F, 1973, **95**(1), 71.
18. T. A. Dow, R. D. Stockwell: 'Experimental Verification of Thermoelastic Instabilities in Sliding Contact', *J. Lubr. Tech. Trans.*, ASME series F, 1977, **99**(33), 359.
19. D. P. H. Hasselman: 'Thermal Stress Resistance Parameters for brittle Refracotry Ceramics: a compendium', *Bull. Am. Ceram. Soc.*, 1970, **49**, 1033.
20. A. W. Paterson, R. Stevens: 'Preferred Orientation of the Transformed Monoclinic phase in fracture surfaces of Y-TZP ceramics', *Int. J. High Technology Ceramics*, 1986, **2**, 135.
21. J. Denape, J. Lamon: 'Sliding Friction of Ceramics: mechanical action of the wear debris', *J. Mat. Sci.*, 1990, **25**, 3592.
22. I. Birkby, P. Harrison and R. Stevens: *The Effect of Surface Transformations on the Wear Behaviour of Zirconia (TZP) Ceramics*, Journal of the European Ceramic Society, 1989, **5**, 37.
23. P. J. Whalen, F. Reidinger, R. F. Antrim: 'Prevention of Low Temperature Surface Transformation by Surface Recrystallization in Yttria-Doped Tetragonal Zirconia', *J. Am. Ceram. Soc.*, 1989, **72**[2], 319.

Author Index

Index

BRITISH CERAMIC PROCEEDINGS 54

CERAMIC FILMS AND COATINGS

Edited by
W. E. Lee

Contributed papers on

Generation and relief of stress in ceramic films; Interphase carbon films in ceramic matrix composites: oxidation behaviour; Conversion/coating of carbon fibres by reaction with SiO(g); The protective properties of alumina coatings, deposited by LP-MOCVD, against high temperature sulphidation of alloys; The influence of fuel gas on the microstructure and wear performance of alumina coatings produced by the high velocity oxyfuel (HVOF) thermal spray process; Fabrication of ceramic coatings using flame assisted vapour deposition; Remote microwave plasma enhanced chemical vapour deposition: application to silicon dioxide and diamond-like carbon layers; The hetrogeneous nature of deposits from hexamethyldisiloxane/oxygen plasmas; The potential of n-propyl silicate ormosils as photonic materials: structural investigation by means of NIR and IR spectroscopy and BET; The strengthening of glass with epoxy resin and ormosil coatings; An empirical and statistical approach for characterising chemical durability within a lead free glass enamel flux region; Wetting and adhesion between float glass and silicon; sol-gel titania surfaces for medical implants, part 1: formation and characterisation; Sol-gel titania surfaces for medical implants, part 2: *in-vitro* evaluation; Crystallisation behaviour of glass-ceramic coating based on barium disilicate; Piezoelectric thin films for SAW devices; Pb (Zr, Ti)O_3 thin films by *in-situ* reactive sputtering on micromachined membranes for micromechanical applications; Microstructual characterisation of Pb (Zr, Ti)O_3 thin films deposited by i*n-situ* reactive sputtering; Characterisation of PZT thin films prepared by a sol-gel route; Effect of microstructure on the optical properties of pyrolytically deposited SnO_2 coatings on float glass; Production of superconducting ceramic $YBa_2Cu_3O_{7-\delta}$ thick films by reaction of eutectic melts with yttria substrates

Book 612 ISBN 0 901716 79 0 246mm x 172mm
EU: £75 / Members £60
Non-EU $150 / Members $120
p&p £3.50 EU / $8.00 Non-EU

Orders to: The Institute of Materials, Accounts Department, 1 Carlton House Terrace, London, SW1Y 5DB   Tel: 0171 839 4071   Fax: 0171 839 2078